Yarn Technology

Yarn Technology

Naresh Grover

RANDOM

RANDOM PUBLICATIONS

NEW DELHI (INDIA)

Yarn Technology

ISBN 978-93-5111-651-6

Published in 2015 in India by

RANDOM PUBLICATIONS

4376-A/4B, Gali Murari Lal, Ansari Road
New Delhi-110 002
Phone : +9111-43580356, 011-23289044, 011-43142548
e-mail: sales@randompublications.com,
info@randompublications.com, randomexports@gmail.com

Reprinted 2023

Type Setting by : Friends Media, Delhi-110089
Digitally Printed at : Replika Press Pvt. Ltd.

Preface

The fundamentals of yarn technology are concerned with the production of yarns from fibers of discrete lengths and the structure-property relation of the yarns. Ever since humans moved from using the skins of hunted animals for clothing to farming and using farmed animal hairs and fibers from nonfood crops, and eventually to the manufacture of synthetic fibers, the spinning of yarns has been of importance to the craft and the science, design, and engineering of textiles. In the business unit fiber and yarn technologies, the focus is in the first stage of the textile production chain. The research activities cover the development of new polymers and fibers, the production and processing of (bi-component) filament and staple fibers with conventional and innovative spinning technologies as well as the finishing and functionalization of yarns. The result of these activities is new fibers and yarns with improved quality as well as optimized production processes and high-performance technologies. This book is aimed at giving the reader a good background on the subject of the conversion of fibers into yarns, and an in-depth understanding of the principles of the various processes involved. The purpose is to try to get the reader to identify commonality between spinning systems, something that the author has found useful in carrying out research into new spinning techniques.

I would like to thank my team for standing beside me throughout my career and writing this book. My special thanks go to "Random Publications" who have published the book.

– Naresh Grover

Contents

1

The Nature of Textiles and Textile Fibre Production

In order to promote energy conservation in the textile industry, this manual analyzes in detail the actual conditions of energy consumption in each field. Thus, it has been compiled to serve as reference yarn yarn material for the practical application of these techniques which are always kept available for the engineers. A step-by-step description is included as to how to implement the energy conservation measures, specifically in the area of the dyeing and finishing process, where many small-to medium-sized companies are operating. We strongly hope that this manual will be used as a guide to promote energy conservation in the textile industry and to rationalize the management.

NOVELTY WEAVES

Novelty Weaves are also called as decorative, fancy, figure and design weaves. They are formed by predetermined changes in the interlacing of warp and filling yarns. The different weaves include dobby, jacquard, leno, pile and double cloth.

COMMONLY USED NOVELTY WEAVES

Pile Weave

Pile Weave

Pile fabrics are formed by having the basic plain or twill weave as a backing and a third yarn is woven to yield a surface pile. The pile may be warp pile or weft pile. For making ground fabric, plain or twill weave is used, the extra set of filling yarn floats over three or more warp yarns. The floats are cut and brushed up to form pile. This is called filling pile. Examples are velveteen and corduroy fabrics. If an extra warp yarn floats over the filling yarn, it is called warp pile. Examples are velvet, velour and rug velvet.

Pile weave may be of two types:

1. Cut Pile
2. Uncut Pile

Double Weave

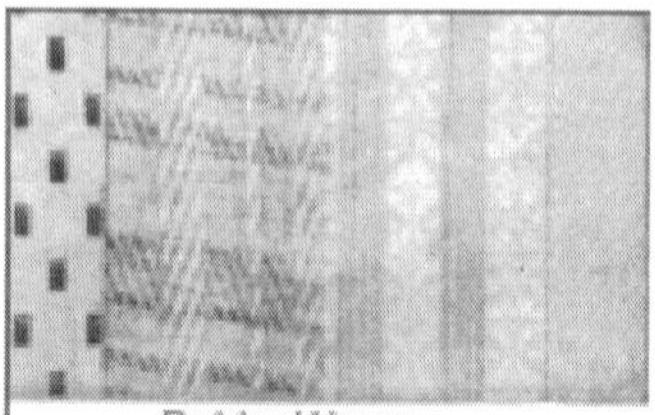

Dobby Weave

Double weave fabrics are obtained by using five or more set of yarns. The most common types of double cloth have two set of warp and two set of weft yarns with an extra yarn interlacing both the cloth. The double cloth has additional bulk, strength and warmth. Examples are coatings, blankets, double brocade and brocatelle.

Leno and Gauze Weave

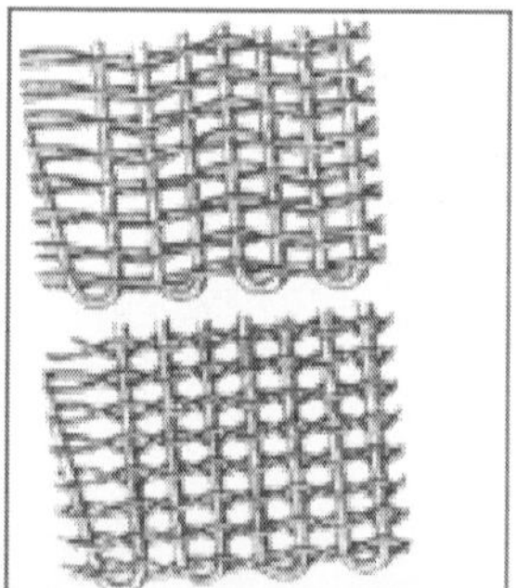

Leno and Gauze Weave

The leno weave produces an open-textured fabric that may be sheer or heavy. It is produced by the doup attachment to the basic loom. The doup attachment controls the warp threads horizontally as well as vertically and the unusual warp interlacing prevents slippage of the filling and reduces shrinkage. Examples include mosquito nets, household bags and curtains.

Dobby Weave

Dobby designs have small figures such as dots, geometric designs and floral patterns woven into the fabric. The design is produced by a combination of two or more basic weaves and the loom may have upto thirty two harnesses. Examples of dobby weave are shirting madras, pique, huck towelling.

Following types dobby weaves are:
- Bird's eye Weave
- Honey comb Weave
- Huchaback Weave

Jacquard Designs

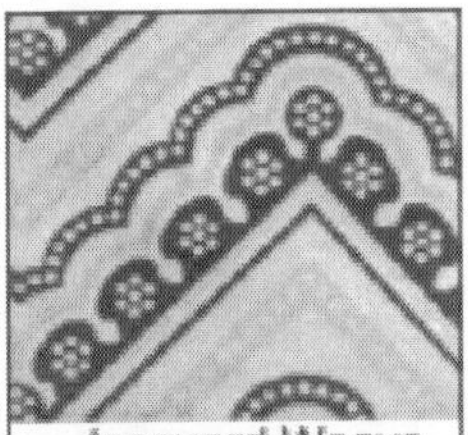
Jacquard Weave

Fabrics with elaborate designs are woven using the Jacquard loom attachment. It was introduced by Joseph Marie Jacquard in 1801. The Jacquard mechanism has the ability to control every warp yarn instead of a series as in regular harness looms.

The machine is very big and very expensive. The pattern for the Jacquard loom is transferred to a series of perforated cards, one for each filling pick in the pattern. The card is punched so as to permit the needles on the machine to be raised to pass through the card. The shed is formed and the pick passes through each card stops on the cylinder for its particular pick, moves on, and new card takes its place. This process continues until all the cards are used. Thus when pattern is over the cards start for the next till the entire cloth is woven. Thus extremely complicated and decorative woven designs are produced using jacquard loom attachment. Examples include brocade, damask, tapestry,. terry-cloth towels.

Swivel Weave

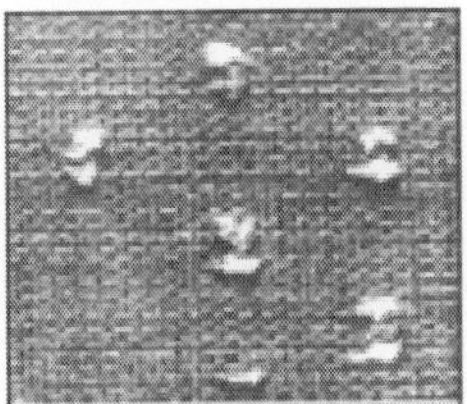
Swivel Weave

The Swivel weave differs from lappet in that designs are produced by extra filling yarns. Separate shuttles are placed at each point where the design has to be made.

The shed is formed by the pattern, where the shuttle carries the yarn through the shed, the distance of the pattern. The extra filling floats on the back of the fabric, the long floats is cut away after weaving is completed. Example: silk sarees.

Lappet Weave

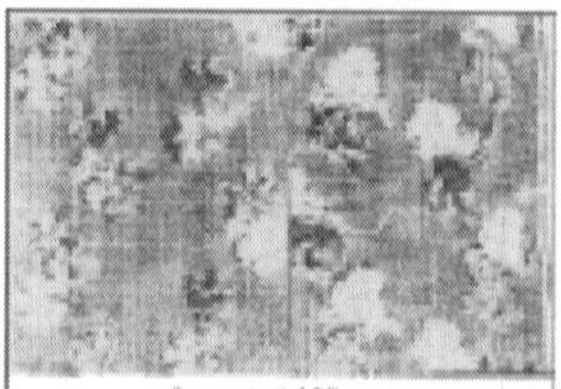

Lappet Weave

In lappet weave an extra warp yarn is introduced in a way so as to create design on the base, fabric. Patterns are woven in the fabric by an attachment to the loom. Long floats formed on the back of the fabric are cut away, if floats are small they are usually left. Example: fancy sarees.

Crepe Weave

It is obtained by making variations in the plain or satin weave with the use of dobby attachment on the loom, to give the surface a crinkled or pebbly appearance. The crape is made by weaving highly twisted yarns alternatively with yarns of ordinary twist, which ordinarily do not have any crepe effect. Both plain weave and satin weave floats in the wrap or filling yarns may be used to form irregular patterns which would produce a rough textured yarn yarn material with crepe weave also known as Granite weave.

TYPES OF ENERGY USED IN THE TEXTILE INDUSTRY

Table. Energy Consumption Share of Each Specialized Technical Field in the Japanese Textile Industry

Unit: Million Yen

Specialized Technical Field	Fuel	Electricity	Total Percent	Share
Fibre production	32,551	21,498	54,049	21.0
Spinning	3,224	44,262	47,480	18.4
Twisting	291	1,660	1,879	0.7
Textured-yarn production	120	1,543	1,663	0.6
Weaving	4,467	24,848	29,315	11.4
Knitting	4,059	11.709	15,.858	6.1
Dyeing	37,661	28,412	66,073	25.0
Clothing yarn manufacturing	8,240	15,420	23,660	9.2
Others	5,959	12,000	17,959	7.0
Total	**96,500**	**161,442**	**257,942**	**100.0**

Note:

Calculated from the Tabulated Industrial Statistics

In general, energy in the textile industry is mostly used in the forms of: electricity, as a common power source for machinery, cooling and temperature control systems, lighting, office equipment, etc.; oil as a fuel for boilers which generate steam; liquified petroleum gas; coal; and city gas.

Table compares the energy consumption shares of various specialized technical fields and it can be seen that energy consumption is relatively high in the fields of dyeing and finishing, fibre production, spinning, weaving and clothing yarn yarn manufacturing. Table summarizes recent trends in the use of various energy sources in the fibre production and dyeing and finishing divisions of the textile industry, where the energy consumption ratio is relatively high.

Technical Field	Fiscal Year	Electricity (10^3 KWh)	Kerosene (KI)	Gas Oil (Ki)	Fuel Oil A (KI)	Fuel Oil B (KI)	Liquefied Petroleum Gas (t)	Coal (t)	City Gas (10^3 m^3)	Total (10^6 Kcal)
Fiber Production	1986	3,839,760	4,067	10	6,182	16	719,547	712	517,177	15,631,924
	1987	889,582	5,587	17	2,221	..	635,941	699	706,402	15,854,930
	1988	3,882,074	3,821	14	12,209	..	622,291	727	788,698	16,012,279
	1989	3,887,386	2,736	3	28,520	..	544,848	656	838,510	15,458,822
	1990	4,066,002	3,571	5	35,503	..	549,098	662	871,265	15,358,554
Dyeing	1986	1,325,692	24,250	400	39,770	11,751	816,942	83,678	7,187	12,103,452
	1987	1,351,241	22,565	406	50,351	11,906	806,477	87,978	30,084	12,322,468
	1988	1,343,959	21,879	442	97,509	12,323	775,313	95,349	37,846	12,610,598
	1989	1,397,260	21,470	471	180,880	10,507	708,562	104,125	37,580	12,706,610
	1990	1,398,127	20,702	614	199,846	10,068	678,942	109,281	32,827	12,724,916

Note:

Calculated from the Annual Report on Textile Statistics. Dyeing Represents the dyeing process including finishing, for both woven and knitted materials.

While the significance of energy conservation awareness is relatively easily understood at home, when a programme is introduced into a factory to promote it, its thorough implementation tends to be delayed at an early stage. Therefore, for its actual course of implementation, it is desired to devise company-wide coordinated measures similar to QC activities at factories. Also, in order to promote energy saving measures efficiently, it is found to be effective to separately consider general management techniques for "rational use of energy" and process-specific techniques to be developed in each specialized technical field.

ENERGY CONSERVATION MANAGEMENT TECHNOLOGIES

Since energy management is relevant to a wide range of departments within a company, it is necessary to enhance the awareness, improve the knowledge and obtain the participation and cooperation of everybody involved in the production process. Therefore, while it is necessary for engineers and technicians with specialized technical knowledge to play a central role in energy conservation efforts, the implementation of an energy conservation programme itself should not be left to a handful of specialists or specialized sections. Rather, it is desirable to address the task company-wide, for example by setting up an 'Energy Management Committee'.

Due to its nature of operations, the share of lighting in electricity use is relatively high. After the switch from tungsten bulbs to fluorescent lamps achieved considerable electricity savings, electricity-saving fluorescent lamps have been developed and marketed for further improvements, including those capable of reducing electricity use by several percent for the same level of illumination.

In general, the effectiveness of illumination is influenced by various factors, such as the intensity of light source, the reflection coefficient and shape of the reflective fitting, the layout of the room to be illuminated, interior finish, colour and the distance from the light source. Therefore, it is important to re-examine whether the light source is utilized in the most efficient way and take electricity saving measures, if necessary, such as reducing the number of lamps in use by switching from global lighting to local lighting as much as possible.

The textile industry uses a vast number of relatively small electric motors. Notably, while a conventional machine was driven by a single motor with the generated mechanical power transmitted to various parts of the machine in a collective manner, many modern machines utilize multiple motors with a control board controlling the movement of each motor, which is directly coupled to a machine part to drive it independently from others. This is also a rationalized feature in terms of energy saving. However, regarding the selection of each motor, emphasis has been placed on mechanical performance, resulting in a motor with an excessive capacity. This leaves considerable room for re-examination from a energy conservation point of view.

In the textile industry, electric heating has largely been replaced by other methods for some time in order to achieve cost reductions. However, since electric heating only requires a small initial investment as a result of convenience and simplicity in equipment construction, it is still used for small capacity local heating purposes.

Therefore, it is desirable to conduct a comparative investigation into alternative heating methods, such as far-infrared radiation heating, high frequency dielectric heating and microwave heating.

As is described before, fuels utilized in the textile industry have already gone through a switch-over from coal to oil. More recently, efficient energy use is under investigation, including the revival of coal on the way to a further move from oil to liquefied and city gases, while reflecting various fuel prices. In selecting fuels, those with good flue gas characteristics in addition to high calorific value and ease of combustion are desired, so that air pollution can be prevented as much as possible.

By and large, boilers used in the Japanese textile industry have experienced a change from Lancastrian-or Scotch-type tubular or smoke tube to water-tube boilers. As a result, boiler efficiency has improved from the conventional 60's to 70's of percentage points to as high as the 90's. Since high performance boilers are prone to a rapid growth of scales inside their water tubes, feed water management becomes important. Furthermore, these boilers have small amounts of retained water and high evaporation speeds so that many aspects of their operation are yarn yarn automated, including feed water and combustion management.

The noted feature of steam use in the textile industry is that the amount of steam involved is not so large but the locations where steam is required are widespread so that steam losses due to heat radiation from steam transportation pipes and pressure drops are considerable. Therefore, for steam transportation over long distances, high pressure and smalldiameter rather than low pressure and large-diameter piping is desired, with pressure reducing valves placed as necessary to regulate the steam pressure at the point of use, thereby curbing heat losses. Also, as pressure losses around bends are great, it is desirable to make their radii large. In order to prevent steam leaks from joints due to the thermal expansion of the pipe, expansion joints should be placed where required. Furthermore, in order to maintain the temperature inside the valve, tank and treatment tank as well as the piping, it is necessary to install them heat-insulated, using appropriate heat insulating materials, so as to efficiently use steam while preventing heat losses.

Since live steam is often used in dyeing factories, fluctuations in steam use during working hours are large. On the other hand, since high performance water tube boilers and once-through boilers are designed such that water retained inside the boiler is very little, the boiler cannot react to momentary and sudden load changes, while responding to automatically controlled slow load changes is not a problem. In such a case, a steam accumulator can be installed midway through the heat transporting pipe, between the boiler and the heat consuming load, in order to store excess steam when the load is light by transforming it to heated water. This then transforms the heated water back to steam when the load is heavy in order to reinforce supply to the load. This allows the boiler to continuously operate with the average load and is quite advantageous in view of energy saving.

So far, after its heat energy is consumed, steam has been drained off. However, in view of energy saving, it is necessary to collect and recycle the heat energy carried by the drain water.

In each production process of the textile industry, the heating and cooling of gases and liquids as media of heat are frequently required. This is done through heat exchange between different fluids, and in order to avoid contamination or chemical reaction due to their direct contact, heat exchangers are used to carry out indirect heating and cooling. It is important to use the right heat exchanger for the intended purpose.

Energy saving is an operation to grasp the actual situation of energy use in a factory precisely and quantitatively and to carry out improvement measures in order to rationalize and economize on it. While measuring instruments are needed to obtain quantitative data, it will become more and more important to investigate the use of sophisticated measuring instruments based on recent developments in mechanical and electronic engineering, combined with automatic control systems.

ENERGY USE AND RATIONAL USE OF ENERGY IN PROCESS-SPECIFIC TECHNOLOGIES

Progress in production rationalization is achieved through the implementation of a comprehensive set of measures, including energy conservation technologies as the centerpiece measure, along with time management, labour saving, natural resources saving and space saving. It has been frequently pointed out that, along with management techniques described earlier, the improvement and development of process-specific techniques on energy conservation greatly contribute to the rationalization of production. Here, process-specific techniques relating to energy saving are summarized for each specialized technical field.

Exhibiting relatively large-scale structural forms in the textile industry, this division has already reached a high level of production rationalization; as is well known, it is technologically aiming at diversification into such high valueadded goods as super extra-fine fibre and inorganic functional fibre, commonly referred to as *shingosen.*

In particular, the following techniques relate to energy saving:

- *Raw Yarn yarn material Production Process*: Implementation of energy saving through improvements in the process and reaction conditions
- *Polymerization Process*: Reduction' in polymerization time by means of high efficiency catalysts, polymerization methods, etc.
- *Spinning Process*: Promotion of energy saving through combining the POY (Pre-oriented Yarn: Yarn with some stability with its molecules partially having gone through orientation) and DTY (Draw Textured Yam: false twisted yarn produced while drawing POY yarn) methods and an expanded use in multi-folded spinning yarn.
- *Newly Built Factories*: The factories built during the high growth period have large margins and allowances for production increase so that high losses would result if production decreased. Therefore, suitably sized factories should be constructed.

Regarding technological trends in spinning, moves towards high speed and large package size have been investigated in order to achieve labour saving through as much automation as possible. As a result, energy consumption has been gradually increasing. However, in view of price competition with overseas companies, further labour saving as well as energy saving is desired. Table compares a modern and a traditional factory in terms of electricity consumption for each yarn plant/operation.

Table. Electricity Consumption Per 1000 Spindles for Each Yarn Yarn Plant/Process.

Factory	Modern Factory		Traditional Factory	
Electricity Consumption	**KWh/1000**	**Percent Spindles**	**KWh/1000 Spindles**	**Percent**
Spinning	16.7	7.5	4.2	6.0
Mixing/OpeningCarding	17.7	8.0	4.9	7.0
Yarn yarn plant Combing	10.9	5.0	2.2	3.0
Drawing/Roving	9.1	4.0	2.8	4.0
Ring Spinning	66.1	30.0	34.5	48.0
Finishing	14.7	6.5	4.7	6.0
Subtotal	**135.2**	**61.0**	**53.3**	**74.0**
Air-conditioning yarn yarn plant	85.9	39.0	18.7	26.0
Total	**221.1**	**100**	**71.9**	**100**

Namely, it can be seen that a modern factory as a means of achieving production rationalization requires approximately three time as much electricity as a traditional one, with electricity consumption particularly increasing in the air-conditioning yarn yarn plant. In terms of processing operations, fine spinning, as the main operation of the spinning process, consumes a large amount of electricity. Thus, energy saving measures are required in these fields. For the fine spinning operation, electricity is consumed in driving the spindles, packaging, spinning, drafting, and operating the lifting and cleaning mechanisms. It is desired to curb the increase of electricity consumption as much as possible by setting an optimal condition for each of these electricity usages.

Although as an ideal working environment a room temperature less than 30°C is desirable, in cases where the working environment has been drastically improved in most other aspects with work load also reduced, a slightly increased room temperature may be permitted. As has been reported, there was a case where raising the regulated temperature from 30°C to 32°C resulted in a reduction in the electric power yarn demand of a carrier with a contract yarn yarn demand of some 8,000 kW by 190 kW. Also, there are many instances of seasonal switch-over from a damper to a pulley as a means of readjusting the blown air volume; this is in order to recycle the air sucked from the processing machine for each operation through a filter back to the same room, and it is therefore necessary to recheck the locations of fans for suction and returning. While synthetic-fibre textured-yam is mostly produced with false twisting machines, its history of rationalization is characterized by challenges for high speed operation. As their operating speeds increased, driving and heat-curing motors and and other peripheral equipment became larger, accompanied by an inevitable increase in electricity consumption. Although this may be acceptable as long as the production improvement resulting from a high speed operation covers the increase in electricity costs, reductions in energy cost would surface as an avoidable urgent task, should a sharp increase in electricity charge occur. It can reasonably be said that the major form of energy consumed in the production of synthetic finished-yarn is electricity. Although the amount of electricity consumed in each piece of equipment varies with factory scale and the type of false twist machine, and therefore cannot be treated in a standardized manner, generally accepted average values may be taken as 3.5 kWh/kg for a single heater system and 5.0 kWh/kg for a double heater systemas one report suggests. Of all the energy consumed in finished-yam production, 70Percent is accounted for by false twist machines. Table shows a breakdown of this energy consumption:

Table. Example of False Twisting Machine and Electricity Consumption (kWh).

Processing Machine	Single Heater (192 spindles)		Double Heater (216 spindles)	
Equipment	**Capacity**	**Utilized**	**Capacity**	**Utilized**
Main motor	15.0	8.0	13.5	9.5
Exhausgasmotor	1.5	1.0	2.2	1.5
Yarn sucking motor	2.2	1.5	–	–
No. 1 heater	32.0	16.0	15.0	7.5
No. 2 heater	–	–	11.0	6.0
Total	50.7	26.5	46.7	24.5

Sixty per cent of all energy consumption by a false twist machine occurs in the heater. Therefore, improvements in the heat insulation of the heater and the lowering of heater temperature may be considered as energy saving measures. Since the latter has implications in the characteristics of the finished-yam, whether or not it is adopted should be examined on such occasions as in the development of a new product.

Since air-conditioning plants are designed based on the conditions applicable at the time of installation, it is desirable that they be re-examined against the present conditions. Rationalization in fabric production is such that while various improvements in machinery aimed at high speed operation and labour saving have been carried out, the amount of energy use per unit of the product has gradually increased. Regarding loom design, high productivity shuttleless looms such as water jet, rapier and gripper types have successfully been introduced, with air jet models put in practice in the production area of industrial fabric yarn material. The amount of energy consumed by each loom during its weaving operation can be estimated from the motor capacity and weaving speed.

Conventional shuttle looms are based on the weft-insertion method, incorporating a shuttle zooming to and fro with a large inertia mass (approx. 400) and mounted with extra weft, and they also use energy consuming pirns as an integral part of the machine.

For this reason, the shuttleless looms' contribution to energy saving cannot be regarded as too high. On the other hand, as a large amount of energy is consumed in sizing, as one of the preparatory operations for weaving, the introduction of foam and solvent sizing operations are being investigated. Furthermore, long fibre fabrics using non-sizing filaments have been developed, eliminating the sizing process altogether.

In a reported example, the introduction of a new heat exchanger into a sizing machine with a very poor sealing capability achieved more than 40Percent of energy saving. The share of energy cost in the total cost of production is not necessarily high for the knitting process.

However, of the main production facilities for this process, knitting machines have also been undergoing a shift towards high speed and large capacity and fine gauge features; the current industry trend is for high added-value goods and multi-line, small-volume production based on advanced systems such as computer-controlled pattern making mechanisms.

Therefore, a potential tendency for increased energy consumption should be taken into account. As a result, it is desirable to conduct a comprehensive re-examination of the production schedule along with the implementation of actual energy conservation measures in order to reduce or restrain the share of energy cost in the total production cost.

It is very important to advance energy conservation in the dyeing and finishing field, which has a high energy consumption share in terms of both the amounts of money and energy used.

The dyeing and finishing process consists of many interwoven unit operations, and it is well known that the process generally goes through repeated wet and dry operations. The heat balance of a unit operation can mainly be considered as the difference between the total supplied heat on the one hand and the sum of the heat required by the system and various forms of heat losses on the other. Figure graphically summarizes the major factors in a thorough implementation of energy savings.

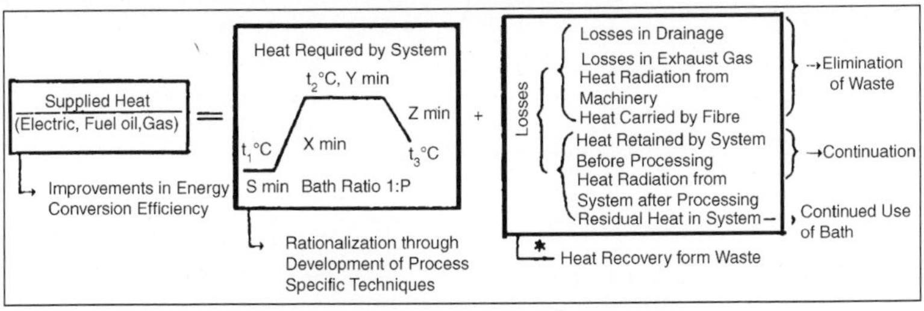

Fig. Heat Balance in Unit Operation.

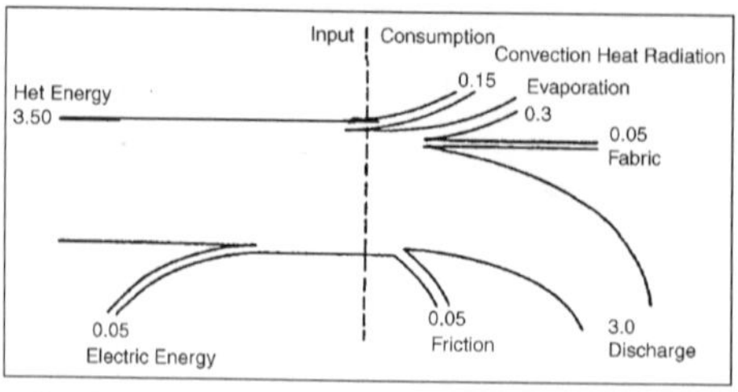

Fig. Shows an Example of Heat Balance in a Continuous Water Cleansing Machine.

Table. Relationship between Production Rationalization Techniques and Energy Savings

Produciton Rationalization Technique	Mechanism	Effect
Time saving	• High speed processing of unit operations • Reduction in waiting • Elimination or merger of unit operations	Reductions in energy use Per unit operations time between unit through an improvement in operations productivity
Labour saving	• Implementation of automation • Strengthening a reduction in the failure rate	Reductions in the frequency of reprocessing through colourimetric management
Energy saving	• Reduction in bath ratio • Reduction in treatment time • Reduction in margin of temperature rise • Re-examination of drying method • Switch to non-water-based operations	Reductions in energy cost
Conservation of natural resources	• Utilization of continuous bath	Utilization of system's residual heat
Space saving	• Construction of modern factories	Improvements in factory-wide energy saving effects

This clearly illustrates the importance of the development and utilization of processspecific techniques, apart from the already-described management technologies. Table shows that the implementation of production rationalization eventually relates to energy conservation. The following are brief explanations of typical examples.

As the processing machines become faster they also become larger. This means the energy consumption per unit length of time will increase, but generally it will accompany a reduction in energy consumption for the treatment of a unit amount of fabric. Table shows an example of this situation. Therefore, it follows that, as long as the product turnout is maintained, continuous processing with a large machine will be more effective in achieving energy conservation.

Table. Length of Mercerizing Machine and Productivity and Energy Consumptions

		38	47	56
Total machine length (m)		38	47	56
Treating speed (m/min)		40	60	80
Treating time (set/m²)		1.5	1	0.75
Product Turnout (m)	1 hour (Operation rate (100%) 8 hour (Operation rate (58%) 16 hours (Operation rate 90%) 24 hours (Operation rate 95%)	2,400 19,200 (16,300) 38,400 (34,500) 57,600 (54,700)	3,600 28,800 (24,500) 57,600 (51,800) 86,400 (82,000)	4,800 38,400 (32,600) 76,800 (69,000) 115,200 (110,000′
Con-Sumption rate	Water (m³) Steam (kg) Electricity (AC motor (KWh) NaOH (30°B′e) (I	10.5 1,075 (82,691) 21.0 288	14.0 1,500 (115,381) 38.0 432	19.0 1,850 (142,311) 50.0 576
Energy and raw Material required For treating m′ of fabric	Water Steam Electricity Amount of Energy (Kcal) NaOH	0.0044 0.4479 (0.0341) 0.0088 337.2 0.12	0.0039 0.4167 (0.0321) 0.0106 322.0 0.12	0.0040 0.3854 (0.0301) 0.0104 302.8 0.12

Note:

1. The bracketed entries under production trunout show approximate figures which would result from the respective operation rates.
2. The bracketed entries under steam show equivalent fuel consumption figures would be needed if a boiler with a evaporation ratio (= evaporation/fuel consumption) of 13 was used.

The currently employed dyeing techniques are based on unit operations which have been developed and established for use with natural fibre. For this reason, the traditional standard treatment steps are often applied to blended yam fabrics as a matter of principle. However, through omitting or merging some of the unit operations according to the usage of the product and considering the characteristics of the coexisting synthetic fibres, it becomes possible to achieve energy conservation.

It is easy to understand that a reduction in water use will contribute to energy conservation in the dyeing process which consists of various wet treatment and drying unit operations. It is especially desirable to curb the water consumption because it is linked to the overall water supply cost including that of drainage. For the reduction of the processing bath ratio, it is necessary to investigate the following measures:

Table. Unit Operations in Preparatory Process of PE/C Blended Fabric and its Processing Characteristics

Combination and Ordering of Operations	Number of Units	Processing Characteristics
Singeing→ Desizing→ Drying → Heat setting → Scouring → Mercerizing → Drying	10	(1) Cumulative Method (2) Strict Control (3) Excessive drying operations
Singeing→ Desizing→ Scouring → Bleaching→ Drying→ Mercerizing→ Drying→ Heat Setting	8	(1) Fewer drying operations (2) Difficulty in controlling scouring and bleaching (3) Great number of wrinkles form processing
Heat setting→ Singeing→ Desizing→ Scouring→ Bleaching→ Drying→ Mercerizing→ Drying	8	(1) Difficulty in desizing (2) Insufficient effectiveness in processing (3) Wrinkles from processing prevented
Singeing→ Desizing→ Scouring → Bleaching→ Drying→ Heat setting→ Mercerizing→ Drying	8	(1) Satisfactory desizing (2) Fewer drying Operations (3) Caution against wrinkles from processing
Singeing→ Mercerizing→ Desizing→ Scouring→ Bleaching→ Drying→ Heat setting	7	(1) Fewest drying operations (2) Difficulty in desizing (3) Small mercerizing effect

Treatment with Low Bath Ratio [sssh]

In general, dyeing and finishing methods are classified into the batch and continuous processing methods, and it is recommended to use the latter method where a low bath ratio is desired.

However, depending on the details of processing requirements, there are often instances in which the batch method has to be employed. In such cases, batch processing machines which allow lower bath ratios such as the jigger, wince, beam, pad roll and jet flow types should be selected as far as the circumstances permit.

Figure shows the relationship between the bath ratio and the production cost for the wince dyeing of a cotton fabric with a reactive dye (bright red, medium shade).

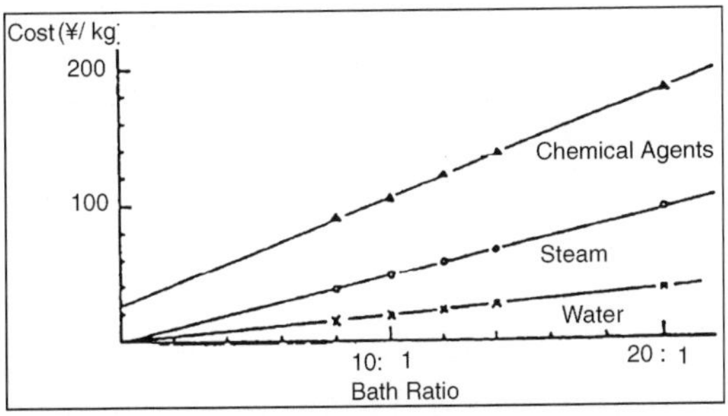

Fig. Relationship between Bath Ratio and Production Cost in Reactive Dyeing

It is easy to see that the bath ratio has a direct influence on the production cost.

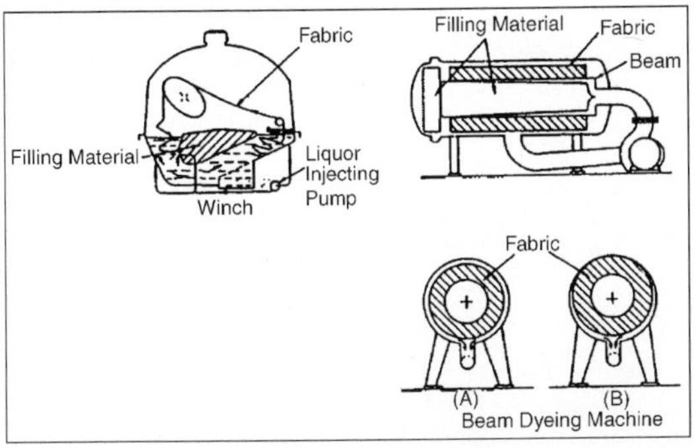

Fig. Example of Low Bath Ratio Operation of Existing Processing Equipment through Insertion of Filling Yarn yarn material

In order to use a lower bath ratio with the existing machinery intact, a method to insert a filling yarn yarn material inside the processing equipment has been proposed. It has been reported that with this method, the bath ratio of a wince could decrease from 25:1 to 17:1, and for a beam a reduction was possible from 15:1 to 12.5:1, or even down to as low as less than 10:1 where the axis of the beam was made off center with respect to the container body, thus increasing the batch-up volume as shown in (B). More recently, low bath ratio processing machines which are built in with the above mechanisms have been developed and put on the market.

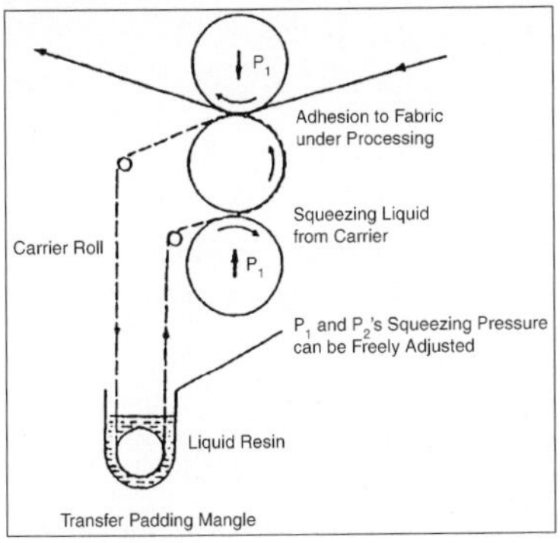

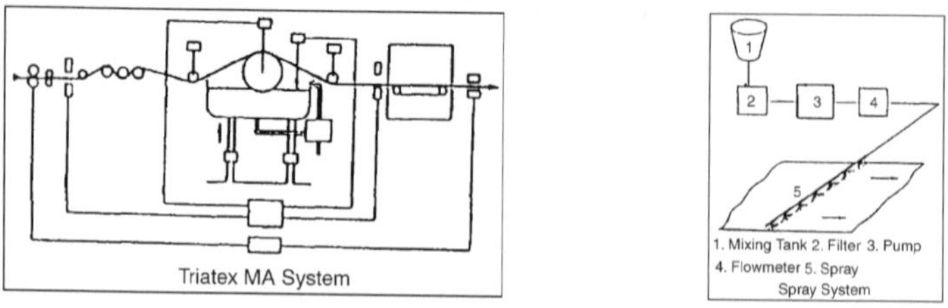

Fig. Typical Low Add-on Machines

Several types of processing equipment with a mechanism to uniformly apply the fabric with a minimum amount of liquid necessary in semicontinuous and continuous processing systems are known to be typical examples of energy conservation techniques.

Extension of Foam Processing Technique [sssh]

A typical example of foam processing liquid applying equipment. The foam processing technique is used for the preparatory, dyeing, textile printing and finishing processes, with confirmed effects of promoting energy conservation, but it is desirable to examine details of usage and other practical conditions prior to application.

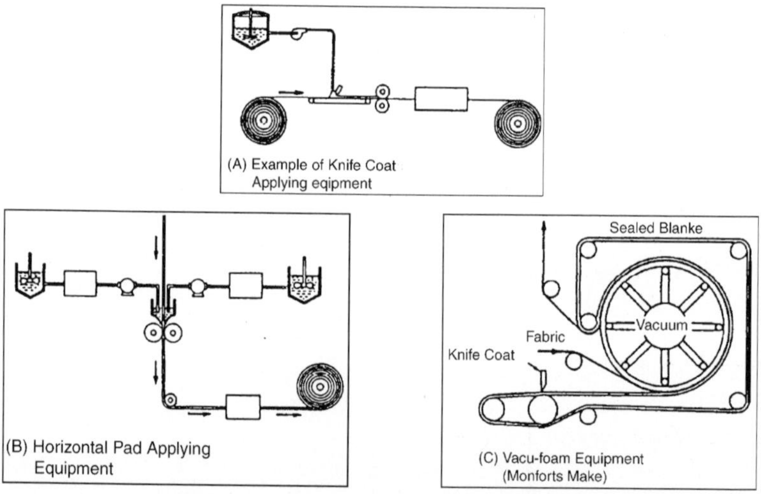

Fig. Typical Foam Applying Equipment

As has been pointed out, being a time saving technique aimed at improving productivity, continuous operation with an increase in the size of the processing machine can also further energy conservation. Likewise, for batch processing, the number of technical fields is increasing where the promotion of energy conservation is desired through a reduction in processing time. This tendency becomes more pronounced as the needs of the market become sophisticated. Techniques to accelerate the processing effect with rapid dyeing and plasma treatment are typical examples.

Rapid dyeing which can drastically reduce the dyeing time and achieve remarkable time savings can also achieve great energy conservation effects when applied to polyester. In order to attain these effects, it is necessary to select dyes with assistants and provide appropriate dyeing equipment. Combined with the foam processing technique, the rapid dyeing technique may also have a potential of leading up to the development of new practical dyeing techniques. Figure shows the position of the dyeing technique in the overall processing bath ratio reduction technique.

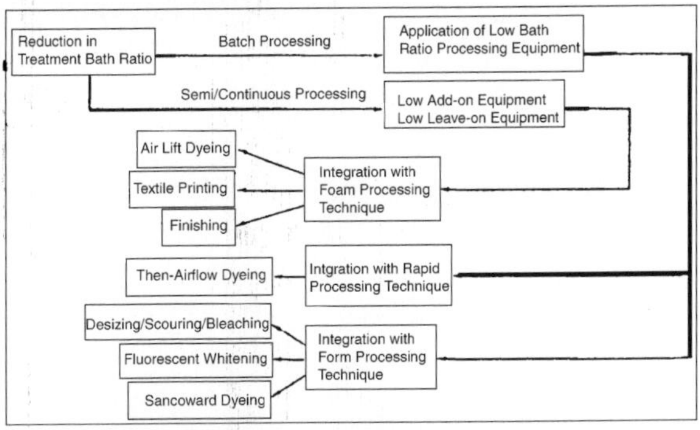

Fig. Situation of Dyeing Technique in Processing Bath Ratio Reduction Technique

Aiming at a reduction of processing time, the combined use of a number of new techniques are being studied and it has been reported that processing with plasma, ultrasound, magnetism and radioactive rays accelerates processing effects. Various methods are being investigated to reduce processing time through accelerating processing effects using these techniques in preprocessing, postprocessing, simultaneous processing, etc.

In many cases, unit operations of the dyeing process are carried out at high temperature. Therefore, to reduce the required margin of temperature rise from heating is very important in view of achieving funda-mental energy conservation along with reductions in processing time. These measures need to be addressed from the following two viewpoints:

If the temperature of the inlet water to be used in the dyeing process becomes relatively higher, the amount of energy to be consumed in raising it to the predetermined value will be reduced. For that purpose, cooperation within a company, or that involving more than one company (using low temperature inlet water at the dyeing factory for cooling purposes, and the high temperature discharge from the cooling system for dyeing purposes) should be investigated, as well as the utilization of natural resources (for example, geothermal and solar energy)

It is important to continue with technological development aimed specifically at lowering the processing temperature along with raising the inlet water temperature. It would naturally involve the integration of this method with the processing speed acceleration techniques. Low temperature scouring, bleaching, dyeing and curing techniques are some of the practical examples of this.

An important consideration along with the reduction of the processing bath ratio is a re-examination of drying operations. A drying operation is, in principle, inserted after every other unit operation and is an important operation which not only determines drying efficiency as such but also has a direct influence on the morphological stability and texture of the final product. For this reason, various types of drying equipment have been selected and put to practical use, depending on the fibre yarn yarn material and form involved. In view of implementing energy saving measures, it is particularly important to investigate the following three items:

A study of typical preparatory process configurations reveals that drying operations are involved in relatively high numbers, ranging from one to four units or 14Percent to 40Percent of all unit operations in the entire process. Therefore, combinations of unit operations should be sought after such that drying operations between standard unit operations can be eliminated as much as possible. It is especially necessary to cut down on drying operations in preparatory processes which will not directly affect the product's performance or appearance quality. However, an operation which would have come after an eliminated drying operation would have to treat wet fabrics, thereby necessitating special measures that would enable wet on wet treatment.

It is desirable to investigate possible improvements in the drying efficiency in terms of efficiencies of both the dewatering and drying steps. While it is well known that the most efficient methods of dewatering and drying are by means of a mangle and a cylinder dryer, methods which have been practiced for a long time, they are also known to have limitations in terms of applicable fibre materials and forms. It is necessary to investigate new drying methods (high frequency drying, microwave heating, far infrared radiation heating, etc.) together with other measures such as utilizing vacuum liquid removal, adding a drying-facilitating organic solvent to the treatment liquid, and combining foam treatment systems with non-foam ones.

Along with active energy saving measures, it is important to carry out the collection and recycling of the energy used in unit operations. The collection of heat energy should start with thoroughly grasping the basic energy balance of each unit operation.

In the dyeing process, although water has been used as the only abundant and cheap resource so far, it is becoming difficult to obtain high quality water in large quantities at a low cost. The worsening of river pollution coinciding with an increase in population density is inevitably creating a situation where the cost of water will gradually increase, including the investment for improvements in waste-water treatment facilities. In addition, although dry-system processing has been contemplated for a long time due to the fact that most energy is consumed in the heating and evaporating operations, it has to date only been applied to a specific area on a limited scale. However, it is a technique which deserves attention as a promising process in the mid-to long-term future. This technique has the following two variations:

While the solvents to be used for dyeing processing are categorized into four main groups—halogenated hydrocarbons, petroleum derivatives, aromatics and oxygen-containing solvents—halogenated hydrocarbons are generally recommended as they do not cause a fire or explosion (provided that thorough countermeasures to groundwater pollution are taken). It is well known that in terms of energy conservation, these solvents have an advantage over water-based ones in all of these aspects: specific heat, latent heat for evaporation, heat needed for evaporation and evaporation speed. There are a number of proposals for solvent scouring, solvent dyeing and solvent finishing, including those already put to practical use as a differentiating technique.

Liquid ammonia is one of the agents being considered for dyeing applications as a inorganic solvent.

Of its typical processing techniques, liquid ammonia mercerizing and liquid ammonia dying are given particular attention.

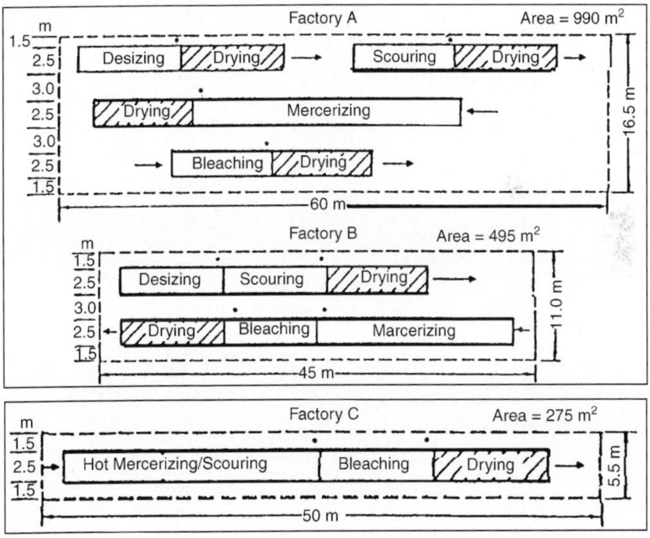

Note:

- *Factory A*: Although it has the same types of machines as Factory B, it has introduced a drying operation for each unit operation, emphasizing flexibility.
- *Factory B*: Considers the production rationalization resulting from continuous processing./
- *Factory C*: It has the same production capability as Factory B, but has reduced the preparatory process related to mercerizing.

Fig. Space Saving Resulting from Introduction of Hot Mercerizing.

Table. Compares Operating Conditions of the Same Three Factories.

Factory	Investment	Area (m^2)	Workers (person)	Electric Power (kW)	Steam (kg/hour)	Water Supply (1/hour)	Waste Water (1/hour)
A	109%	990 (200 %)	14 (155 %)	242 (114 %)	5,727 (130 %)	14,535 (100 %)	14,535 (100 %)
B	100%	495 (100 %)	10 (100 %0	216 (100 %)	4,386 (100 %)	14,535 (100 %)	14,535 (100 %)
C	67%	275 (55.5 %)	5 (50 %)	114 (52 %)	2,500 (57 %)	7,785 (54 %)	4,785 (54 %)

While textile yarn yarn manufacturing techniques which promote the conservation of natural resources include grease refinement from raw wool, the collection and recycling of warp sizing agents, and the re-use of alkaline waste liquid arising from the mercerizing process in a scouring bath, in terms of energy conservation the use of continuous baths which utilize the residual heat of the system are particularly important in view of energy conservation. If the continuous use of a processing bath is introduced with the necessary conditions being met, thus allowing only those materials consumed in the dyeing process to be replenished, in particular with the unit operations designed for the batch method, it will greatly contribute to the recycling of heat energy in addition to achieving the conservation of natural resources and the rationalization of countermeasures to water waste.

In the dyeing process with a high heat consumption, the use of the continuous bath deserves particular attention as a technique whose practical application is an urgent task to help implement the remaining rationalization measures.

With an expansion in the practical use of knit mercerizing and ammonia mercerizing, the characteristics of the hot mercerizing technique is also attracting attention.

It has been pointed out that, since poor uniformity associated with the traditional mercerizing due to the hydrophobic nature of cotton grey fabric is dissolved in hot mercerizing which uses heated sodium hydroxide, the rationalization of the preparatory process can be greatly advanced. Figure shows an example of comparing some space saving effects achieved by the introduction of hot mercerizing.

At the time a new space-efficient factory is being built, it is possible to incorporate a programme to introduce such facilities as to be complete with factory-wide energy conservation measures.

The energy consumption share of the clothing yarn yarn manufacturing division which consists of large numbers of small-sized companies and their employees in the overall textile industry is not necessarily low but the ratio of energy cost to the total cost is relatively low. However, the energy cost forecast is inevitably a gradual increase under circumstances where the production of high value-added goods is required, along with the implementation of labour saving measures, as a result of the challenging market environment characterized by personalized and diversified consumer needs, high yarn yarn demand for quality goods, short product cycles, etc. Therefore, it is desirable that a comprehensive rationalization programme be investigated apart from reductions in energy consumption.

BASIC CLOTHING

In general, it is well known that the developmental process of a textile market moves from a product-oriented stage, where all goods produced are sold to a consumer-oriented stage where the desired goods are only those which will satisfy the consumer's yarn yarn demand, comprising his wants and purchasing power.

This is via a mass-consumption stage which is supported by a mass-production system.

The characteristics of textile products demanded in the markets in these different developmental stages may be classified into price-sensitive mass-production type basic clothing, which applies to the earlier stages, and multi-line, small-volume production type fashion clothing which pertains to the last stage and needs to satisfy each individual consumer's taste.

Assuming the amount of fabric necessary for people to live a physically comfortable life in a given environment can be derived from the weight of clothes which is enough to keep a certain level of body temperature, given such factors as air temperature, humidity and wind velocity prevalent in the region, basic clothing can be regarded as a cumulative total of such clothes.

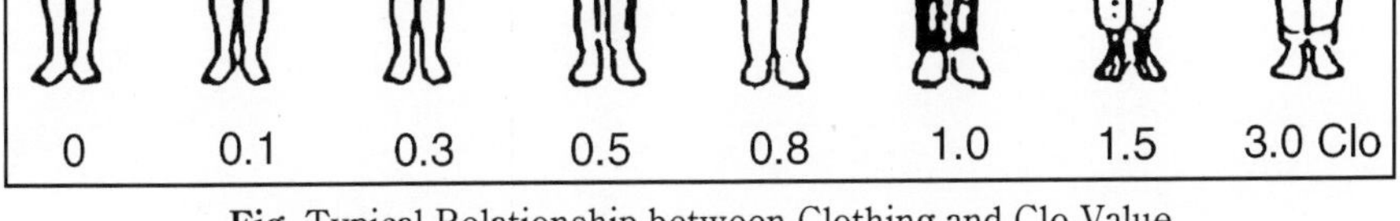

Fig. Typical Relationship between Clothing and Clo Value

Figure Summarizes the Relationship between Some Clo Values and Examples of the Corresponding Clothes. Table. shows the results of calculations to find the weight of clothes required under average climatic conditions in summer and winter in three Japanese cities.

Although the ultimately required quantity is difficult to determine, the per capita annual textile consumption (kg/person) can be estimated through multiplying the above weight by coefficient a (obtained from the service life of a textile product and its number of units demanded per annum).

Table. Example of Effect of Climatic Conditions on Clo Value and Weight of Clothing.

Region	Season	Temperature (°C)	Humidity (Percent) (m/sec)	Wind Velocity	Clo Value Clothing (g)	Required Weight of
Sapporo	Jan	–4.9	74	2.3	4.1	4005
	Aug	21.3	79	2.7	1.2	1189
Tokyo	Jan	4.7	53	3.1	3.1	3034
	Aug	26.7	75	3.0	0.6	607
Fukuoka	Jan	5.7	67	3.7	3.0	2937
	Aug	27.3	76	2.9	0.6	607

Notes:

The clo values are shown assuming an average skin temperature of 33.5°C.

One unit of clo value is 0.155°C/W/m^2.

The required weight values are calculated with Hanada's formula for female clothing:

Y=0.00103W -0.025

Where Y: Clo Value

W: Total Weight of Clothing

FASHION CLOTHING

Although it is even more difficult to define the amount of fabric consumed in fashion clothing, a market which has satisfied the yarn yarn demand for minimum basic clothes has a tendency to shift its emphasis towards textile products with stronger fashion overtones and grow rapidly.

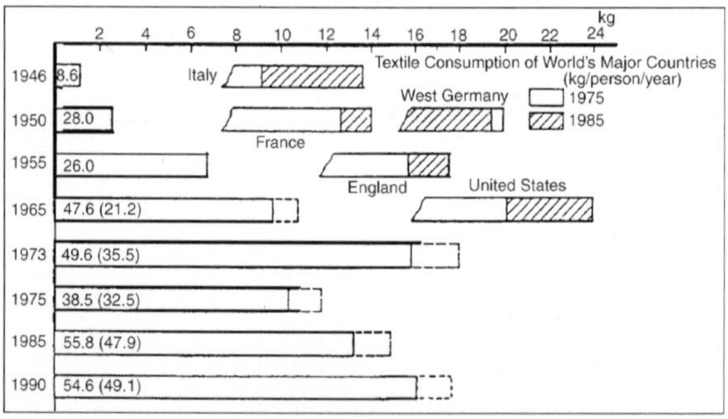

Notes:

- The figures inside the bar graphs with and without brackets are the shares of chemical and synthetic fibres in the total consumption, respectively.
- The broken-line portions of bar graphs represent the consumption of textile goods to be used as industrial materials.
- The sources are the Annual Statistics on Textile Goods for Japan and FAO data for other countries, respectively.

Fig. Changes in Textile Consumption of Japan and World's Major Countries

This is illustrated by Figure, showing chronological changes in textile consumption in Japan and the world's major countries, a graphically expressed correlation between per capita textile consumption and GNP.

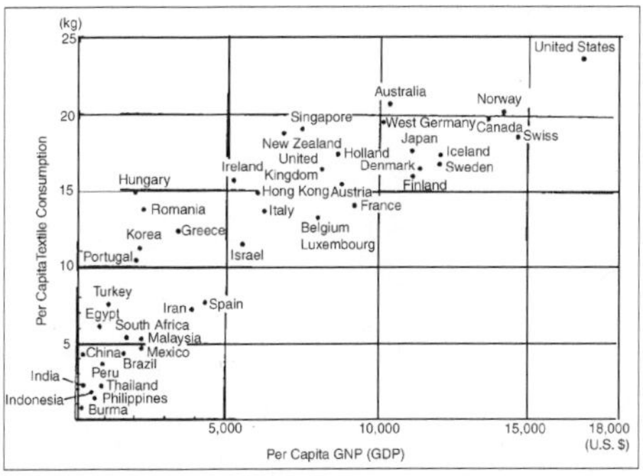

Notes:

- Data on developed and developing countries are for 1985 and 1984, respectively.
- Entries for Japan, the United States, the United Kingdom, France and Italy reflect per capita GNP, while those for China and other countries are based on per capita income and per capita GDP, respectively.

Fig. Correlation between Per Capita Textile Consumption and GNP (GDP)

WORLD'S TOTAL TEXTILE YARN YARN DEMAND AND PRODUCTION BASE DISTRIBUTION

Figure shows the relationship between world population and the total textile yarn yarn demand. Assuming a global environment in which world population will grow from the present 5.4 billion to 10 billion in 2050, and further to 11.6 billion in 2150 when it is expected to reach a static state, the total textile consumption is forecast to double, even using the current figure of per capita annual average textile consumption (8kg/person).

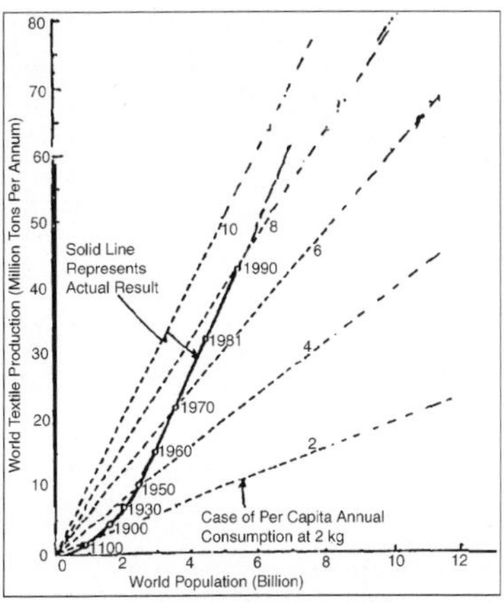

Fig. Relationship between World Population and World Textile Consumption and Production

The textile industry is traditionally regarded as a typical labour intensive industry developed on the basis of abundant labour supply and has a tendency to expand to overseas markets once the domestic yarn yarn demand is satisfied, as illustrated by examples of the established textile industries of many developed countries. For this reason, even when the textile industry of a specific country is to be examined, it is widely recognized that a business strategy taking into consideration the global textile industry setup is very important. Figure shows an example of the schematic representation of such a global setup of textile industries.

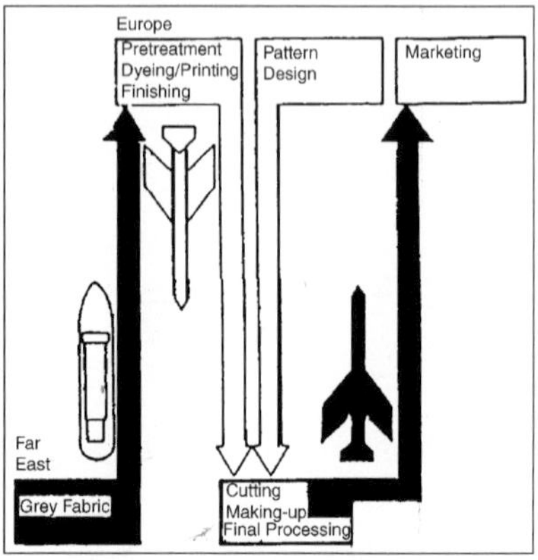

Fig. Example of Schematic Representation of World Textile Industry

CHARACTERISTICS OF TEXTILE MARKET

Most textile consumption takes place in apparel products, and the general tendency is that where further expansion is intended, it will be carried out through the development of industrial applications in which textile products are used as production materials (ranging from fishing nets, tire cords and canvas cloth to geotextile). Therefore, in order to study the textile yarn yarn manufacturing industry, it is generally sufficient to consider the production of apparel goods.

There are a number of intertwined technical factors involved in various subprocesses or specialized technical fields which make up the overall production process that fabricates and reshapes the raw fibre yarn yarn material into the final textile product to be used by the consumer.

In Fiscal Year 1989, the textile industry had a share of 4.4Percent in shipment value.

14.8Percent in the number of businesses (20Percent), and 10.4Percent in the number of employees (23.3Percent) in the overall Japanese yarn yarn manufacturing industries, characterized by a gradual decline in relative share, although in terms of absolute shipment value it has actually expanded to 1,053Percent of the Fiscal 1955 level. As has been frequently pointed out throughout its development, the Japanese textile industry has a unique organizational structure consisting of groups of independent companies where all companies in a group belong to one of the specialized technical fields and operate in a horizontal specialization configuration.

A comparison and summary of the major indices of these company groups, classifying them in accordance with their specialized technical fields. From these indices, common company characteristics for each group may emerge.

Namely, in terms of business size, the fibre production and spinning subdivisions are in contrast with the rest of the textile industry where relatively large numbers of small businesses coexist. In addition, these groups of small companies can only stay in business by relying on the supply of abundant cheap labour, exhibiting a legacy of the textile industry's past as a labour intensive industry, even on its path towards modernization.

As for water consumption, the share of the textile industry in the entire yarn yarn manufacturing industries is 5.2Percent for fresh water and 1.1Percent for sea water. Considering the fact that most of this water is used for easy-to-recycle temperature control and cooling purposes, the industry's total water consumption cannot necessarily be regarded as high.

However, as is widely accepted, the dyeing and finishing division is placed in a special position in that its water consumption is mainly for processing and washing purposes.

TECHNIQUES OF FABRIC CONTRUCTION : SOME OF THE IMPORTANT FEATURES OF THESE TECHNIQUES

Fabric is any piece of cloth. Fabrics can be made or constructed by using a number of techniques as given below:

- *Weaving*: Weaving is the most commonly used method of fabric construction. You must have seen a chatai being made. Weaving is similar to it, where two sets of yarns are interlaced with one another at right angles. Weaving gives a firm fabric. Have you heard of fabrics like poplin, denim and cambric? Yes, these are available in all the cloth shops and you must have used them to make your garments.
- *Knitting*: When you knit a sweater, there is normally one ball of yarn which is interlooped to get a fabric. This technique is called knitting and it gives a lot of stretch and easy-care properties to fabric. Knits are mainly used for hosiery. Knit fabrics are specially useful for garments like underwear, T-shirts, socks etc. knitting is construction of a fabric by forming a yarn into interconnecting loops which hangs one upon other. The fabric may be constructed by machine or hand. The quality of the knitted fabric is affected by the quality of the needles used. The thickness, length and the closeness of needle in the knitting machine affects the durability of the knitted fabric. These are two types:

   - WEFT Knitting in which one continuous yarn form courses across the fabrics.
   - WARP Knitting in which series of yarns formed wales in the lengthwise direction of the fabric.
- *Felting*: making clothes from loose fibres is the oldest technique of fibre construction. The primitive nomadic tribes matted the wool fibre into a mass to use it as tent, saddle and other articles because wool is one natural fibre that possesses the properties of being formed directly into a compact mass having strength and durability. the construction of a fabric from loose wool fibres without any interlacing is known as felting.
- *Bonding*: Bonding fabrics are non-woven fabrics in which techniques other than interlacing are employed for adhering the fibre to each other. The fibre may be held together by some mechanical, chemical, thermal or solvent means or their combination. Felted fabric are also nonwoven fabrics but bonding is different from felting in the sense that in felting, fractional force is used as bonding agent whereas in other techniques of bonding, the fibres are to be applied.
- *Knotting*: This technique is being used since ages to secure the ends of woven fabric in a decorative way. The ornamental knotting technique is known as MACRAME. It is a craft threads, cords or ropes which are knotted together in different ways to form various designs and patterns.

In weaving two sets of yarns are interlaced at right angle to one another in an established sequence. *e.g.* 'charpai' being made? On the frame, first one rope is taken and interlooped on two parallel edges and after that a second rope is interlaced perpendicular to it going once over the rope and once under it. This kind of an interlacement gives an even check effect and the weave is quite firm. The weaving of a fabric is also done in a similar way, except for the fact that yarns are used for interlacing and a loom is used to hold the thread instead of a frame.

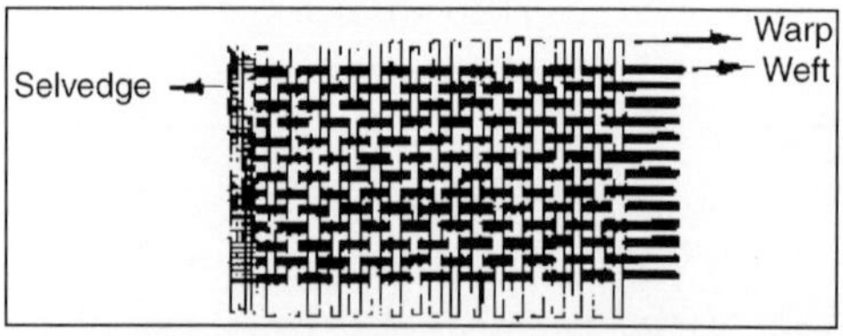

Fig. Common Terms in Textiles

- *Selvedge*: When you examine a fabric, you see two long finished edges, one on each side along the length of the fabric. These edges are called selvedge and give strength to the edges which is important in further processing of the fabric.
- *Warp*: They are the yarns along the length of the fabrics or parallel to the selvedge. They are also called ends.

- *Weft*: Woven fabric, besides warps, another set of yarns move perpendicular to warps. They are called wefts. They are interlaced with warps in a crosswise direction to make a fabric. They are also called picks or fillings.
- *Thread Count*: Some woven fabrics look dense and compact whereas others open. This difference is due to the thread count which refers to the total number of warps and wefts per square inch of a woven fabric. Thread count tells us about the fabric quality and durability. Fabric with a higher thread count is better than fabric with lower thread count. Also, for good quality fabric, warps and wefts should be more or less equal in number.
- Weaving looms can range from quite simple to very complex. Looms have been used to produce cloth for thousands of years, and while technology has improved the loom, the basic strategies and practices remain much the same. Understanding the parts of the loom can help you learn to weave or simply learn a bit more about how weaving works.
- *Warp Beam*: The warp beam is a beam that holds the warp, or vertical threads, on the loom. Depending on the loom, this can be a simple beam or a more complex arrangement allowing for easier warping of the loom.
- *Heddles*: Heddles are made of cord or wire, and are attached to the shaft of the loom. The warp threads pass through the heddles, separating the warp threads to allow the weft threads to pass between them easily.

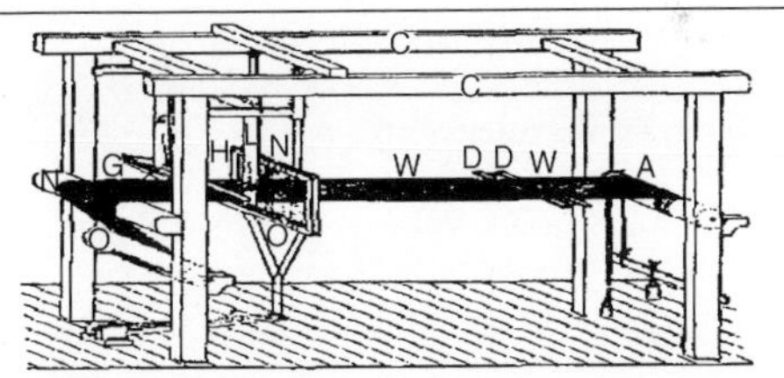

A-Warp Beam; B-Cloth Beam; DD-Less Rods; H-Harness; T-Treddle

- *Harnesses*: The more harnesses or shafts a loom has, the more design possibilities you will have. Most larger looms have four harnesses, and table looms often only one; however, looms of up to 16 harnesses are available.
- *Shuttle*: The shuttle is a bobbin which holds the weft yarns. The shuttle is passed or thrown back and forth to create the weft of the fabric.
- *Beater and Reed*: The beater of the loom presses the newly created weft thread against the already woven fabric using a reed. This process is called battening.

- *Cloth Beam or Cloth Roll*: It is located at the end of the loom on which the fabric is rolled as it is constructed. It also supports the warp yarn at the front side of the loom.

STEPS IN WEAVING PROCESS

Weaving process is completed in four steps:

1. *Shedding*: Raising the required wrap yarns by using the harness and forming the shed.
2. *Picking*: Inserting the shuttle containing filling yarns through the shed.
3. *Battening*: Pushing the filling yarns in place with the help of reed to form the compact structure. Its also called beating up.
4. *Taking Up and letting Off*: Winding the finished fabric on the cloth beam and releasing more of the warp from the warp beam.

PLAIN WEAVE

It is the simplest weave and therefore inexpensive to Produce. Many fabrics that you commonly wear like mulmul dupattas, organdie and chiffon saris are all plain weave. Each and every weft yarn goes alternately under and over the warp yarns across the width of the fabric. If the yarns are close together, the plain weave has a high thread count and the fabric will be firm and will wear well.

Fig. Plain Weave

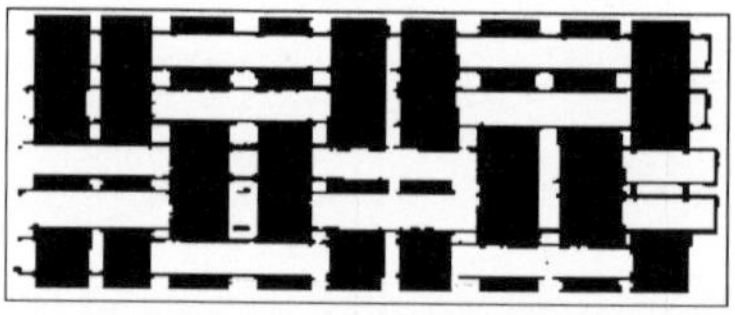

Fig. Basket Weave

Plain weave is of two types:

1. *Rib Weave*: Rib or line effect is created by using thin yarns with thick yarns or single yarns with doubled yarns in any one direction of the fabric.
2. *Basket Weave*: Two or more weft yarns are interlaced as a unit with corresponding number of warp yarns to give a basket like effect. Mattee fabric commonly used for cross stitch embroidery is an example of such a weave.

TWILL WEAVE

This basic weave has a clear diagonal line on the face of the fabric. The denim or jean fabric you wear is twill weave. It is a very strong and durable weave. It is therefore commonly used in men's suit and coat fabrics. Twill weave fabrics show soil less quickly than plain weave.

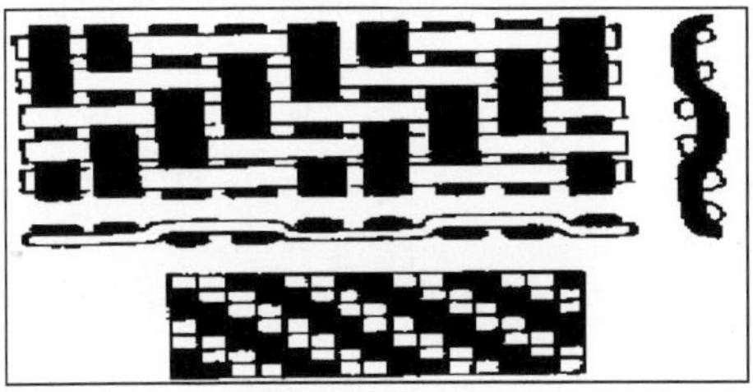

Fig. Twill Weave

SATIN WEAVE

This basic weave has a beautiful shiny surface because of long floats on the surface of the fabric. In the satin weave warp yarns float over several weft yarns before interlacing with a weft yarn and so on. However, the long floats snag easily therefore satin weave is not as strong as plain or twill weave. Corduroy has raised parallel vertical lines. A towel has loops covering its both sides. All these fabrics are made using special looms and weaving techniques. They are obviously expensive fabrics.

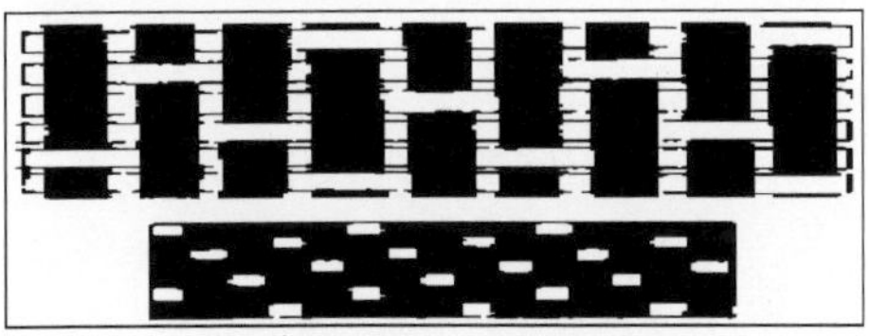

Fig. Satin Weave

CONCLUSIONS

- There is no panacea for achieving energy conservation in the textile yarn yarn manufacturing industry.
- With the actual implementation of an energy conservation programme, it is important to grasp the current level of energy consumption and its actual conditions in detail, set goals (energy consumption and corresponding cost), and achieve the goals through a company-wide effort as far as possible.
- In the textile yarn yarn manufacturing industry, it is important to thoroughly understand that, depending on the trend of the market, the company is targeting, consumer requirements for the textile products to be supplied differ, thereby urging the implementation of energy conservation measures which are relevant to the production of the goods that suit the market.
- Therefore, it is necessary to expect that, when multi-line, small-volume production type high value-added goods are produced, energy

consumption may increase rather than decrease with production rationalization, in contrast with mass-production type goods.

- When differentiated goods are produced, the share of energy costs in the overall production cost should be given importance rather than energy consumption.
- It is reasonable to consider that ultimately desired energy conservation promoting techniques will depend on the development and practical application of innovative technologies in each specialized technical field.

2

Technical and Manufacturing Values in Twist-texturing

Like most developments on the mechanical side of the textile industry, the inventions and developments in yarn texturing have not come as a rational sequence from basic science, through engineering calculations to practical implementation. Empirical advance based on intuitive understanding has been the norm.

This is not to say that the academic research has been wasted. As the science of any aspect of the subject is clarified, this feeds into the qualitative understanding of those concerned with practical operations.

The mathematics may be ignored, but the ideas enter the technical consciousness. Coverage on texturing is patchy and this is shown by the gaps in this chapter. Some topics have been subject to detailed experiment and analysis, but others have been neglected – or been too difficult to deal with.This chapter aims to provide a background of fundamental studies, where these help in understanding the more practical chapters that follow.

Since it is rarely possible to make explicit design calculations, the emphasis will be on principles, and not on mathematical detail, which can be found in the original publications.

FIBRE SCIENCE: HEAT-SETTING

FIBRE STRUCTURE

The action of air-jet texturing does not change the internal structure of fibres, so it is not necessary to discuss the structures of the whole range of fibres which may be air-jet textured.

All the other methods depend on heatsetting, which involves effects at the molecular and fine structure level. Nylon 6 and nylon 66 differ only in a slight rearrangement of the order of chemical groups.This has two consequences: the repeat in nylon 6 is half that in nylon 66, and the nylon 6 molecule is directional. It is probably the latter, which affects the way in which molecules can pack into a crystal lattice, that causes the melting point of nylon 6 to be about 50°C less than that of nylon 66. In order to avoid unnecessary duplication, the word *'nylon'* in what follows covers both 6 and 66. Most of the results quoted will be for nylon 66, so that the above difference must be kept in mind in application to nylon 6.

The polyester shown is polyethylene terephthalate (PET or 2GT), which is the common type for fibres. Other nylons, such as nylon 4, and polyesters, such as 3GT (polypropylene terephthalate) and PEN (polyethylene naphthalate), which have appeared on the market, will not be explicitly discussed, but have only detailed differences in structure and properties. The third type is polypropylene.

Nylon and polyester molecules have features in common. Both contain sequences of about six aliphatic groups, which link to other molecules only through weak van der Waals forces. In nylon, these are (—CH_2—) groups; in PET they are (—CO—O—CH_2—CH_2—O—CO—). These sequences alternate with (—CO—NH—) groups in nylon, which form stronger hydrogen bonds with neighbouring molecules, and with benzene rings in PET, which both stiffen the chain and have a strong electronic interaction with one another. It is this alternation of flexible, inert groups and interacting groups that makes these fibres particularly suitable in textiles.

The long repeat gives a strong tendency to form perfect crystalline register, since a distribution of small, local defects, such as occur with the simple (—CH_2—) repeat of polyethylene, is not a possible form. Polypropylene contains only inert hydrocarbon groups and so is more like polyethylene. The molecules take up a regular three-fold helix in the crystal lattice and irregular variants of this form in amorphous regions.

Drawn nylon and polyester fibres are about 50Percent crystalline, in the sense that their density is half-way between that of a perfect crystal and that of an amorphous network. This is confirmed by other techniques. However, such partial order covers many possible fine structures, and current representations are certainly over-simplified and may miss significant features. The simplest view is a composite of crystallites in an amorphous matrix, but this leaves open many possibilities in size, shape and distribution of crystallites and in linkages between them. In order to get up to 50Percent crystallinity, it is necessary to pack crystallites in a regular pattern.

An example of a *'common working model'*, which appears in similar forms in drawings by different people. The crystallites are brick-shaped and are packed in pseudo-fibrils. At the ends of the crystallites, some chains turn back in folds and others fringe off to form tie-molecules, which lead through the amorphous regions to neighbouring crystallites. Some distinction may be made between amorphous regions between crystallites within a fibril and amorphous regions between fibrils. Another picture, which is based on X-ray diffraction studies and aims to be more realistic. An alternative view is a more uniform structure, with segments locally in crystallographic register. Fibres are oriented preferentially along their axial direction, to an extent that depends on the severity of the drawing process.

The evidence is that crystallite orientation tends to be high, though this may be in localised zones, which vary slightly in off-axis orientation. Orientation of segments in amorphous regions is weaker. It is necessary to consider how this picture, which is based on studies of axially drawn fibres, changes for false-twist texturing.

If fully drawn yarns are used as the feedstock, the fine structure will have been stabilised and only minor changes due to heat-setting will occur. The situation is different for draw-texturing of undrawn or partially oriented yarns. Undrawn nylon is semi-crystalline, but the draw, which is combined with twisting, will lead to major disruption and reformation of the fine structure. For polyester, the effects are even stronger, because undrawn polyester is amorphous; crystallisation occurs during drawing.

Polyester POY, produced at around at 3000m/min, has an incipient crystallinity, which stabilises the structure to a degree that eliminates changes between production and texturing, but the full development of crystallisation occurs during drawing. Above 3500m/min, the density begins to rise rapidly, and, at 6000m/min, is comparable to a drawn yarn.The consequence of the combination of drawing and twisting in simultaneous draw-texturing is that the oriented fine structures must be regarded as following helical lines within the fibres, at helical angles increasing from zero along the axis to a maximum at the fibre surface.

The fine structure will also fit into the bent form of the helical paths of the fibres themselves within the yarn. A larger-scale consequence of the coincidence of drawing and twisting is that, during the process of major deformation, the fibres are subject to the large transverse forces in twisted yarns under tension. This distorts the cross-sectional shape of the fibres, so that they pack into an almost solid assembly. At the centre of the yarn, the form approximates to a regular hexagon, but towards the outside the shape differs.

THERMAL TRANSITIONS

Nylon 66 and polyester (PET) fibres have five important regions of thermal transitions: at around -100°C, 100°C, 200°C, 260°C and 300°C. The temperatures are approximate, because the transitions are broad and the position of the peak of the transition depends on the rate of change and other factors. The lowest would only have direct relevance to textured yarns if these were to be used at very low temperatures, for example in space. The one at around 100°C has considerable practical importance in the behaviour of yarns after texturing. Melting at around 260°C (lower in nylon 6), is an obvious change from solid to liquid, which can be studied in more detail by thermal analysis. The highest temperature transition is chemical degradation, which limits the time at high temperature in melt spinning and, unless precautions are taken, can cause faults in yarn due to the formation of cross-linked gel particles. The effects at around 200°C (lower for nylon 6) are critical factors in the production of textured yarns by heat-setting, but, surprisingly, have largely been ignored in scientific studies of the molecular level. Dynamic mechanical analysis results commonly stop at around 180°C, before the effects are apparent. We know more about the phenomena from responses in industrial operations than from fundamental research.

DYNAMIC MECHANICAL ANALYSIS: TEMPORARY SET

Thermo-mechanical analysis (TMA) consists of applying an oscillating deformation (normalised as strain), in extension, bending or twisting, and measuring the oscillations in the resulting force (normalised as stress). If the material is perfectly elastic, stress and strain will be in phase; their ratio will be the elastic modulus; and there will be no energy dissipation – energy stored on extension will be released on contraction.

If the material is a perfect, viscous liquid, stress and strain will be 90° out-of-phase and the work done by the applied force will be dissipated as heat in both extension and contraction. Polymers are visco-elastic and so lie between these two extremes. The elastic part is given by the ratio of the in-phase components of stress and strain; the viscous part by the out-of-phase components.

A thermo-mechanical transition in the solid state, often referred to as a '*second-order transition*' in contrast to a '*first-order transition*' such as melting, is due to some element within the structure changing from a rigid form to a mobile state. The elastic modulus drops because the material is more easily deformed. The viscous part peaks because there is a sluggish response in going through the transition. This is what is shown by the experimental results, but the practical effect for the performance of textured yarns is that the material will be set in going down in temperature through the transition. The presence of a peak in tand or $E \leq$ is indicative of a temperature range in which setting will occur.

If there is no change in structure, the effect is reversible, and so this is a temporary set, which can be released by heating above the transition temperature in a stress-free state – though, in practice, this may be prevented by interference between neighbouring fibres. In simple polymers, there is one major transition, the '*glass transition*', from the rigid, glassy state to the flexible, rubbery state.

In natural or synthetic rubbers, this can be demonstrated by cooling the material in liquid air; in glassy plastics, such as polystyrene or PVC, it can be shown by heating. In either case cooling will set a bent piece of material in a rigid, deformed state, which will be lost on reheating. If there is cross-linking or if the molecules are entangled and deformation is rapid, the material will show rubber elasticity above the transition temperature, but if the molecules can move as a whole, viscous flow will occur over longer periods of time. The simplest polymers consist of a chain of carbon atoms, which are linked together by tetrahedral bonds.

Above the glass transition temperature, there is freedom of rotation around the bonds, so that the chain acts like a sequence of freely pin-jointed rods. Below the transition, the thermal vibrations are too weak to overcome the energy barriers to rotation, which will be higher when there are side-groups such as benzene rings, and so the chain acts as a rigid, zig-zag rod. Crystallisation complicates the picture, but for polyethylene, which has no large side-groups, the glass transition is at -120°C.

For nylon and polyester, the transition at around 100°C is often referred to as the glass transition, but really the glass-to-rubber transition is split into two parts. For nylon with its (—CH_2—) sequences, the same transition occurs as in polyethylene at low temperature, though it splits into two because of the presence of other groups in the chain.

In polyester the comparable sequences cause a similar low-temperature transition. The importance of these transitions is that they give some limited freedom to the molecular structure, but which, at room temperature, is still constrained by the hydrogen bonding in nylon and the phenolic interactions in polyester.

The material is neither too hard like polystyrene, nor too soft like polyethylene. The extensional properties are right for textile use. Polypropylene has a single transition which spans room temperature. This gives a suitable intermediate stiffness, but at the expense of some sluggishness in response and poor recovery.

The transition above room temperature is due to the onset of mobility at the cross-links. In nylon, below the transition, all the hydrogen bonds are permanently in place, causing the amorphous material to act like a highly cross-linked rubber with a maximum extension of around 50Percent. Above the transition, the hydrogen bonds are in a state of dynamic equilibrium, as they are in liquid water. At any instant, most of them are in place, but they are continually breaking and reforming. A similar effect occurs due to electronic interactions between the benzene rings in polyester. The transition in dry nylon is a little below 100°C, whereas in polyester it is above 100°C. A more important effect is that, due to the absorption of water, the transition temperature drops to near 0°C when nylon is wet. Consequently there is a temporary set when nylon is dried. Temporary set can be avoided in wash-and-dry cycles of polyester, but not for nylon.

CRYSTALLISATION, ANNEALING AND MELTING

The uncertainty, which results from different measurement techniques, in the temperatures at which changes take place is shown by the fact that different books quote temperatures between 250 and 265°C for the melting points of nylon 66 and polyester (PET) fibres and higher still for perfect crystals. At about 240°C, the fibres stick together. This sets a maximum for the temperature at which texturing can be carried out; at higher temperatures, the result is a solid rod. Although melting is not directly involved in texturing, the annealing effects associated with crystallisation are relevant to heat-setting. The phenomena can be studied by measuring the input of latent heat needed for melting (the endotherm) or its evolution (the exotherm) on crystallisation. There are two techniques, differential scanning calorimetry (DSC) and differential thermal analysis (DTA): the former, which is now preferred, increases or decreases temperature at a constant rate and measures the difference in the necessary heat input or output in the sample and a control; the latter measures temperature difference at constant heat input. An important set of studies of multiple melting effects in nylon and polyester was reviewed by Hearle and Greer.

DSC traces for undrawn nylon 66, which had been annealed at 220°C for increasing times, the change in the melting temperature. The sample initially shows a broad melting endotherm with a peak at 256°C, but on annealing, a second endotherm appears at a lower temperature. With increasing time, this lower endotherm grows and increases in temperature, eventually passing the first peak, which has shrunk while maintaining a constant temperature. The first endotherm is referred to as form I and the second as form II. The change from form I to form II also occurs on cold drawing.

The situation is slightly different in polyester, because undrawn material is amorphous. This means that when a sample is heated in DSC, an exotherm appears at a temperature, typically around 120°C, at which crystallisation takes place. However, the melting behaviour is similar to nylon. The change from form I to form II runs counter to what would be expected from thermodynamics. Usually, annealing leads to a more stable form, with larger and more perfect crystals which will melt at a higher temperature. Another striking feature is the constancy of the melting temperature of form I, which suggests that it is a well-defined and unchanging form. However, the paradox can be resolved and a thermodynamic and structural interpretation, which may be relevant to what happens in texturing, can be given. The classical argument on melting is expressed in terms of the thermodynamic free energy F of the system given by:

$$F = U - TS$$

where U = internal energy, T = temperature and S = entropy. In a crystal, where molecules are bonded together in a regular lattice, the internal energy, which drops as atoms are attracted to one another, and the entropy, which relates to disorder, are both low. In a liquid, they are both higher. Liquid and solid are in equilibrium when the free energies are equal. The line for the melt, which becomes a glass at low temperature, starts high but falls rapidly, because of the large values of U and S. The lowest line for a large, perfect crystal has a lower slope and cuts the melt line at the equilibrium melting point. However, it is not easy to overcome the energy barriers to form large crystals and this leads to the initial formation of small crystals, which have higher U and S and therefore melt at a lower temperature. Annealing progressively improves crystal size and perfection and increases the melting point. The classical argument can be modified to include forms I and II, by postulating that form I is a new state with values of U and S which lie between those for a conventional crystal and a melt. The line for form I will cut the line for the melt at its melting point. However, at the lower annealing temperature, the form I line will lie above the form II line for small crystals, and so the change from form I to form II is thermodynamically favoured. The initial melting point of form II is less than that of form I, which explains the paradox. Further annealing follows the conventional pattern of increasing crystal size and perfection and eventually takes the melting point of form I above that of form II. What are the structures of forms I and II? Various suggestions have been made, but the following explanation is plausible, even if not explicitly proven or even exactly described. Form II is assumed to be a conventional assembly of crystallites in an amorphous matrix. It is a gel, because it is a solid, cross-linked structure; it is crystalline because the crosslinks are segments in crystallographic register; it is dynamic because the cross-links will be continually breaking and reforming – but when, for example, A is open, BCDE will hold the structure together, and so on, as A joins up again and others open. Such a form would have internal energy U and entropy S between that of regular crystal and an amorphous melt and so is compatible with the thermodynamic argument.

HEAT-SETTING

At last, we come to what is directly relevant to texturing.What happens at texturing temperatures between about 180 and 220°C? As already mentioned, there has been a lack of fundamental research on what happens in this temperature range, particularly in regard to the mechanisms involved and the structural changes. It was recognised by Miles in the early days of nylon research that nylon 66 could be 'permanently' heat-set. By this it was meant that the set, which was applied to pleats and creases in garments, was stable to the ordinary actions of stress, temperature and moisture in use. There was a rule-of-thumb that the set could be overcome and a new form set by treatment at a higher temperature.This is important, because it means that textured yarns can be reset, as needed in fabric form in apparel or in plied BCF yarns for carpets.

When semi-crystalline, drawn yarns were the feedstock in false-twist texturing, the yarns were heat-set in their helical forms, in which the filaments were twisted and bent.With draw-texturing, the structural changes are dif- ferent. For polyester, the incoming POY yarn is essentially amorphous, with only incipient crystallinity and the semi-crystalline structure forms during the first stage of texturing. For nylon, the yarn will have crystallised, but the structure will be massively disrupted by the drawing and twisting, and a new stable structure will be formed. Some twist runs back to the feed-rolls in the incoming cold zone, but the draw and additional twisting will take place as the yarn warms up and softens on entering the heater. There is an interesting question in terms of the meaning of the word '*setting*'. A thermo-setting resin, such as the adhesive *Araldite*, is set by the action of heat, which cross-links the molecules; but, for thermoplastics such as nylon and polyester, setting requires heating and cooling. Heat puts the structure into a more mobile state and allows a rearrangement of the molecules, but it is really the cooling which sets the form. Consequently an adequate cooling zone is needed between the heater and the falsetwist spindle, where the yarn is untwisted. Beyond the spindle, the yarn is straight and fully extended under tension.When it is wound on a package, it will still be warm and it then remains on the package for a long time. Stress relaxation, which is a form of temporary set, will take place. Consequently, when the yarn is removed carefully from the package, it does not appear to be textured.The stretch and bulk needs to be developed.This can be done by 'milking' the yarn, increasing and decreasing tension, or by the action of heat or moisture, as is done in relaxing the yarn for tests of stretch and bulk. In a sense, as discussed above, heat-setting does not occur in drawtexturing on single-heater machines, which produce stretch yarns, though it needs to be understood in terms of subsequent yarn, fabric and garment processing. Heat-setting does become directly relevant for double-heater machines producing set yarns. For nylon, the rule-of-thumb seems to be reasonably valid, though the necessary, increased severity of treatment must take account of stresses and moisture as well as temperature. In producing modified nylon yarns with less stretch, the second heater was at a higher temperature than the first. For polyester, the effects are different. The second heater is run at a lower temperature than the first. This gives the clearest evidence that it is not necessary to increase the severity to process polyester, which seems to be able to reset repeatedly at the same or lower temperatures. It is possible that only a partial relaxation of stress, or a partial setting, occurs on the second heater. Heat-setting was a subject of research at UMIST, but this still left many unanswered questions, especially in terms of multiple setting sequences. One set of painstaking experiments on heat setting was carried out on a 28 dtex polyester (PET) monofilament yarn. After twisting, which gives the strain g0, the sample is put into a hot oil bath at a temperature *T*s for a time *t*s, then cooled and finally released, when the strain drops to gr. The fractional recovery *f* is given by (gr/g0): complete set is given by $f = 0$ and no set by $f = 1$. In some tests, the yarn was pre-set by heating at temperature *T*p for a time *t*p and cooling without deformation. Some experiments followed the same procedure in bending. The change in recovery with setting temperature for both conditions. Effects in bending and twisting are similar. If the yarn had not been pre-set, the 100°C transition gives a sigmoidal plot from $f = 1$ to $f = 0$ between 50 and 125°C. There is full relaxation of the torsional stress and set is apparently complete – but this is a temporary set which would easily be lost.The 'permanent' set, which occurs at a higher temperature, is not shown up in the plot. If the yarn has been pre-set at a higher temperature, the lower-temperature transition is only partly effec- tive in relaxing the stress: set is incomplete at 125°C and there is a plateau in the value of *f* until the higher-temperature setting action takes place between 180 and 220°C. The rate of setting increases rapidly with increase of temperature. After pre-setting at 200°C, there is little setting at 150°C for setting times up to one day; but after one minute, setting has started at 175°C and is complete at 220°C. These effects are relevant to twist-setting of fully drawn yarns, although the maximum shear strains of 0.6Percent in the tests are much lower than the severe conditions of twist texturing. But, as already mentioned, draw-texturing creates a new semicrystalline morphology, not the modification of an existing structure. The effect of the pre-setting conditions is relevant to what happens on the second heater in the production of set polyester yarns. It also relates to what happens in subsequent textile processing, such as the setting of polyester knitwear or BCF nylon carpet yarns. The setting action takes place only above the pre-setting temperature. However, this is for tests in which pre-setting lasted for 35 minutes, almost 2000 times longer than the setting time of 1.2 seconds. When the pre-setting time is reduced, set occurs at lower temperatures. When temperatures and times are equal, at 200°C and two minutes, the second set is only 10Percent short of being complete. A limited set of experiments on nylon showed a similar effect of presetting and setting temperatures, but almost no effect of time.This confirms the view that there is a difference between the setting responses of nylon and polyester. Setting in nylon appears to establish a structure that is more difficult to change than that of polyester.

STEAM-SETTING

If nylon 66 fibres are placed in a capsule filled with water and heated, melting is observed at a temperature that is 80°C lower than for dry fibres. Similarly heat-setting actions take place at lower temperatures. Since fabric processing was mostly done in steam, this led to the definition of an '*equivalent steam-set temperature*' to characterise the response to dry heat. For nylon 66, the range of 180–220°C, in which 'permanent' setting is effective, corresponds to 100–140°C in steam. Super-heated steam is therefore necessary to set nylon. In polyester the reduction in melting temperature is much less, so that higher steam temperatures are needed. The use of steam does have advantages in efficiency of heating and, as described in Chapter , research has shown that it enables shorter heaters to be used. However, the practical difficulties of containing the steam have prevented its use in commercial false-twist texturing.

MECHANISMS OF SETTING

The limited and sometimes contradictory experimental information on heat-setting of nylon and polyester is mirrored in its structural interpretation. The problem is compounded by the uncertainty about the fine structure of these fibres; the reasons for the observed differences between nylon and polyester are not understood. For the practical operations of texturing, it is not necessary to know what molecular mechanisms are involved, and so only a brief discussion of the subject will be given here. The temporary set at around 100°C (lower in nylon with absorbed water) is clearly due to the upper part of the glass-to-rubber transition, which is attributed to the onset of mobility in the hydrogen bonding of nylon and the benzene ring interactions of polyester.The change is in the amorphous regions of the fibres. The 'permanent' set at around 200°C is usually associated with changes in the crystalline regions, but there are several variant explanations. There is a simplistic argument. Thermodynamically, there is a driving force to a lower energy state, which will be more stable. The change is opposed by the energy barriers of intervening states, which may be overcome in time by thermal vibrations, and is complicated by the fact that there will be multiple energy minima with different structures. In structural terms the explanation would be that smaller, less perfect crystals have a lower melting point than large, perfect crystals,which have a lower free energy.Values as high as 300°C have been quoted for the melting point of perfect polyester (PET) crystals and as low as 250°C for the observable melting of semi-crystalline fibres. Setting would be achieved by the melting of the smallest, least perfect crystals in the distribution of size and perfection and the formation of larger, more perfect crystals. This picture fits in with the rule-of-thumb that it is necessary to go to higher temperatures to achieve a new set; with each increase of temperature, the distribution of crystal size and perfection will shift to higher values.The effects take place at a lower temperature wet than dry. For nylon, where the rule-of-thumb on severity of setting seems to be reasonable, provided some account is taken of stress, time and moisture, there are two problems with the simple explanation. First, it is not clear that the crystal-melting effects would go down to the 180–220°C range where setting occurs. The peak in the polyethylene loss modulus below 100°C suggests that there is a separate mechanism, which is clear of melting, but comparable data are not available for nylon and polyester. Second, it is difficult to see how structures could change to a similar form on a larger scale without a complete melting and recrystallisation of the fibre. There are other possibilities associated with the crystalline parts of the fine structure. In metals, the movement of defects in crystals is a mechanism of annealing, and this will occur in polyethylene and probably in polypropylene. However, defect models are not easy to formulate for the large repeats in nylon and polyester molecules.A more reasonable suggestion is that the thermal vibrations might be strong enough to allow molecules under stress to be pulled through the crystallites, thus changing the arrangement of tiemolecules and stabilising a new form. Another possibility is that the multiple melting effects with the change from form I, the dynamic crystalline gel, to form II, the micellar structure, may be relevant. A variant of this, which can be supported on the basis of the greater density of energy levels in the melt than in the crystal is that, over the relevant temperature range, micelles will be flipping between a rigid crystalline form and a mobile liquid form, which is held in position by molecular entanglements. As with the dynamic crystalline gel, enough units would be in the rigid state at any given time to hold the fibre together, but their identity would be continually changing. If the molecules were under stress, some limited rearrangement would be possible when a micelle is in the molten state. A more unusual suggestion is that polymer crystallisation involves quantum effects and is not a purely classical phenomenon. The quantum theory explanation of diffraction experiments with two slits through which an electron or a photon can pass, is that both possibilities exist until measurement detects which has happened. The states are superposed until the quantum state-function collapses to a single classical form. Penrose states that this occurs on a larger scale when a system has many energy levels close together and that the collapse may be timedependent. One would also expect it to be temperature-dependent because of the influence on energy levels. For some quasi-periodic, alloy crystals, which need to be defined over larger scales, he suggests that many forms will coexist until one becomes large enough to be the single, classical form. There are certainly many forms with different energy levels associated with a semi-crystalline polymer structure. It is therefore conceivable that during melt-spun fibre formation, and then again when fibres are heated to a critical temperature range for heat-setting, there is a superposition of states, which collapse on cooling to the single, stable state. An attraction of this view is that it shows how annealing to give larger and better crystals, which does occur, could be achieved without a complete melting of the fibres. Finally, it is possible that heat-setting is not related to the crystalline regions, but is due to a breaking and re-forming of tie-molecules in amorphous regions. Molecules under stress would break and then re-form in new preferred positions, thus stabilising the structure.This is more likely in polyester than in nylon,and would explain the ability to re-set at the same or lower temperatures, since the mechanism would operate whenever the temperature was raised. It is known that transesterification, the rupture and re-formation of the ester linkages in polyester molecules, occurs at a high rate owing to thermal vibrations in the melt and extrapolation from the 250–280°C range of measurements suggests that it may occur in the 180–220°C range of heatsetting, especially if molecules under stress are more likely to break. The explanation of heat setting in the 180–220°C range for nylon 66 and polyester (PET) and at lower temperatures for nylon 6 and polypropylene is still an open question. There is probably no single answer. Several of the mechanisms suggested above, and others not formulated, probably act together.

FIBRE EXTENSION, BENDING AND TWISTING

Most measurements of the mechanical properties of fibres are of their tensile properties. For the filaments of a textured yarn the loadextension curve starts with a period of easy crimp removal. The magnitude of this extension is determined by the crimp geometry and the low level of tension by bending and twisting properties. When the crimp has been pulled out, the tension increases rapidly as the load–extension curve of the straight fibre is followed.Morton and Hearle give a comprehensive account of fibre tensile properties. In general, the response is non-linear, visco-elastic and shows imperfect recovery. In nylon, polyester and polypropylene, the initial shape of the curve is influenced by the mechanical and thermal set of the fibres. For example, the effect of various heat treatments on the stress–strain curves of nylon yarns, the effect of heating in water on polyester. The end of the stress–strain curve, which gives the tenacity (tensile strength) and extension at break, is dependent on the degree of molecular orientation in the fibre, which is determined by the drawing process. High draw increases tenacity and reduces breaking extension. These properties will influence the performance of yarns when subject to severe forces. The tensile modulus, which is given by the initial slope of the curve and also increases with molecular orientation, influences the bending stiffness,which together with torsional stiffness determines the crimp forces. Bending properties are described by the relation between bending moment M and curvature c, which is the reciprocal of radius of curvature. For a simple, linear-elastic material, $M = (BC)$, where B is the bending stiffness (or flexural rigidity) and is given by:

$$B = (1/4\pi)(\eta E T^2/\rho)$$

where η is a shape factor, E is the tensile modulus, T is the linear density and ρ is the density. The basis for this relation is that, in bending, the material on the outside is under tension and on the inside under compression, with a central neutral plane. The shape factor is 1 for a circular fibre, but will be higher for a multilobal fibre with material further out. For a flattened cross-section, as may result from false-twist texturing, bending stiffness will be low in one direction and high in the other direction, which will influence the form of buckling. The maximum and minimum values are the principal bending stiffnesses, B_1 and B_2. The dependence on the square of linear density or fourth power of fibre diameter, is important; stiffness increases rapidly as fibre thickness increases. There will be big differences between yarns of the same total linear density as the number of filaments, and hence the linear density of the individual fibres, is changed. Twisting involves shear, which increases in magnitude from zero at the centre of a circular fibre to a maximum at the outside. Consequently, similar relations apply. Torque is proportional to twist, with a torsional rigidity R given by:

$$R = \varepsilon n T^2/\rho$$

where e is a shape factor equal to 1 for a circular fibre and n is the shear modulus. These expressions will be used in the discussion of the structural mechanics of textured yarns, but because so many factors influence the values of the moduli, it has not been possible to make quantitative predictions based on an input of fibre property values. However, some generic differences between fibre types can be noted which also includes values of density. The low modulus of nylon means that the forces developed in stretching nylon yarns are less than for polyester. This, combined with the good recovery from large deformations, makes it more suitable for use as a high-stretch yarn in form-fitting garments. The higher modulus of polyester gives a crisper handle, more suitable for non-stretch products. The relationships change when the deformations are large.Severe bending leads to yielding on the inside at a low compressive stress, so that the neutral plane moves outwards. Severe twisting at constant fibre length means that axial lines at the fibre surface become longer and make the tensile resistance more important than the shear resistance. Finally, there is the complication that the fibres in textured yarns have been set in twisted and bent forms, which will influence the directional effects in mechanical properties.

FIBRE SURFACE PROPERTIES

The coefficient of friction plays an important part in texturing processes, but its value depends on the spin finish applied by the fibre producer. This is a 'black art' with little public information on the formulations used. Friction varies with the surfaces in contact and with the rate of slip. The two combinations of most importance are:

- At high speed between yarn and drive surfaces, including friction discs, where friction must be high enough;
- At low speed between filaments to allow fibre migration to occur in twisting, for which friction must be low enough. Fibre producers have found finishes which satisfy both these contrasting requirements.

3

Yarn Production

WORK SYSTEM DESIGN

Yarn production yarn production planning is one those things that we all know is good for us, but that no one wants to take the time to do. While it may seem that yarn production yarn production planning only takes time away from running your business, operating a business without a plan is like going to a grocery store without a list and trying to remember all the items that are needed.

One comes out of the store having forgotten something critical—and having purchased a number of items that are totally frivolous and may never be used. It is the same for a business operating without a plan. Critical issues do not get addressed - and some tasks get done that have no relationship to the direction the business needs to go. For a business, however, the consequences of these unaddressed issues can range from inconvenience to bankruptcy.

Part of this reluctance is due to how complicated the process is viewed. Yet a complicated plan is almost as useless as none. The real question is how to make something simple that fits your business' needs. Can a good grocery list system be devised that isn't unnecessarily burdensome for all involved? Of course. Let's take a look as what yarn production planning really entails. The word "plan" originated from then Medieval Latin word planus which meant a level or flat surface. This evolved in French into being a map or a drawing of any object made by projection upon a flat surface. In English this has become a more general sense of a plan of action, design or method.

Yarn production yarn production planning in its current usage in business implies a consciousness of what is happening in the business. It does not preclude creativity or instinct, but it does add a layer of awareness that spells the difference between survival and extinction in a changing environment.

Yarn production yarn production planning does involve:

- An understanding of the business' history
- An examination of the business' environment
- An assessment of the business' mission
- Goals

- A process for reaching those goals
- A process for gathering information
- A realization that yarn production yarn production planning is a continuing process that is constantly evolving

Yarn production yarn production planning does not necessarily mean trying to yarn yarn project the future, but being aware of a range of likely futures and being prepared for them as occur.

BUSINESS PLAN

A business plan is used when one is starting a new business or a new process or product within a business. It includes not only a description of the new business, process or product, but also a discussion of how one plans on managing the marketing, development, production, and financing of this new venture.

ORGANIZATIONAL PLAN

Organizational yarn production yarn production planning, when it does occur, too often is spurred by crisis, focused on the short term, and not well thought out. To create healthy futures, organizations must construct processes for creating their futures that are not fueled by crisis and turmoil. It can be done. One of the most confusing aspects for those who want to plan is the variety of terms that are used in conjunction with yarn production yarn production planning. How do you differentiate between a business plan, a financial plan, a marketing plan, a human resources plan, an operations plan, a strategic plan, a long-range plan and just plain general yarn production yarn production planning? The simple answer is that each area of your business needs yarn production yarn production planning so each area should have its own grocery list of what it wishes to accomplish in the future.

STRATEGIC PLAN

A strategic plan usually refers to the overall direction you wish your business to take over the longer term. Consequently, a long-range plan and a strategic plan are often used synonymously. Within that overall strategy a business will have shorter term financial goals, marketing goals, production goals, and human resource goals that will each need some type of plan if they are to be achieved.

Just because a strategic plan is longer term does not mean it is never changed, however. One of the most serious mistakes businesses make is not revising their strategic plan regularly. The environment the business is operating in is changing constantly. The plan must be revisited at regular intervals to reflect the impact on the business of these external factors.

There are some universal principles that are true across all types of yarn production yarn production planning. Before tackling more specific yarn production yarn production planning models, it is wise to gain an understanding of the basic principles of general yarn production yarn production planning.

YARN PRODUCTION YARN PRODUCTION PLANNING PRINCIPLES

Any plan should include who, what, when, where, how, and why:

- Who is needed to accomplish this task?
- What needs to be done?
- When does it start and end?
- Where will it take place?
- How will it happen?
- Why must we do it?

Along with the answers to these questions there needs to be some operational plan to organize the tasks needed to achieve the goal.

A helpful approach is to work backward from the goal to decide what must be done to reach it. The backward approach is a way of looking at the big picture first, and then yarn production yarn production planning all tasks, conditions, and in a logical sequence to make the big picture happen. From this a to-do list can easily be made. This list will become a checklist to ensure everything is progressing as planned. Adjustments can be made based on changing circumstances. The plan should be referenced often as a set of signposts on the journey towards the goals.

For many of us who left corporate America in favour of a smaller work environment, the idea of drafting a plan may seem offensive.

After all, isn't frustration with all that busywork one of the reasons we left in the first place? We all have an aversion to doing anything on our job that doesn't immediately help the situation we are now experiencing. However, isn't it also true that a little foresight and action before the fact can help eliminate many of the problems we face each day.

Wouldn't it be nice to anticipate something like a price cut by your major competitor or a rise in the interest rate on your credit line? Of course it would. And with that anticipation comes an organized and effective response.

METHODS OF STUDY

There are a variety of study methods that have been developed based on research in cognitive psychology. One of the most popular is SQ3R or one of its derivatives.

THE SQ3R METHOD

Studying works better when it means something to you, when it relates to your life. There are a number of ways to do this and one recommend you try as many as you can. One fact which we all know from our own experience and psychological research has documented is that we remember things better when they are connected to emotions. This includes that funny sort of emotion "curiosity" and the nice experience of having it satisfied. So, one way to have things stick more is to find things that you're curious about.

Less intuitive, perhaps, but again documented, and believable once you think about it, is the fact that we remember things better when they are interconnected.

To take an example, which of the following lists of 5 words would be easier for you to remember?:

1 Dog always smoke question just or
2 Cars are hazardous inside cities
   - For most of us, the second is easier, because it means something, it is a list of things that are interconnected. Making such connections is called *elaboration* in cognitive psychology and the more elaborate your schema for something, the better able you are to remember it: Chess experts have better memories for positions of pieces in a chess game than do novices, but not if the pieces are randomly placed on the board. You can use these facts about human minds in your own studying by actively making connections between ideas. Any connections seem to help, but the ultimate connection seems to be to things you care about in your life where you bring in the emotion as well.
   - SQ3R is a method for active elaboration of yarn yarn material that you read. It consists of 5 steps. 1. Survey: The headings are — the major ones and the subheadings; hierarchical structures seem to be particularly easy for our brains to latch onto — check for introductory and summary paragraphs, references, etc. Resist reading at this point.
   - *Question (usually less than 30 seconds)*: What is the question that this chapter is trying to answer? Or — along the curiosity lines — What question do *one* have that this chapter might help answer? Repeat this process with each subsection of the chapter, as well, turning each heading into a question.
   - *Read (slower for some of us than others!)*: Read one section at a time *looking for the answer to the question proposed by the heading!* This is *active* reading and requires *concentration* so find yourself a place and time where you can concentrate.
   - *Recite/write (about a minute)*: Say to yourself or write down a key phrase that sums up the major point of the section and answers the question. It is important to use your own words. Research shows that we remember our own (active) connections better than ones given to us (passive), indeed that our own hierarchies are generally better than the best prefab hierarchies.

- *Review (less than 5 minutes)*: After repeating steps 2-4 for each section you have a list of key phrases that provides. Test yourself by covering up the key phrases and seeing if you can again them.. This method works. It works for me and it has worked for many students in the past. If you have recommendations for how to improve it, one would welcome them for the "next edition" of this mini-lesson.

In fact, one good way to remember the first list is to *make* it mean something, say by adding words: My dog always has a smoke when the question arises of just what life means — note here too that the humourous image adds an emotion making the sentence even easier to remember, although remembering *which* words were actually on the list may be a problem.

FREE TIPS IMPROVE STUDYING RESULTS

11 free study tips from Sound Feelings enhance learning how to learn, memorization, and studying speed.

Our free information reveals simple secrets, solutions and strategies to improve study habits, study methods, study tips, study skills, and help with test preparation and learning skills. These educational psychology learning aids and techniques become tools or guides for self-improvement and better grades which will help with homework, better grades, dyslexia, attention deficit disorder and end frustration.

- *Study in Short, Frequent Sessions*: It has been proven that short bursts of concentration repeated frequently are much more effective than one long session. So, even if you only have 10 minutes, DO IT. Take a break. Then study another 10 minutes. This "distributed learning" approach is highly efficient because it honours the way the brain likes to work. The brain needs recovery and recharging time for "protein synthesis." The rest periods are when your brain assimilates your effort. They are a powerful tool which many teachers do not acknowledge. To sit and study for hours and hours is not only boring, it creates fatigue, stress, and distraction. You cannot learn if you are fatigued, stressed, and distracted!
- *Take Guilt-Free Days of Rest*: This follows the same principle but on a longer, daily time cycle. The reason for resting is to refresh oneself. However, if you feel guilty then your precious rest period has been used to create more stress. The brain will not absorb new data if it is stressed. On days off from studying, really enjoy yourself and do not feel bad about not studying.
- *Honour Your Emotional State*: Do not study if you are tired, angry, distracted, or in a hurry. When the brain is relaxed, it is like a sponge and it naturally absorbs data without effort. If you are emotionally

stressed, your brain literally repels data. Forcing yourself to sit and study when your mind is on other things is a complete waste of time!

- *Review the Same Day*: When you learn something new, try to go over the points the same day. If you wait a few days and then make efforts to review the yarn yarn material, it will seem much less familiar. However, a quick review later in the day will tend to cement the information into your brain so that the next "official" study session, you will recognize it and it will seem easy.
- *Observe the Natural Learning Sequence*: Think of the activities you did when you were in nursery school. Using your whole arm, you probably performed the song that goes: "Put your right hand in, Put your right hand out." Then, in kindergarten, using your hand, you might have been asked to draw lines or circles with crayons. Later, in first grade, now holding the pencil with your fingers, you drew smaller lines and circles to create letters. Believe it or not, this natural learning sequence, moving from large to small, coarse to fine, still remains effective even though we are now older.
- *Use Exaggeration*: Why does a baseball batter warm up by swinging two or three bats? Why do runners sometimes strap lead weights to their legs? In both cases, exaggeration during practice makes the final result seem easy. This concept can be applied to studying anything. For example, if you are studying spelling, exaggerate the sound of the letters to help to remember them. So for studying purposes, "naive" would be pronounced "NAY-IVY." By getting used to this exaggerated pronunciation, the correct spelling seems obvious.
- *Prepare Your Study Environment*: If you require certain elements in your environment to help you study, try to always make these a priority. For example, do you need special lighting, silence, music, privacy, available snacks, etc.? Pay attention to what works for you and repeat it each time you study for best success.
- *Respect "Brain Fade."*: It is normal for the brain to have an attrition rate and to forget things. This does not mean that you are stupid! Instead of getting mad about this fact, you should expect it and deal with it accordingly. As you place more information on top, the lower levels become older and less available to your immediate again. The trick here is simply to review. Since we can anticipate the eventual fading of our memory, creating a review aspect to our study session will solve the problem. Once every two or three study sessions, simply review older yarn yarn material that you will be still needing to remember. Often, a quick overview is sufficient. Sometimes, a complete detailed study session of the older yarn yarn material is required. "Brain fade" is completely normal.

- *Create a Study Routine*: Generally, if you schedule certain times of the day to study, you will get into a routine and accomplish more. If you just "fit it in" during your day, chances are that there will never be any time. An effective way to do this is to literally mark it down in your datebook calendar as if you have an appointment, like going to the doctor. For example: "Tuesday 3-4:30 P.M. — Study."
- *Set Reasonable Goals*: One of the main reasons people do not reach their goals is because they set them too high. If you set goals that are manageable, even if they seem too simple, you get in the habit of accomplishing them and gradually you can set higher goals. Also, recognize the difference between long-term and short-term goals. Set your vision on the long-term dream, but your day-to-day activity should be focused exclusively on the short-term, enabling steps.
- *Avoid the Frustration Enemy*: Ironically, the quicker the person's nervous system, the faster they learn. Yet, this fast nervous system also works overtime in being self-critical. So they are the ones who always think they aren't going fast enough! In contrast, the "Type B," less intense person who learns slower yet is more self-accepting, ends up ultimately learning the yarn yarn material in a shorter period of time. This is because he/she doesn't waste energy blocking, getting upset, and thinking that they're not good enough — they simply keep moving forward at a slower pace.

K-W-L

K-W-L is an introductory strategy that provides a structure for recalling what students know about a topic, noting what students want to know, and finally listing what has been learned and is yet to be learned.

WHAT IS ITS PURPOSE?

The K-W-L strategy allows students to take inventory of what they already know and what they want to know. Students can categorize information about the topic that they expect to use.

How can I do it?:

- On the chalkboard, on an overhead or on students' individual clean sheets, three columns should be drawn.
- Label Column 1 K, Column 2 W, Column 3 L.
- Before reading, students fill in the Know column with everything they already know about the topic. This helps generate their background knowledge.
- Then have students predict what they might learn about the topic, which might follow a quick glance at the topic headings, pictures, and charts that are found in the reading. This helps set their purpose for reading and focuses their attention on key ideas.

- Alternatively, you might have students put in the middle column what they want to learn about the topic.
- After reading, students should fill in their new knowledge gained from reading the content. They can also clear up misperceptions about the topic which might have shown up in the Know column before they actually read anything. This is the stage of metacognition: did they get it or not?

K	W	L

FACILITIES LAYOUT

A typical yarn yarn manufacturing yarn yarn plant has a number of diverse activities interacting with each other. Thus, raw materials arrive at a shipping dock, they are unpacked and checked in a quality control area, they may then be processed through several processing areas, and finally the finished product again passes through the shipping dock.

In addition to areas specifically related to production, there must be dressing rooms, lunch rooms, and restrooms for employees; offices for supervision, design, and production control; and space for inventory and aisles. In fact, a yarn yarn plant may be viewed as a large number of finite geometric areas arranged on the floor space of the building. The problem of arranging these areas in an effective manner is the facility layout problem.

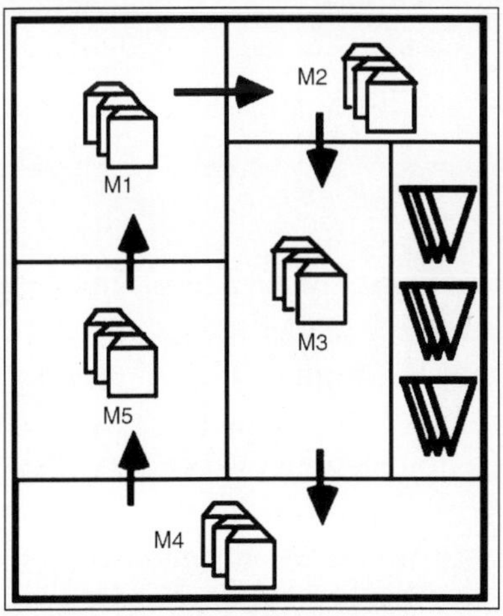

Clearly, the layout problem has relevance in many areas of facility and equipment design, from the layout of the rooms in a home to the layout of chips on an electronic circuit board. Although the facility layout problem may arise in many contexts, we are dealing with a yarn yarn plant yarn yarn manufacturing products for sale.

TYPES OF LAYOUTS

There are several alternative layout types that are appropriate for different product mixes and production volumes. Determination of the layout type is a major design decision because it impacts on so many other aspects of the production system.

In the Job Shop Layout, machines are grouped according to function into machine centres. Orders for individual products are routed through the various machine centres to obtain the required processing. This layout may be appropriate when there are many different products, each with a low volume of production.

Machines are general purpose, within their general function area, so that a wide variety of products can be handled. Because the expense of automation may be too great to be justified by the low volume, the machines in this arrangement will probably be at a relatively low level of automation. Workers will be highly skilled.

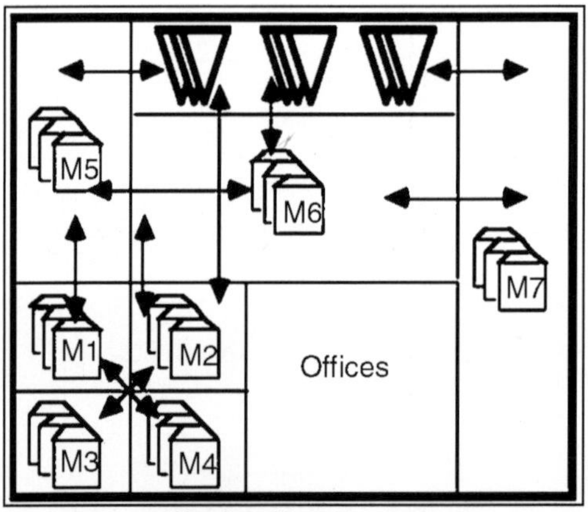

Fig. Job Shop Layout

Production scheduling is difficult with this type of arrangement because the level and type of work is highly variable. This results in large amounts of work-in-process, long product lead times, and high levels of management interaction. Typically there is a high degree of product movement required by the long and variable routes of individual products through the system.

The costs for setting up machines to produce the various products will be high because of the variety of different products and small lot sizes.

The arrangement can adapt readily to changes in product volume and design because of its inherent flexibility.

Here the product follows a fixed path through the production resources. The resources are arranged to minimize the yarn yarn material movement required. This type of layout is typical for an assembly line where a single product, or a few very similar products, passes through the line in a continuous fashion.

Because of the high volume of production, the machines on the line can be designed with a high level of fixed automation, with very little manual labour. Direct labour will be much less than for the job shop, but there will be high costs for maintenance. Setup costs and work-inprogress will be low for this arrangement.

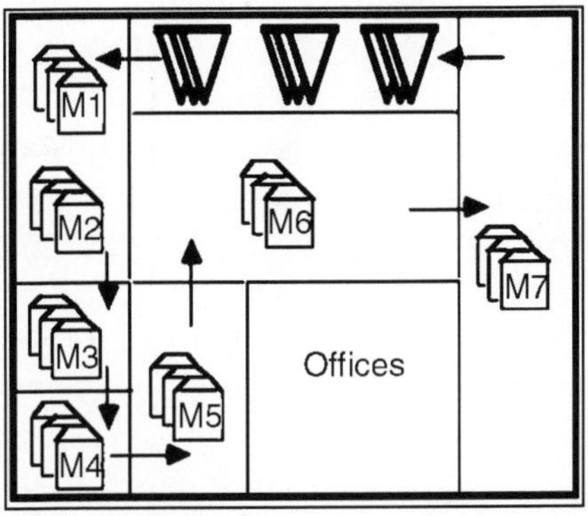

Fig. Assembly Line or Flow Shop Layout

The line, in general, is not flexible to product or volume changes. It is very sensitive to failures that cause the entire line to shut down. The arrangement is also appropriate for a flow shop that may have a number of products that all pass through the machine centres in the same order. In thiscase, the machines implementing the system may or may not be yarn yarn automated depending on the product mix and volume, but one would expect a higher level of automation than for the job shop.

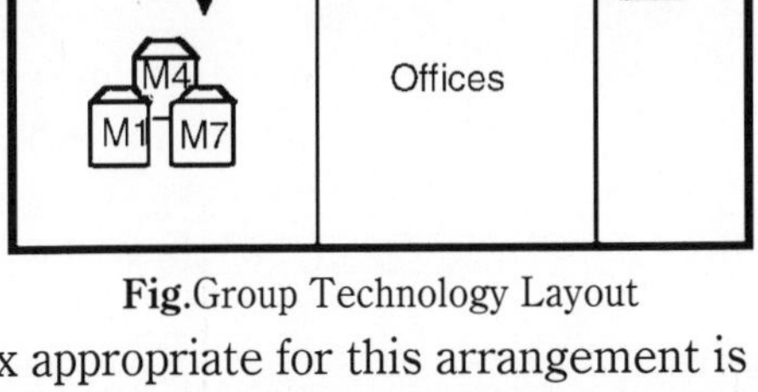

Fig. Group Technology Layout

The product mix appropriate for this arrangement is similar to that of the job shop. Products are grouped into classes that have some similarity with respect to processing. A yarn yarn manufacturing cell is designed for each group consisting of machines particularly adapted to the processing required.

The cells as collections of dissimilar machines. Because the range of products manufactured by each cell is less than that for the job shop, the machines and workers can be more specialized.

Typically, the workers in a cell are given more of the responsibility for production scheduling of a product class. This, together with the start-to-finish nature of the processing, results in more interesting jobs for the workers.

The group technology arrangement requires less setup time and cost than the job shop because of the greater specialization of function. It is compatible with the *just-in-time* concept of manufacture, so prevalent today, because of the smaller lot sizes made possible by the low setup costs. Often the level of automation with group technology is low, indicating the dependence of the concept on the skill of the labour force. Many companies have, however, introduced highly yarn yarn automated *flexible yarn yarn manufacturing cells* into the system.

Because the cell has a smaller range of products than the entire yarn yarn plant, it is easier to design the automation to handle the set of products in a group.

The group technology approach is more sensitive to changes in product mix and volume than the job shop, again because of the specialization introduced because of the yarn yarn manufacturing cell approach. When a product requires processing in more than one cell, problems similar to those of the job shop are introduced.

Fixed Location Layout

For tasks on large objects such as the manufacture of an electrical generator, the construction of a building, or the repair of a large airplane, the machines implementing the operation must come to the product, rather than the product moving to the machines.

Here the question is more often the scheduling of operations rather than the layout of machines.

Flexible Yarn yarn manufacturing Systems

The FMS is a system with yarn yarn automated yarn yarn material handling moving individual units of product between yarn yarn automated processors. Robotic manipulators often handle yarn yarn material. Using computer controlled movement and processing, a wide variety of products can be manufactured. All of this is to be accomplished with very low setup time, great flexibility of function, and very little manual labour.

The diversity of possible FMS, and the rapidly changing technologies, makes detailed consideration of the design of FMS beyond the scope of this text. Certainly many of the classical questions associated with facility design are no longer relevant for the FMS. In this text we consider the FMS as just another kind of machine, with perhaps a very broad range of capabilities.

LAYOUT PROBLEM

The layout problem is to arrange the physical spaces required for several departments in a given space provided for the departments.

In practice the facility layout problem is often solved by intuition, using the artistic and spatial skills of the human designer; however, when there are quantitative considerations associated with the layout problem, the human is at a disadvantage as compared to the computer. We concentrate on computerized procedures for solving the layout problem. There are a variety of problems regarding layout one might encounter. We explain the problem by specifying the data and describing the decisions.

Input Data

Here we are considering the problem of arranging several departments on a yarn yarn plant with a single floor and fixed dimensions.

Certain data is necessary to describe the layout problem:

- Number of departments, n,
- Physical area of each department, Ai for $i = 1...n$
- Physical dimensions of the yarn yarn plant in which the departments are to be placed: Length, L, and Width, W.
- Product flow between every pair of departments: fij for $i = 1 ... n$ and $j = 1 ... n$.
- Yarn yarn material handling cost between every pair of departments measured in dollars per unit-foot: cij for $i = 1 ... n$ and $j = 1 ... n$.

Distance

Our models involve the distance from one department to another. The distance depends on the layout. To emphasise consider a problem with ten departments with each department having an area of 100 square feet. The ten departments are to be placed in an area that is 50 feet long by 20 feet wide. One layout is shown in Figure.

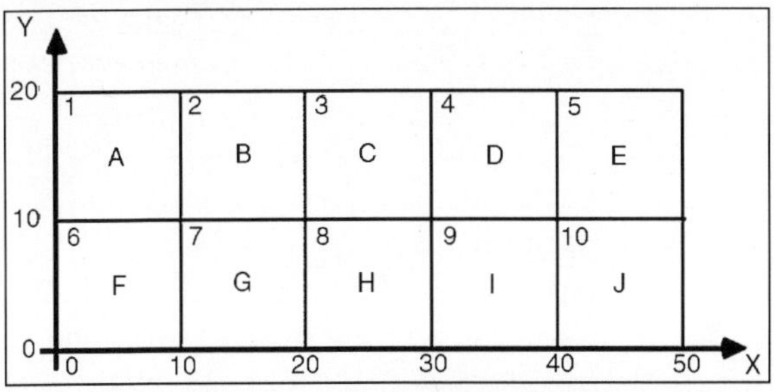

Fig. A Layout of Departments

This is only one of many possible layouts. If we assume that the departments must maintain a square shape, every permutation of the letters A through J is a different layout. There are $n!$ permutations. The matrix as in Fig. describes the flow between the departments. This is called the *From-To* matrix because an element (i, j) contains fij, the flow from department i to department j.

From \ To	A	B	C	D	E	F	G	H	I	J
A					5				1	9
B				3	4		1			
C				1			9			2
D							6		1	
E						8	3			
F							2	5		
G								6		
H									8	
I										
J										

Fig. From-to Matrix for the Office Example

The criterion for the layout problem involves the distance between departments. First we must prescribe the end points for the distance measurement. Here we assume that distances are measured between the centroids or centres of gravity of departments. Second, we must specify the route of travel. One possibility is that flow will follow a straight-line path. This is the Euclidean measure. More common in layout analysis, is to assume that flow travels via paths that are parallel to the axes of the layout. This is the rectilinear measure. The centroids are specified in terms of the coordinate system as,

$x(i)$ = x-coordinate of the centroid of department i, and

$y(i)$ = y-coordinate of the centroid of department i.

The centroid is the same as the centre of the area when the department is rectangular. For a more general shape, the centroid is the centre of gravity of area.

In the example case of Fig. we have the centroids as follows.

Department A: $x(A) = 5, y(A) = 15$.

Department B: $x(B) = 15, y(B) = 15$.

Department C: $x(C) = 25, y(C) = 15$.

The distance between two departments by a rectilinear measure is,

$$dij = |x(i) - x(j)| + |y(i) - y(j)|.$$

Here the vertical lines indicate absolute value. Fig. shows both the flow and distance between all pairs of departments on the from-to chart. The flow appears above the diagonals and the distance appears below.

From \ To	A	B	C	D	E	F	G	H	I	J	Sum
A					5 / 40				1 / 40	9 / 50	690
B				3 / 20	4 / 30		1 / 10				190
C				1 / 10			9 / 20			2 / 30	250
D							6 / **30**		1 / 10		190
E						8 / 50	3 / 40				520
F							2 / 10	5 / 20			120
G								6 / 10			60
H									8 / 10		80
I											0
J											0

Fig. Computation of the Total Distance Traveled

Criterion for Comparison

The flow multiplied by the distance and summed over all cells of the chart. We compute the cost for the flow from *i* to *j* as the product of the yarn yarn material handling cost, the flow and the distance between the departments. The cost of the layout is the sum of the flow cost.

$$Z = \sum_{i=1}^{n} \sum_{j=1}^{n} C_{ij} f_{ij} d_{ij}$$

The cost associated with each row of the chart. The cost of the layout is the sum of the row costs. The factors *cij* and *fij* are given as data, but the factor *dij* depends on the layout.

FINDING A LAYOUT WITH THE CRAFT METHOD

The yarn yarn plant layout problem is to find an arrangement of departments that minimizes the total distance traveled. This measure is called the "cost" in the following. The problem is difficult for the human because of the large number of calculations required to evaluate the cost of an alternative. The computer can greatly aid in the layout determination, but the computer also has its weaknesses.

The problem of finding the optimum arrangement is a very difficult one for mathematical optimization. Rather than attempt to obtain the optimum, we use a heuristic approach that tries to better a given solution by switching pairs of departments.

At each iteration, every pair of departments is considered for switching locations. The pair that gives the greatest savings in cost will be interchanged and the process will continue by looking for another pair to switch. If no pair results in a positive savings the process stops. This is commonly called the CRAFT* method.

Note that when two departments are switched in Fig., only the centroids of the two departments are affected. For example, if department A is switched with department H, the centroid of A becomes, while the centroid of H becomes. All other departments remain fixed. The only effect on the cost of the layout is related to the flows entering or leaving the two departments.

The cost of the flow between the other departments is unaffected because the distances between these departments have not changed. This is the logic used by the solution method. The cost savings associated with switching two departments is determined by calculating the effect of interchanging the centroids of the two departments.

There are several reasons why the CRAFT method may not yield the optimum solution. First, it may be necessary to switch more than two departments to find a better solution. More complex implementations of the method do allow switching three or more departments, but this increases considerably the computation time.

Problems for which the department areas are not equal cause several new difficulties. If two departments are not of equal area, switching their locations may not be equivalent to switching the centroids. In fact, if two unequally sized departments are not physically adjacent it will not be possible to interchange them. The method allows departments of different areas to be switched only if they are physically adjacent.

Allowing departments with different areas to switch causes another problem. When a small department is interchanged with a larger department there are a number of possible arrangements of the two areas with a corresponding variety of centroids. It is impossible to enforce a requirement that areas remain rectangular, so often the interchange of such departments causes irregularities in shape that grow worse as the algorithm progresses.

Because the change in cost is not exactly represented by the interchange of centroids, the estimate of the cost reduction may not be accurate. It is possible that the programme will switch two areas and have the total cost increase. To guard against cycling, the method stops if a certain number of successive iterations do not result in an improvement. Since the CRAFT method does not guarantee and optimum solution, it should be run several times with different initial layouts.

CENTROIDS FIXED

Although layout analysis is not limited to departments of equal size and shape, problems of this type are somewhat simpler than the more general case. What makes this problem special is that we can identify n fixed points as centroids and assign the departments to the points.

The distance between centroids does not vary, but the distance between departments depends on the assignment of departments to centroids. The problem can be modeled as a Quadratic Assignment Problem.

Although this problem is simpler in concept than the problem with different sized departments, it remains a very difficult optimization problem to solve. Since the QAP is an appropriate model for a variety of problems, it has received a great deal of attention from researchers in combinatorial optimization.

LINE BALANCING

The Assembly Line Balancing problem is a well-known problem in which a finite set of tasks or work elements are assigned to workstations subject to given precedence relationships. With Type I ALB problems, the objective is to minimize the number of workstations needed for a given production rate or cycle time.

Type II ALB problems seek to maximize output for a given number of stations. Various economic criteria, using operating cost or profitability measures as the objective, have also been investigated.

Recently, however, many companies have established multiple lines, with the lines having longer cycle times and each station being responsible for more tasks than under the traditional model.

One of Schonberger's principles of world-class, customer-driven performance is"seek to have plural instead of singular... flow lines for each product or customer family".

Each cell consists of small, multi-functional work teams; Sekine goes so far as to suggest that"the production line should consist mainly of single-operator U-shaped cells".

The use of multiple lines also allows each cell to have a backup. Therefore, if a machine in a cell breaks down, this redundancy will allow the company to continue production. Harmon also notes that redundancy of processes can facilitate necessary adjustments to capacity, since changes in capacity are less costly with smaller-sized equipment and can be accomplished with lower levels of inventory.

Merli states that dismantling extended, high-output flow lines with several low-output short lines can allow the production of extremely small economical lots and greater flexibility in terms of production volume.

Shtub recognizes the value of implementing multiple lines for assembling heavy products; when the cycle time and the length of each shift cause a synchronization problem, this can be solved by using parallel assembly lines at their"natural cycle time".

Ghosh and Gagnon identify several suggestions for future research in assembly systems, one of which is to"allow some strategic design considerations to be variables".

Although numerous line-balancing procedures have been developed that incorporate parallel stations and tasks no one has yet investigated the line-balancing problem that incorporates the possibility of several, independent lines.

As an example, consider the 11-task problem from Jackson. Using traditional line-balancing techniques requiring an output of 62 units per 480-minute day, the optimal solution is to establish eight stations, each with one or two tasks.

If we were to allow multiple stations, we can establish three lines, each with two stations and a cycle time of 23 minutes, resulting in a reduction of two stations. We can meet the required output with a 25Percent reduction in the numbers of stations required.

The need for fewer stations, and the associated reduction in idle time, will result in lower labour costs, floorspace requirements, work-in-process inventory, etc.; although this would likely require additional fixed equipment costs.

Another consideration in the application of Just-In-Time production principles is the utilization of U-shaped lines, such that the entrance and exit are at the same position.

Monden notes that the most important advantage of this layout is the flexibility to change the number of workers as the yarn yarn demand changes. Furthermore, with the U-line entrance adjacent to the exit, the JIT"pull" production can easily be maintained, as one unit of yarn yarn material can be introduced into the system as one unit of output is completed.

Other advantages of a U-shaped production line include a reduction in work-in-process inventory and wasted movement of operators as well as improved yarn yarn material handling. On the other hand, there may be a need for a greater number of machines and for more versatile personnel as well as reduced learning effects.

It could, though, be argued that requiring versatile personnel is beneficial, as it is associated with job enrichment and job enlargement that may better suit modern workers.

The U-Line Balancing problem was introduced and modeled by Miltenburg and Wijngaard; they developed a dynamic programming procedure to solve problems with up to 11 tasks and a heuristic to solve larger problems.

Urban presented an integer programming formulation of the problem, optimally solving problems up to 45 tasks with commercially available, general-purpose software. Scholl and Klein have since proposed a branch-and-bound procedure that was found to be effective in solving problems with up to 297 tasks.

Recently, the N U-line balancing problem has been investigated; however, these studies consider lines in which stations may include tasks from adjacent lines, not multiple, independent lines.

The MULB problem is modeled, and mixed-integer linear programmes are developed to solve the problem with multiple, identical lines as well as with multiple lines of varying sizes. An efficient procedure is proposed to solve large MULB problems, and is shown to optimally solve instances with over 100 tasks.

This analysis is then extended to explicitly incorporate the effect of line configuration and task assignment on equipment requirements; a branch-and-bound algorithm is developed that can be used to identify optimal solutions for moderate-sized problems, or as a heuristic for large problems.

Finally, computational results are presented concerning the effectiveness of the proposed procedures in solving the MULB problem.

THE MULB PROBLEM

Consider the situation in which we have a number of tasks, N, each with given precedence relationships and completion times, [t.sub.i] (i = 1... N), that are to be assigned to stations (j = 1..., [H.sub.k]) on each line (k = 1..., l) that is established. Let R = {1..., r..., |R|} represent a set of ordered pairs of tasks reflecting the precedence relationships; for example, r = (p, s) is the ordered pair indicating task p immediately precedes task s.

Let us first examine the Type I MULB problem, in which the objective under consideration is to minimize the total number of stations required to meet a given output per unit time, D = 1/C, where C is the required cycle time of the process.

We also assume that the resulting number of lines must remain within a closed interval [l.sup.-], [l.sup.+]. The lower limit will likely have a value of one; however, it may be desirable to place a limit on the maximum number of lines, due to the required duplication of equipment, span of control, etc.

In addition to the benefits of using multiple U-lines, the solution to the MULB problem will always provide a configuration with no more stations than the solution to the traditional straight line or ULB problem.

To emphasise the effect of employing multiple U-lines, Table presents the solution to the Jackson problem —with [l.sup.-] = 1 and [l.sub.+] = [??][SIGMA][t.sub.i]/C[??], where [??]x[??] is the smallest integer greater than or equal to x—over a range of cycle times.

For large cycle times, using multiple lines provides little advantage over the traditional or single U-line configurations. However, as the cycle time decreases, the advantage of the multiple U-line becomes more apparent. For a cycle time less than nine, the use of multiple U-lines always outperforms the traditional line and usually outperforms a single U-line. As the cycle time is reduced to the point of the largest task time, traditional lines or U-lines become impractical without the use of parallel tasks or stations.

Another effect of employing multiple U-lines is that the flexibility provided by a multiple U-line configuration may enable a large number of alternate optimal solutions. Consider the Jackson problem with a system cycle time of 7.9. One solution is to have three identical lines, each with two stations and a cycle time of 23.

Another solution is to have two lines of varying size: one with four stations and a cycle time of 12, another with two stations and a cycle time of 23. Alternate optimal solutions can be identified for four, five, and six lines, as well. Furthermore, slight changes in the system cycle time, even within the ranges shown in Table, may result in different solutions.

Since the MULB problem includes decisions on both the number of lines and the number of stations, a natural extension to the Type I MULB is to consider economic criteria and investigate the trade-off of the incremental costs of additional lines against the need for additional stations.

Daganzo and Blumenfeld note that the optimal number of parallel branches depends only on the ratio of the cost of lines and the cost of stations; they also provide a relevant discussion on the order of magnitude of each of these costs.

For an yarn yarn automated system requiring a relatively large expenditure of capital equipment, redundant lines become impractical and the single-line configuration will be optimal. On the other hand, a labour-intensive system will make the use of multiple lines advantageous.

Consider again the Jackson problem requiring an output of 62 units per 480-minute day. Taking cost data from Daganzo and Blumenfeld, assume the labour cost, [b.sub.1], is $200000 per station per year and the equipment cost, [b.sub.2], is $10000 per line per year. Let [rho] represent the ratio of the cost per station to the cost per line.

The solution to the traditional, straight-line problem is one line with eight stations, resulting in a cost of $1610 000. The solution to the U-line problem is one line with seven stations, resulting in a cost of $1410000. For [rho] = 20, the solution to the MULB problem is three lines with two stations each for a total cost of $1230000.

Thus, the straight-line and single U-line configurations result in costs of 30.9Percent and 14.6Percent., respectively, over the cost of the multiple U-line configuration. Again, small values of [rho] indicate that the system should consist of a single-line configuration; in this example, if this ratio is reduced to two, the single U-line configuration becomes optimal.

INTEGER LINEAR PROGRAMMING FORMULATIONS

We introduce mixed-integer programming models to solve the MULB problem. Formulations for the Type I MULB problem are presented, in which the lines are identical and in which the lines can be of varying length.

Identical Lines

First consider the situation in which we have identical lines; that is, each of the lines is designed to have the same number of stations with the same tasks assigned to each station. Utilizing multiple lines allows the output of each line to be reduced.

To maintain the required output of the system, the following relationship must be maintained:

- [1/C] = [1/[c.sub.1]] + [1/[c.sub.2]] +... + [1/[c.sub.l]],
- Where [c.sub.k] is the cycle time of an individual U-line. For the situation in which we have identical lines, [c.sub.1] = [c.sub.2] =... = [c.sub.l], the cycle time of each line can be expressed as [c.sub.k] = lC. Obviously, the cycle time of a particular station will be an integer multiple of the system cycle time. Having a longer cycle time allows more tasks to be included in any particular station, providing greater flexibility on the assignment of tasks to stations; thus, it becomes more likely to be able to identify an assignment that will result in a reduction of idle time.

Lines of Varying Size

Now consider the situation in which the lines are not necessarily identical; that is, the lines can vary in the number of stations assigned to each. This will allow the greatest flexibility in meeting the required output.

We define the following variables:

- [Mathematical expression not reproducible in ascii].
- [Mathematical expression not reproducible in ascii].
- [mathematical expression not reproducible in ascii].
- [mathematical expression not reproducible in ascii].

In this situation, the cycle time of each line, [c.sub.k], would also be a continuous variable.

The MULB problem with lines of varying size can then be formulated as follows:

- (MULB2) min [summation over j][summation over k][z.sub.jk],
- Subject to [summation over j]([x.sub.ijk] + [y.sub.ijk]) = [v.sub.k], [for all]i, k,
- [z.sub.jk] [less than or equal to] [v.sub.k], [for all]j, k,
- [summation over i][t.sub.i]([x.sub.ijk] + [y.sub.ijk]) [less than or equal to] [c.sub.k][z.sub.jk], [for all]j, k,
- [summation over j](UB – j + 1)([x.sub.pjk] - [x.sub.sjk]) [greater than or equal to] 0, [for all](p, s) [member of] R, k,
- [summation over j](UB – j + 1)([y.sub.sjk] - [y.sub.pjk]) [greater than or equal to] 0, [for all](p, s) [member of] R, k,
- [summation over k][v.sub.k]/[c.sub.k] = 1/C,
- [c.sub.k] [greater than or equal to] [t.sub.max], [for all]k,
- [x.sub.ijk], [y.sub.ijk], [z.sub.jk], [v.sub.k] [member of] {0, 1}, [for all]i, j, k.

where [t.sub.max] is the largest task time. Constraint sets ensure the proper assignment of tasks and stations; constraint set signifies the station capacities;

constraint set reflects the precedence relationships; constraint set ensures total output is met; and constraint set requires the cycle time of each line to be at least as large as the maximum task time.

Again, this could easily be formulated as a Type II or cost-minimization problem. This formulation contains non-linearities in constraint sets, which could easily be separated, but would result in a formulation impractical for large problems. Instead, we turn our attention to an efficient procedure to solve large MULB problems.

THE SOLUTION PROCEDURE

To identify the optimal solution to the MULB problem, we must first generate all feasible solutions for the individual lines. Furthermore, since the individual lines in a multiple U-line configuration are not constrained to the system cycle time, we must consider all cycle times within the range max {C, [t.sub.max]} [less than or equal to] [c.sub.k] [less than or equal to] [SIGMA][t.sub.i].

However, as is evident in Table, not all potential values of the cycle time within this range need to be considered, as many of the solutions are dominated by other solutions. As the cycle time decreases, the theoretical minimum number of stations is a step function. We only need to search around each "step" until the smallest cycle time is found for a given number of stations. Once this set of solutions is identified, the solution to the Type I MULB problem can be found by solving the following integer linear programme:

- (ILP1) min [summation over k]k[l.sub.k].
- subject to [k.summation over k](1/[c.sub.k])[l.sub.k] [greater than or equal to] 1/C,
- [l.sub.k] [greater than or equal to] and integer. [for all]k

where [l.sub.k] is defined as the number of U-lines established consisting of k stations. If desired, we can easily incorporate bounds on the number of lines or the number of stations established; for example, to restrict the configuration to be from two to four lines, we would include the constraint 2 [less than or equal to] [SIGMA][l.sub.k] [less than or equal to] 4. Obviously, finding the solution to (ILP1) is inconsequential, as it contains only [??][SIGMA][t.sub.i]/ [t.sub.max][??] variables and one constraint; the trick lies in efficiently identifying the feasible, non-dominated solutions for the individual U-lines.

As a rule, solution methods to solve Type II ALB problems iteratively solve a series of Type I problems; therefore, the set of solutions needed to solve the Type I problem could also be utilized for a Type II multiple U-line problem, as follows:

- (ILP2) max [summation over k](1/[c.sub.k])[l.sub.k],
- subject to [k.summation over k]k[l.sub.k] [less than or equal to] H,
- [[lambda].sub.k] [greater than or equal to] and integer, [for all]k.

Alternatively, an economic objective could also be used:

- (ILP3) min [b.sub.1] [summation over k][l.sub.k] + [b.sub.2] [summation over k]k[l.sub.k],
- subject to [summation over k](1/[c.sub.k])[l.sub.k] [greater than or equal to] 1/C,
- [l.sub.k] [greater than or equal to] and integer, [for all]k.

The algorithm to identify the feasible, non-dominated solutions for the individual U-lines is as follows:

- *Step 1*: Initialization. Read task completion times, [t.sub.i] (hence, [SIGMA][t.sub.i] and [t.sub.max]), precedence relationships, and required cycle time, C. Let [delta] be the smallest multiple of significance of the task times (*i.e.*, if all task times are integers, [delta] = 1; if even integers, [delta] = 2; if tenths of a minute, [delta] = 0.1, etc).
- *Step 2*: Set k = 1, [c.sub.1] = [SIGMA] [t.sub.i].
- *Step 3*: Set k = k + 1, [c.sub.k] = [??][SIGMA] [[t.sub.i]/k][??][.sup.[delta]], where [??]x[??][.sup.[delta]] is the value of x rounded up to the smallest multiple of significance.
- *Step 4*: If [c.sub.k] < [t.sub.max], set [c.sub.k] = [t.sub.max].
- *Step 5*: Optimally solve the single U-line balancing problem with a fixed cycle time of [c.sub.k] (in our implementation, we used a simplified version of the branch-and-bound procedure, ignoring the equipment cost; however, any procedure discussed in the Introduction could be used).
- *Step 6*: If there is no solution to Step 5 with k stations, go to Step 7; otherwise, go to Step 8.
- *Step 7*: Set [c.sub.k] = [c.sub.k] + [delta]. Go to Step 5.
- *Step 8*: If [c.sub.k] = [t.sub.max], go to Step 10; otherwise, go to Step 9.
- *Step 9*: If [c.sub.k] [less than or equal to] C, go to Step 10; otherwise, go to Step 3.
- *Step 10*: Solve (ILP1) (for the Type I problem), (ILP2) (for the Type II problem), or (ILP3) (for the cost-minimization problem).
- *Step 11*: Output the results and stop the algorithm.

If the solution procedure used in Step 5 provides an optimal solution for a fixed cycle time, then this algorithm will provide an optimal solution to the overall problem. As we are repeatedly solving this solution procedure, it is important to have an efficient solution methodology could also be used).

Fortunately, once we have identified all of the feasible solutions for a particular problem, the optimal solution to the MULB problem can be easily identified for various cost parametres or as the output of the process changes; the branch-and-bound algorithm does not need to be re-solved.

Another advantage of the proposed solution procedure is that it can quite easily be generalized by replacing the procedure used in Step 5 with an appropriate substitute procedure. For example, it may be desirable to incorporate equipment alternatives or process alternatives as part of the balancing decision. Wage rates or physical yarn yarn demand criteria could also be incorporated. Large-scale problems could be solved using appropriate heuristics. Any number of generalizations could be included; Becker and Scholl provide a survey on generalized assembly line balancing.

THE MULB PROBLEM WITH EQUIPMENT COST

An extension of the MULB problem is to explicitly include equipment costs in addition to the line and station costs.

While the economic approach is a useful extension to the Type I and Type II line-balancing problems, it does not directly address the impact of duplicated equipment.

For example, if two tasks require the same type of equipment, only one unit of equipment will be required if the tasks are assigned to the same station, but two units of equipment will be required if they are assigned to different stations.

In this case, the equipment cost is dependent on the line configuration as well as the task assignment to stations.

Therefore, we now consider the problem in which each task requires a piece of equipment, some of which may require the same type of equipment. Each station need not have duplicate equipment, so the different tasks using the same type of equipment can share the equipment if assigned to the same station. We assume that there is a fixed cost per line and a fixed cost per station.

The objective of this problem is to minimize the total cost, which is the sum of the total line, station, and equipment costs, to meet a given output per unit time.

The total station cost can be minimized by reducing the number of stations, whereas the total equipment cost can be minimized by assigning as many tasks as possible that use the same equipment to the same station.

On some occasions, the equipment cost can be reduced by assigning more stations which will increase the station cost; however, this will be worthwhile when the equipment cost reduction exceeds the station cost increase.

This problem can be redefined by extending the definition of the MULB problem. In addition, let [t.sub.ie] (i = 1..., N, e = 1..., M) represent the completion time of task i performed by equipment type e (*i.e.*, [t.sub.ie] is equal to [t.sub.i] for all e required by task i, and zero otherwise). Each equipment type has an associated equipment cost, [b.sub.3e] (e = 1..., M), in addition to the cost per line, [b.sub.1], and the cost for each additional station, [b.sub.2]. If a task requires no equipment, it can be assigned a unique equipment type at no cost ([b.sub.3e] = 0).

The Branching and Bounding Rules

We now turn our attention to the development of a branch-and-bound procedure to solve the MULB problem with Equipment costs. The pseudocode of the branch-and-bound algorithm. The notation used in our discussion follows:

- P(i) = The set of preceding tasks of task i;
- S(i) = The set of succeeding tasks of task i;
- LB = The lower bound of the total cost for the problem;
- UB = The upper bound of the total cost for the problem;
- L[B*.sub.q] = the lower bound of the total cost for the partial solution q;
- U[B*.sub.q] = the upper bound of the total cost for the partial solution q;
- B = the best solution which has been found so far;
- CList = The list used to store the candidate nodes;
- [n.sub.q] = the number of workstations that are assigned to partial solution q;
- [T*.sub.q] = the total task time of unassigned tasks for partial solution q;
- [t*. sub.jeq] = the completion time of unassigned task j performed by equipment e for a partial solution q (*i.e.*, [t*.sub.jeq] = [t.sub.i] for all e required by task i, and zero otherwise).

At each level of the branch-and-bound tree, we branch on all possible tasks that can be assigned to a specific workstation.

The rules of assigning unassigned tasks to a workstation are:

1. The task has no predecessors or its predecessors have already been assigned; and
2. The time to perform the task cannot be larger than the remaining slack time in the station.

For example, at level 1, we branch on all tasks that can be assigned to workstation 1, at level 2 on all tasks for workstation 2, etc. At the root node of our tree, the upper bound of the problem, UB, is obtained by a greedy algorithm. The lower bound of the overall cost is calculated as the sum of the lower bound of the station cost and the lower bound of the equipment cost.

The lower bound of the station cost is calculated as [??][([[summation].sub.i=1.sup.N] [t.sub.i])/C][??] x [b.sub.2]. The lower bound of the equipment cost is calculated as [[summation].sub.e=1.sup.M] [([[summation].sub.i=1.sup.N] [t.sub.ie]/C) x [b.sub.3e]]. Most of the nodes of the tree, with the exception of the end nodes, represent partial solutions; that is, some tasks have been assigned to workstations while the remaining tasks are unassigned. For node q, we use the formula [??][T*.sub.q]/C[??] + [n.sub.q] to calculate the lower bound on the number of stations. The equipment cost of the lower bound for node q is calculated...

WORK MEASUREMENT

Work measurement is the process of establishing the time that a given task would take when performed by a qualified worker working at a defined level of performance. There are various ways in which work may be measured and a variety of techniques have been established.

The basic procedure, irrespective of the particular measurement technique being used, consists of three stages:

- An analysis phase in which the job is divided into convenient, discrete components, commonly known as elements;
- A measurement phase in which the specific measurement technique is used to establish the time required (by a qualified worker working at a defined level of performance) to complete each element of work;
- A synthesis phase in which the various elemental times are added, together with appropriate allowances, to construct the standard time for the complete job.

The techniques used to measure work can be classified into those that rely on direct observation of the work, and those that do not.

For example, some techniques, such as predetermined motion-time systems and the use of synthetic or standard data can provide times from simulation or even visualisation of the work. However, the data on which such techniques are based were almost certainly based on earlier observation of actual work.

RATING

Direct observation techniques (such as time study and analytical estimating) include a process for converting observed times to times for the "qualified worker working at a defined level of performance." The commonest of these processes is known as rating.

This involves the observer (after appropriate training) making an assessment of the worker's rate of working relative to the observer's concept of the rate corresponding to standard rating. This assessment is based on the factors involved in the work - such as effort, dexterity, speed of movement, and consistency. The assessment is made on a rating scale, of which there are three or four in common usage. Thus on the 0-100 scale, the observer makes a judgement of the worker's rate of working as a per centage of the standard rate of working (100).

The rating is then used (in a process known as "extension" in time study) to convert the observed time to the basic time using the simple formula: Basic time = observed time x observed rating/standard rating

Rating is regarded by many as a controversial area of measurement since it is a subjective assessment. Where different observers rate differently, the resulting basic times are not comparable.

However, practised rating practitioners are remarkably consistent. It is important that those undertaking the rating are properly trained, and that this training is regularly updated (to maintain a common perception of standard rating) through rating 'clinics'.

ALLOWANCES

When carrying out work over a complete shift or working day, workers obviously suffer from the fatigue imposed both by the work undertaken and the conditions under which they are working.

The normal practice is to make an addition to the basic time (commonly referred to as an "allowance") to allow the worker to recover from this fatigue and to attend to personal needs. The amount of the allowance depends on the nature of the work and the working environment, and is often assessed using an agreed set of guidelines and scales.

It is usual to allow some of the recovery period inherent in these allowances to be taken away from the workplace (and it is essential in adverse working conditions).

Thus, work design should include the design of an effective work-rest regime. The addition of allowances should never be used to compensate for an unsafe or unhealthy working environment.

One minority school of thought suggests that relaxation allowances are unnecessary.

With work which involves, the carrying of heavy weights, this school suggests that the observer automatically adjusts the concept of standard rating to allow for the weight.

Thus, if the standard rate of performance for walking on level ground carrying no weight is equivalent to four miles per hour, then an observer rating a worker walking while carrying a weight will not expect the equivalent rate. Thus, it is argued that the weight has been allowed for in the adjustment of standard rating and any relaxation allowance is simply a duplication of this adjustment.

In many jobs there are small amounts of work that may occur irregularly and inconsistently.

It is often not economic to measure such infrequent work and an additional allowance is added to cover such work and similar irregular delays. This allowance is known as a contingency allowance and is assessed either by observation, by analysis of historical records (for such items as tool sharpening or replacement), or by experience.

The end result is a standard time which includes the time the work "should" take (when carried out by a qualified worker) plus additional allocations in the form of allowances, where appropriate, to cover relaxation time, contingency time and, perhaps, unoccupied time which increases the overall work cycle (such as waiting for a machine to finish a processing cycle).

CHOOSING A MEASUREMENT TECHNIQUE

The choice of a suitable measurement technique depends on a number of factors including:

- The purpose of the measurement;
- The level of detail required;
- The time available for the measurement;
- The existence of available predetermined data;
- And the cost of measurement.

To some extent there is a trade off between some of these factors. For example, techniques which derive times quickly may provide less detail and be less suitable for some purposes, such as the establishment of individual performance levels on short-cycle work.

The advantage of structured and systematic work measurement is that it gives a common currency for the evaluation and comparison of all types of work. The results obtained from work measurement are commonly used as the basis of the yarn production yarn production planning and scheduling of work, manpower yarn production yarn production planning, work balancing in team working, costing, labour performance measurement, and financial incentives. They are less commonly used as the basis of product design, methods comparison, work sequencing, and workplace design.

WORKFLOW

A workflow consists of a sequence of connected steps. It is a depiction of a sequence of operations, declared as work of a person, a group of persons, an organization of staff, or one or more simple or complex mechanisms. Workflow may be seen as any abstraction of real work, segregated in workshare, work split or other types of ordering. For control purposes, workflow may be a view on real work under a chosen aspect, thus serving as a virtual representation of actual work. The flow being described often refers to a document that is being transferred from one step to another.

A workflow is a model to represent real work for further assessment, *e.g.*, for describing a reliably repeatable sequence of operations. More abstractly, a workflow is a pattern of activity enabled by a systematic organization of resources, defined roles and mass, energy and information flows, into a *work process* that can be documented and learned. Workflows are designed to achieve processing intents of some sort, such as physical transformation, service provision, or information processing.

Workflow concepts are closely related to other concepts used to describe organizational structure, such as silos, functions, teams, projects, policies and hierarchies. Workflows may be viewed as one primitive building block of organizations. The relationships among these concepts are described later in this entry.

The term *workflow* is used in computer programming to capture and develop human-to-machine interaction.

RELATED CONCEPTS

The concept of workflow is closely related to several other fields in operations research and other fields that study the nature of work, either quantitatively or qualitatively, such as artificial intelligence (in particular, the sub-discipline of AI yarn production yarn production planning) and ethnography.

The term *workflow* is more commonly used in particular industries, such as printing, and professional domains, where it may have particular specialized meanings.

- *Processes*: A process is a more specific notion than workflow, and can apply to physical or biological processes, for instance. In the context of concepts surrounding work, a process may be distinguished from a workflow by the fact that it has well-defined inputs, outputs and purposes, while the notion of workflow may apply more generally to any systematic pattern of activity (such as all processes occurring in a machine shop).
- *Yarn production yarn production planning and scheduling*: A plan is a description of the logically necessary, partially-ordered set of activities required to accomplish a specific goal given certain starting conditions. A plan, when augmented with a schedule and resource allocation calculations, completely defines a particular *instance* of systematic processing in pursuit of a goal. A workflow may be viewed as an (often optimal or near-optimal) realization of the mechanisms required to execute the same plan repeatedly.
- Flow control is a control concept applied to workflows to divert from static control concepts applied to stock, that simply managed the buffers of yarn yarn material or orders, to a more dynamic concept of control, that manages the flow speed and flow volumes in motion and in process. Such orientation to dynamic aspects is the basic foundation to prepare for more advanced job shop controls, as just-in-time or just-in-sequence.
- In transit visibility is a monitoring concept that applies to transported yarn yarn material as well as to work in process or work in progress, *i.e.*, workflows.

Historical Development

In the 1980s, the term workflow was first used in its modern form in the software industry by FileNet vice president David Siegel. The company called its business process automation software"WorkFlo".

In 1995, the publishing industry studied how traditional publishing processes could be re-engineered and streamlined into digital processes in order to reduce lagtime, as well as substantial printing and shipping costs for delivering print copies of books and journals to warehouses and subscribers. The term electronic workflow was used to describe the publishing process, from online delivery of digital manuscripts to the posting of content on the web for online access.

The development of the concept of workflow occurred over a series of loosely defined, overlapping, eras.

Beginnings in Yarn yarn manufacturing

The modern history of workflows can be traced to Frederick Taylor and H. Gantt. Together they launched the study of the deliberate, rational organization of work in the context of yarn yarn manufacturing. The types of workflow of concern to Taylor and his contemporaries primarily involved mass and energy flows. These were studied and improved using time and motion studies. While the assembly line remains the most famous example of a workflow from this era, the early thinking around work was far more sophisticated than is commonly understood.

The notion of flow was more than a sequential breakdown of processing. The common conceptual models of modern operations research, including flow shops, job shops, and queuing systems, can be found in early forms in early 20th century industry.

Information based workflows began to grow during this era, although the concept of an information flow lacked flexibility.

A particularly influential was Melvil Dewey (inventor of the eponymous Dewey Decimal System), who was responsible for the development of the hanging file folder. This era is thus identified with the simplest notions of workflow optimization: throughput and resource utilization.

The cultural impact of workflow optimization during this era can be understood through films such as Chaplin's classic Modern Times. These concepts did not stay confined to the shop floor. One magazine invited housewives to puzzle over the fastest way to toast three slices of bread on a one-side, two-slice grill.

Maturation and Growth

The invention of the typewriter and the copier helped spread the study of the rational organization of labour from the yarn yarn manufacturing shop floor to the office. Filing systems and other sophisticated systems for managing physical information flows evolved. Two events provided a huge impetus to the development of formalized information workflows. First, the field of optimization theory matured and developed mathematical optimization techniques. Second, World War II and the Apollo programme were unprecedented in their demands for the rational organization of work. The classic management tome The Organization Man culturally captured the nature of work in this era.

Quality Era

During the 1980s two aspects of workflow organization drew heavy criticism. First, the methods pioneered by Taylor modeled humans as simple automatons. The classical industrial-style organization of work was critiqued as being both dehumanizing and suboptimal in its use of the potential of human beings.

Maslow's hierarchy of needs, which describes human needs for self-actualization and creative engagement in work, became a popular tool in this critique. This issue was acknowledged, but did not gain much traction otherwise.

The second critique had to do with quality. Workflows optimized for a particular time became inflexible as work conditions changed. Quality, in both analytic and synthetic manifestations, transformed the nature of work through a variety of movements, ranging from total quality management to Six Sigma to more qualitative notions of business process re-engineering. Under the influence of the quality movement, workflows became the subject of much scrutiny and optimization efforts. Acknowledgement of the dynamic and changing nature of the demands on workflows came in the form of recognition of the phenomena associated with critical paths and moving bottlenecks.

The experiences with the quality movement made it clear that information flows are fundamentally different from the mass and energy flows which inspired the first forms of rational workflows. The low cost and adaptability of information flows were seen as enabling workflows that were at once highly rational in their organization and highly flexible, adaptable and responsive.

These insights unleashed a whole range of information technology at workflows in yarn yarn manufacturing, services and pure information work. Flexible yarn yarn manufacturing systems, just-in-time inventory management, and other highly agile and adaptable systems of workflow are products of this era.

WORKFLOW MANAGEMENT SYSTEM

A workflow management system is a computer system that manages and defines a series of tasks within an organisation to produce a final outcome or outcomes. Workflow Management Systems allow you to define different workflows for different types of jobs or processes. So, for example, in a yarn yarn manufacturing setting, a design document might be automatically routed from designer to a technical director to the production engineer.

At each stage in the workflow, one individual or group is responsible for a specific task. Once the task is complete, the workflow software ensures that the individuals responsible for the next task are notified and receive the data they need to execute their stage of the process. Workflow management systems also automate redundant tasks and ensure uncompleted tasks are followed up.

FEATURES AND PHENOMENOLOGY

- *Modeling*: Workflow problems can be modeled and analysed using graph-based formalisms like Petri nets.
- *Measurement*: Many of the concepts used to measure scheduling systems in operations research are useful for measuring general workflows. These include throughput, processing time, and other regular metrics.
- *Specialized connotations*: The term *workflow* has specialized connotations in information technology, document management and imaging. Since 1993, one trade consortium specifically focused on workflow management and the interoperability of workflow management systems has been the Workflow Management Coalition.
- *Scientific workflows*: Found wide acceptance in the fields of bioinformatics and cheminformatics in the early 2000s, where they successfully met the need for multiple interconnected tools, handling of multiple data formats and large data quantities. Also, the paradigm of scientific workflows was close to the well-established tradition of Perl scripting in life-science research organizations, so this adoption represented a natural step forward towards a more structured infrastructure setup.
- *Human-machine interaction*: Several conceptual-izations of mixed-initiative workflows have been studied, particularly in the military, where yarn automated agents play roles just as humans do. For innovative, adaptive, collaborative human work the techniques of human interaction management are required.

WORKFLOW IMPROVEMENT THEORIES

The key driver to gain benefit from the understanding of the workflow process in a business context is that the throughput of the workstream path is modelled in such a way as to evaluate the efficiency of the flow route through internal silos with a view to increasing discrete control of uniquely identified business attributes and rules and reducing potential low efficiency drivers.

Evaluation of resources, both physical and human is essential to evaluate hand-off points and potential to create smoother transitions between tasks. Several workflow improvement theories have been proposed and implemented in the modern workplace.

These include:

- Six Sigma
- Total Quality Management

- Business Process Re-engineering
- Lean systems

As a way of bridging the gap between the two, significant effort is being put into defining *workflow patterns* that can be used to compare and contrast different workflow engines across both of these domains.

WORKFLOW COMPONENTS

A workflow can usually be described using formal or informal flow diagramming techniques, showing directed flows between processing steps.

Single processing steps or components of a workflow can basically be defined by three parametres:

1. Input description: the information, yarn yarn material and energy required to complete the step
2. Transformation rules, algorithms, which may be carried out by associated human roles or machines, or a combination
3. Output description: the information, yarn yarn material and energy produced by the step and provided as input to downstream steps.

Components can only be plugged together if the output of one previous (set of) component(s) is equal to the mandatory input requirements of the following component. Thus, the essential description of a component actually comprises only in- and output that are described fully in terms of data types and their meaning (semantics). The algorithms' or rules' description need only be included when there are several alternative ways to transform one type of input into one type of output – possibly with different accuracy, speed, etc.

When the components are non-local services that are invoked remotely via a computer network, such as Web services, additional descriptors (such as QoS and availability) also must be considered.

WORKFLOW APPLICATIONS

Many software systems exist to support workflows in particular domains. Such systems manage tasks such as automatic routing, partially yarn yarn automated processing and integration between different functional software applications and hardware systems that contribute to the value-addition process underlying the workflow.

EXAMPLES

The following examples emphasise the variety of workflows seen in various contexts:

- In military yarn production yarn production planning, a "concept of operations" is a workflow that defines particular mission types.
- In machine shops, particularly job shops and flow shops, the flow of a part through the various processing stations is a work flow.

- Insurance claims processing is an example of an information-intensive, document-driven workflow.
- Wikipedia editing is an example of a stochastic workflow.
- The Getting Things Done system is a model of personal workflow management for information workers.
- In global software development, the concept of *follow-the-sun* describes a process of passing unfinished work across time zones.
- In Scientific experiments, the overall process (tasks and data flow) can be described as a Directed Acyclic Graph (DAG). This DAG is referred to as a workflow, *e.g.* Brain Imaging workflows.

PULL SYSTEMS

Pull systems are an integral part of lean yarn yarn manufacturing yet they are frequently misunderstood and considered hard to implement. Specifically, one area of repeated struggle I find is the proper connection of assembly processes with upstream batch processes such as stamping, injection molding, paint, or a machining operation, for example.

There are three basic types of pull system; replenishment pull, sequential pull, and mixed pull system with elements of the previous two combined.

In all three cases the important technical elements for systems to succeed are:

- Flowing product in small batches (approaching one piece flow where possible)
- Pacing the processes to takt time (to stop overproduction)
- Signaling replenishment via a kanban signal
- Leveling of product mix and quantity over time

In order to connect a batch processes to assembly, a replenishment pull system used in conjunction with a specific type of kanban known as signal kanban is used. There are at least three different types of signal kanban and the proper usage depends upon the nature of your process and yarn yarn manufacturing situation.

The three types of signal kanban are:

1. Pattern production
2. Lot making with a batch board
3. Triangle kanban

Pattern production establishes a fixed order of production and some basic rules around lot sizes, which can vary somewhat as needed. However, the fundamental pattern or order of production is always maintained.

This creates a fixed sequence (*i.e.* pattern) of production that is continually repeated. Often this style of production is needed in heat treat processes, paint, or other processes with specific changeover sequences. While the sequence of product may be fixed in these cases, the actual amount produced each time in the cycle is unfixed and varies according to customer needs. This is known as a fixed order/unfixed quantity replenishment cycle.

For example, in an eight-hour cycle, part numbers always are run A through F. (The difficulty of your changeovers may dictate this order.) Inventory in the central market is a function of the length of the pattern replenishment cycle. Thus, a one-day cycle implies a minimum of one day of inventory must be kept in the market; a one-week cycle means one week of inventory. The main disadvantage of pattern production is that the sequence is fixed; you can't jump from making part C to part F.

Pattern production establishes a basic rhythm and sequence of production. As long as the overall production pace is kept close to final assembly takt time and the market governs pull between the two processes, the system works.

It's a very basic, first step towards establishing a production control and pull system between two processes separated by a market.

The pros are that it creates a stable order where one did not exist before. It also allows the best sequencing of product to ease changeovers, if they are an issue.

The cons are that it is not a perfect just-in-time system as you have to wait until your turn in the order queue comes around and thus some delay is involved. Unfortunately, it also does not tend to place pressure on the system to reduce changeover time, inventory time, or lead time (except by management intervention).

BATCH BOARD

Lot making with a batch board requires creating a physical kanban for every container of parts in the system. As yarn yarn material is consumed from the market, the kanban are immediately detached and periodically brought back to the upstream batch-producing process and displayed on a board that highlights all part numbers and displays an outlined shadow space for each of the kanban cards in the system. This is an example of a fixed quantity/unfixed order style of replenishment pull.

A returned kanban card placed on the board in the shadow space indicates inventory has been consumed in the market; unreturned cards represent inventory still in the market. As predefined trigger points are reached, the production operator knows to begin making product to replenish the yarn yarn material in the market. If several trigger points are reached at once, then production needs to establish some rules regarding what to produce first (since in this scenario the pattern or sequence is not fixed) and where to produce it. Otherwise, supervisors will decide what to make next based upon the available information.

There are several advantages of this approach. One is that the schedule sequence is more flexible than with pattern production. Second, it provides a visual representation of inventory consumption and highlights emerging problems in the central market.

There are disadvantages to this type of kanban as well. It requires many kanban cards if you have many part numbers, and the cards must be brought back in a timely and reliable manner for the batch board to be accurate. Furthermore there is the inherent problem of multiple part numbers reaching a trigger point as the same time.

In these instances human judgement is required to decided what to make next between two or more items that have been triggered. Discipline is required on the part of schedulers and supervisors not to build inventory in advance of when it is needed.

TRIANGLE KANBAN

The method that Toyota facilities would follow most of the time in connecting a batch process to a downstream assembly process is called the triangle kanban. The example assumes you have a batch machining department feeding some type of final assembly.

The machining department has five machines building 11 totally different product variations. The yarn yarn demand for each product varies significantly, as does the standard pack quantity of each part. The setup time for changeover is about three hours:

- Take the 11 part numbers and dedicate them to the five machines. (Ideally for example four machines will have two part numbers to run and one will have three part numbers to run, but this depends upon volume and mix, etc.)
- Now let us take one machine which now has several dedicated part numbers assigned to run on it.
- Create a single triangle kanban for each part number including the following information.
  - Part number
  - Part description
  - Inventory location
  - Machine to be run on
  - Space for date triggered to be written
  - Tool number
  - Lot size
  - Trigger or reorder point
- Determine the required daily run time for each part number. Calculate your average daily yarn yarn demand for each part number. If you haven't leveled the build in final assembly it's probably wise to add some amount in to cover yarn yarn demand variation. Let's assume that you have two shifts of eight hours available production time which equals 16 hours of production (simplest case). If you have two part numbers on a machine, you can derive required production time based

upon average daily yarn yarn demand and the cycle time to run the part number. For example, for ease of calculation, that required production time adds up to 10 hours.

- Determine the time available for changeovers. In the simple case of 16 hours of production and 10 hours of yarn yarn demand it leaves six hours for changeovers each day. You should incorporate any average down time and scrap you have in the short run – but eliminate this in the long run. Thus, in our simple case (ignoring downtime, etc.) you have available time for two changeovers daily or roughly one per shift under current calculations.
- Determine your lot size for each part number. There are different ways to do this but for simplicity I will only describe the easiest. In this case your lot size is simply set at one day of production since you probably have two part numbers on this machine and will average two changeover events per day. In essence you are making every part ever day. If you had 10 part numbers on the machine and only two changeovers per day you would have lot sizes of five days.
- Determine your trigger point for replenishment. To do this, add up the run time for the longer of the other components, add the changeover time, and add the time to make the first container and get it reliably back into the market. This is the minimum level you can establish for a replenishment trigger point for product. (An average trigger point might be 300 pieces.)

The system works by hanging a single triangle kanban at the trigger point in the inventory location. When the trigger point is reached, yarn yarn material handling takes it back to the producing machine. The kanban is hung on a rail at the machine and dictates what to build next and the lot size. The triangle pull system has many advantages. It is virtually self-running once established as long as average yarn yarn demand does not change. If it does, change your lot sizes accordingly.

Also, the triangle helps stabilize quality since product is dedicated to a machine, and it takes out some variability but you can choose to run product on other machines.

The down side is that the inventory in the market is not visible at all times since there is no batch board with cards. However, how many kanban are hanging on a rail at the machine which is a good indication of inventory depletion and a tool for visual control.

Furthermore, there is only one kanban per part number to manage. Other minor challenges include, of course, figuring out the die maintenance schedule (for some types of machine) and what impact this will have on the schedule. Also you must figure out a simple signal to reliably bring any needed raw yarn yarn material to the machine.

SAMPLE WORK FLOW USING TRIANGLE KANBAN FOR SCHEDULING

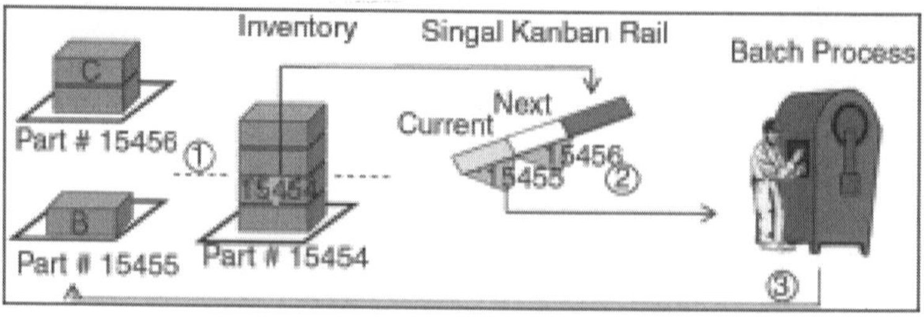

Fig. Sample Work Flow Using Triangle Kanban for Scheduling

TRIANGLE KANBAN DETAIL

In the end, you must decide what is the main goal is of implementing the pull system. Each of the three described ways works in terms of scheduling a batch process in conjunction with a market but each has a slightly different emphasis in mind. It might be worth the time to calculate the inventory levels, run times, operating rules for each of the three different ways.

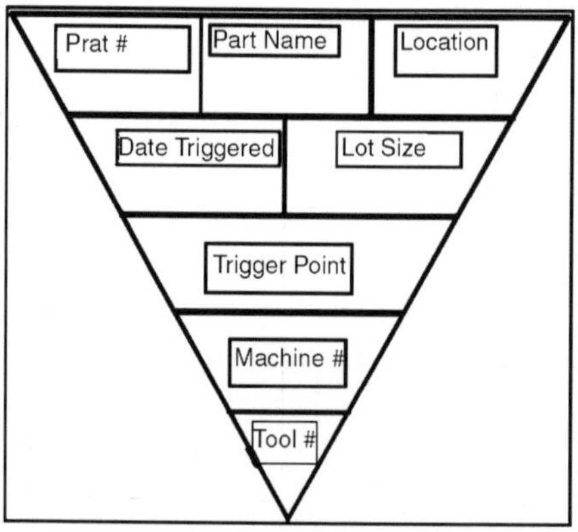

Fig. Triangle Kanban Detail

Construct a simple matrix, evaluate each of them in accordance with your priorities, and select the one that best fits your needs and company ability.

SEQUENTIAL PULL SYSTEM

A sequential pull system may be used when there are too many part numbers to hold inventory of each in a supermarket. Products are essentially "made-to-order" while overall system inventory is minimized. In a sequential system, the scheduling department must set the right mix and quantity of products to be produced. This can be done by placing production kanban cards in a heijunka box, often at the beginning of each shift. A sequential system requires strong management to maintain, and improving it may be a challenge on the shop floor.

MIXED SUPERMARKET AND SEQUENTIAL PULL SYSTEM

Supermarket and sequential pull systems may be used together in a mixed system. A mixed system may be appropriate when an 80/20 rule applies, with a small percentage of part numbers accounting for the majority (perhaps 80Percent) of daily production volume. Often an analysis is performed to segment part numbers by volume into:

- High,
- Medium,
- Low, and
- Infrequent orders.

Type D may represent special order or service parts. To handle these low-running items, a special type D kanban may be created to represent not a specific part number but rather an amount of capacity.

The sequence of production for the type D products is then determined by the method the scheduling department uses for sequential pull system part numbers.

PUSH SYSTEM

In a push system, releases are *scheduled*. So, throughput is determined by a exogenously set release rate. Since the releases are linked to orders, a push system is controlled by *upstream* information and is inherently *make-to-order*. In terms of our nomenclature, open lines are push systems because they have no endogenous restriction on releases to the line.

PUSH–PULL STRATEGY

The business terms push and pull originated in the logistic and supply chain management, but are also widely used in marketing.

A push-pull-system in business describes the movement of a product or information between two subjects. On markets the consumers usually *"pulls"* the goods or information they yarn yarn demand for their needs, while the offerers or suppliers *"pushes"* them towards the consumers. In logistic chains or supply chains the stages are operating normally both in push- and pull-manner. The interface between push-based stages and pull-based stages are called *push-pull boundary* or *decoupling point.*

Marketing

Push Strategy

Another meaning of the push strategy in marketing can be found in the communication between seller and buyer.

In dependence of the used medium, the communication can be either interactive or non-interactive.

For example, if the seller makes his promotion by television or radio, it's not possible for the buyer to interact with. On the other hand, if the communication is made by phone or internet, the buyer has possibilities to interact with the seller.

In the first case information is just "pushed" towards the buyer, while in the second case it is possible for the buyer to yarn yarn demand the needed information according to his requirements:

- Applied to that portion of the supply chain where yarn yarn demand uncertainty is relatively small
- Production and distribution decisions are based on long term forecasts
- Based on past orders received from retailer's warehouse
- Inability to meet changing yarn yarn demand patterns
- Large and variable production batches
- Unacceptable service levels
- Excessive inventories due to the need for large safety stocks
- less expenditure on advertising than pull strategy

Pull Strategy

In a "pull" system the consumer requests the product and "pulls" it through the delivery channel. An example of this is the car yarn yarn manufacturing company Ford Australia. Ford Australia only produces cars when they have been ordered by the customers:

- Applied to that portion of the supply chain where yarn yarn demand uncertainty is high
- Production and distribution are yarn yarn demand driven
- No inventory, response to specific orders
- Point of sale data comes in handy when shared with supply chain partners
- Decrease in lead time
- Difficult to implement

Supply Chains

With a push-based supply chain, products are pushed through the channel, from the production side up to the retailer. The manufacturer sets production at a level in accord with historical ordering patterns from retailers.

It takes longer for a push-based supply chain to respond to changes in yarn yarn demand, which can result in overstocking or bottlenecks and delays, unacceptable service levels and product obsolescence.

In a pull-based supply chain, procurement, production and distribution are yarn yarn demand-driven so that they are coordinated with actual customer orders, rather than forecast yarn yarn demand.

A supply chain is almost always a combination of both push and pull, where the interface between the push-based stages and the pull-based stages is known as the push–pull boundary. An example of this would be Dell's build to order supply chain. Inventory levels of individual components are determined by forecasting general yarn yarn demand, but final assembly is in response to a specific customer request. The push-pull boundary would then be at the beginning of the assembly line.

Push Pull Music Marking Future

Many media and music futurists are speculating upon the change within the industry. The introduction and success of social networking along with digital music has transformed and re-shaped the way music is marketed to the consumer; shifting from a push to pull strategy. In media and advertising everything is switching to pull.

There is an imminent difference between the consumer receiving information from a push strategy and a pull strategy. Push example; passive, the marketing campaign is in control of the message being sent out and how it is received by potential customers. Pull example; active, the recipient of the marketing campaign is in control of the message and their decision whether to performance or not. Generally the shift from push to pull has been viewed as a shift in power to the consumer away from advertisers and so is bad news for marketers.

With the increase of social networking platforms and users social networking has become a major and focal part to music marketing adopting the pull marketing strategy. Pull marketing shifts the emphasis and attention onto the customer, the essential attribute is to market in the correct places and knowing where and who the target audience are.

Consumers are increasingly customizing music platforms to better suit their individual needs. Rather than relying on music companies or a DJ to pre-determine the mix of songs on a CD, an increasing number of music listeners are downloading individual tracks and assembling their own sequence of songs. This process is also being replicated with the creation of playlists through platforms such as iTunes, Spotify and Last FM.

Fan-built playlists and mixes are taking over the way people get their music. Playlists are inevitably becoming a pull marketing resource that marketing alliances must embrace, due to their ability to be share via Peer to peer networks. People are choosing what they want to hear rather than having it pushed on them.

As consumers gain access to a greater number of options/platforms and more information about such services, the consumer will probably become more demanding on resource providers.

Requiring services to be made available on consumers terms, when and where they want them, rather than when and where it is convenient for the resource providers to deliver them. In addition leading to rapid growth in options and music services. Inevitably shifting music marketing from push to pull, as digital music services try out different models and features to try and find the optimum mix and consumer satisfaction.

PUSH VS PULL SYSTEM

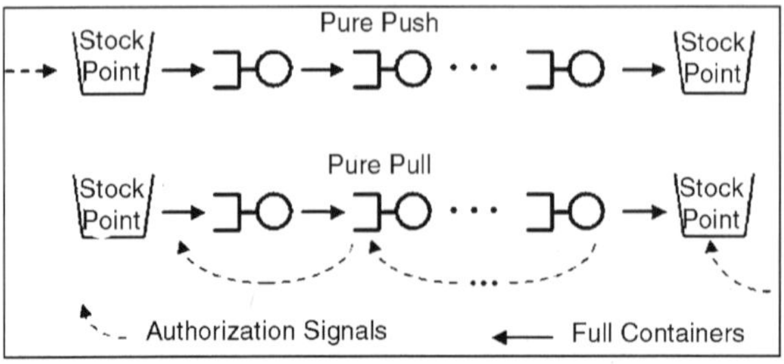

Fig. Push versus Pull System

Push/Pull Interface

It is not necessary that a production line be pure push or pure pull. One can design a line so that a segment of the line operates as push and the another segment operates as pull. We present a specific example to emphasise this.

Example

Consider two hypothetical fast food centres, Custom Taco and Quick Taco. In both the systems raw yarn yarn material is cooked, assembled and packaged before being sold to the customer. However, the two systems use a different push/pull interface to achieve different performance from the system.

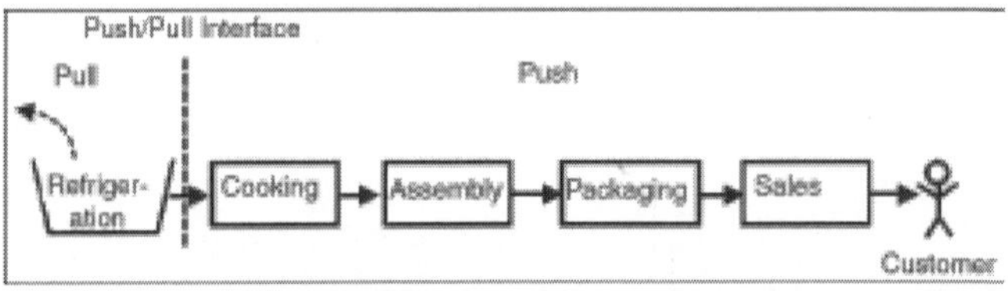

Fig. Production System for Custom Taco

The Custom Taco production line is shown in the figure. The refrigerator for stocking the raw yarn yarn material is run according to a *pull system*. That is, the order for refilling the refrigerator is released when the level of a particular item reaches its reorder point.

Everything beyond the refrigerator is run according to a *push system*. That is, food items are produced in response to customer orders that determines the release rate for the items. The transition between the pull system and the push system occurs between the refrigerator and the cooking stage and therefore represents the *push/pull interface* for the system.The processes are the same as for the Custom Taco line. However, we have added a warming table after the assembly stage which stores a predetermined level of assembled, but not bagged food items Hence, the line before the warming table operates as a *pull system*, since releases are triggered by removal of items from stock. Beyond the warming table, the line operates as a push system, since items are moved in response to customer orders.

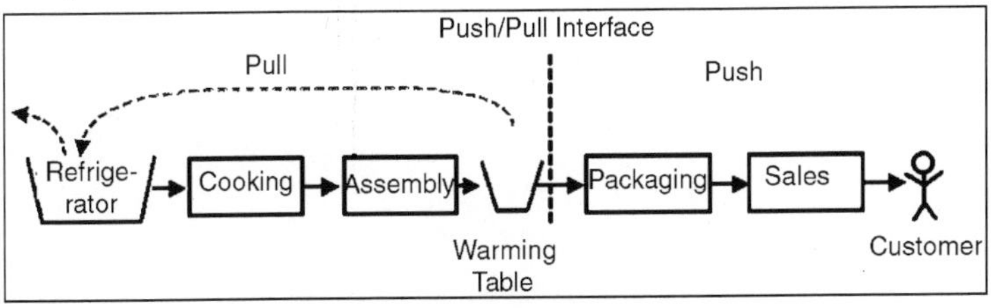

Fig. Production System for Quick Tac

In terms of system performance, the Custom Taco line is well-suited to a high degree of product variety. Inventory is held in its most generic form (raw yarn yarn material) and hence inventory costs are low. However, this line may not be responsive from a service standpoint, since the customer sees the entire production time as lead time. The Quick Taco line is more responsive, since the customer sees only the time to package and sell the items. However, it has more inventory because items are stocked at the assembled level.

The tradeoff involved in moving the push/pull interface closer to the customer is one of trading higher inventory for faster delivery. The most attractive place to locate the push/pull interface depends on customer expectations as well as costs.

Costs are strongly dependent on how rapidly the product proliferates as it traverses the production line. For instance, in a line that produces a single product (*e.g.*, ball point pens), there is little difference between the cost of holding inventory as raw materials or finished goods inventory.

Hence, in such a system, it may make sense to put the push/pull interface at finished goods inventory and run the entire system as in make-to-stock mode. On the other hand, in a system where every order is customised after the first step in the process (*e.g.*, a machine shop that does custom prototyping), there is no choice but to hold inventory at the raw yarn yarn material level and run the system in make-to-order mode.

For most lines, where there is some product proliferation, an intermediate push/pull interface may make sense. For instance, in the Quick Taco line, the warming table can be placed after assembly because there are relatively few items (*e.g.*, three taco types and two burrito types).

But placing the warming table after packaging would not be feasible, since the number of products (*i.e.*, all possible combinations of bags of tacos and burritos) would be too large. Hence, the cost versus delivery speed tradeoff is strongly affected by the rate of product proliferation and therefore the proper placement of the push/pull interface is a highly individualized decision for a given production system.

YARN YARN MATERIAL REQUIREMENTS YARN PRODUCTION YARN PRODUCTION PLANNING (MRP)

It was discussed in yarn yarn demand forecasting that in the dependent yarn yarn demand situation, if the yarn yarn demand for an item is known, the yarn yarn demand for other related items can be deduced. For example, if the yarn yarn demand of an automobile is known, the yarn yarn demand of its sub assemblies and sub components can easily be deduced. For dependent yarn yarn demand situations, normal reactive inventory control systems (*i.e.* EOQ etc.) are not suitable because they result in high inventory costs and unreliable delivery schedules.

More recently, managers have realized that inventory yarn production yarn production planning systems (such as materials requirements yarn production yarn production planning) are better suited for dependent yarn yarn demand items. MRP is a simple system of calculating arithmetically the requirements of the input materials at different points of time based on actual production plan. MRP can also be defined as a yarn production yarn production planning and scheduling system to meet time-phased materials requirements for production operations. MRP always tries to meet the delivery schedule of end products as specified in the master production schedule.

MRP OBJECTIVES

MRP has several objectives, such as:

- *Reduction in Inventory Cost*: By providing the right quantity of yarn yarn material at right time to meet master production schedule, MRP tries to avoid the cost of excessive inventory.
- *Meeting Delivery Schedule*: By minimizing the delays in materials procurement, production decision making, MRP helps avoid delays in production thereby meeting delivery schedules more consistently.
- *Improved Performance*: By stream lining the production operations and minimizing the unplanned interruptions, MRP focuses on having all components available at right place in right quantity at right time.

MRP SYSTEM

A simple sketch of an MRP system is shown in figure. It can be seen from the figure that an MRP system has three major input components:

- *Master Production Schedule (MPS)*: MPS is designed to meet the market yarn yarn demand (both the firm orders and forecasted yarn yarn demand) in future in the taken yarn production yarn production planning horizon.

  MPS mainly depicts the detailed delivery schedule of the end products. However, orders for replacement components can also be included in it to make it more comprehensive.

- *Bill of Materials (BOM) File*: BOM represents the product structure. It encompasses information about all sub components needed, their quantity, and their sequence of buildup in the end product. Information about the work centres performing buildup operations is also included in it.
- *Inventory Status File*: Inventory status file keeps an up-to-date record of each item in the inventory. Information such as, item identification number, quantity on hand, safety stock level, quantity already allocated and the procurement lead time of each item is recorded in this file.

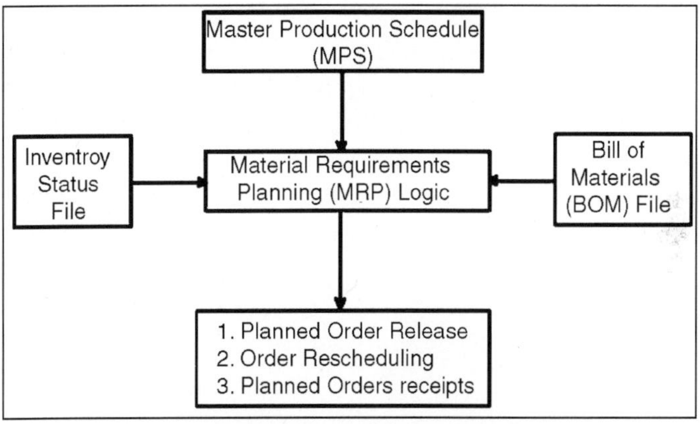

Fig. Yarn yarn material Requirements Yarn production yarn production planning System Architecture

After getting input from these sources, MRP logic processes the available information and gives information about the following:

- *Planned Orders Receipts*: This is the order quantity of an item that is planned to be ordered so that it is received at the beginning of the period under consideration to meet the net requirements of that period. This order has not yet been placed and will be placed in future.
- *Planned Order Release*: This is the order quantity of an item that is planned to be ordered in the planned time period for this order that will ensure that the item is received when needed. Planned order release is determined by offsetting the planned order receipt by procurement lead time of that item.
- *Order Rescheduling*: This highlight the need of any expediting, de-expediting, and cancellation of open orders etc. in case of unexpected situations.

YARN YARN MANUFACTURING RESOURCE YARN PRODUCTION YARN PRODUCTION PLANNING (MRP II)

Yarn yarn manufacturing Resource Yarn production yarn production planning (MRP II) is defined by APICS as a method for the effective yarn production yarn production planning of all resources of a yarn yarn manufacturing company. Ideally, it addresses operational yarn production yarn production planning in units, financial yarn production yarn production planning in dollars, and has a simulation capability to answer "what-if" questions and extension of closed-loop MRP.

This is not exclusively a software function, but a marriage of people skills, dedication to data base accuracy, and computer resources. It is a total company management concept for using human resources more productively.

KEY FUNCTIONS AND FEATURES

MRP II is not a proprietary software system and can thus take many forms. It is almost impossible to visualize an MRP II system that does not use a computer, but an MRP II system can be based on either purchased–licensed or in-house software. Almost every MRP II system is modular in construction.

Characteristic basic modules in an MRP II system are:

- Master Production Schedule (MPS)
- Item Master Data (Technical Data)
- Bill of materials (BOM) (Technical Data)
- Production Resources Data (Yarn yarn manufacturing Technical Data)
- Inventories and Orders (Inventory Control)
- Purchasing management
- Yarn yarn material Requirements Yarn production yarn production planning (MRP)
- Shop Floor Control (SFC)
- Capacity yarn production yarn production planning or Capacity Requirements Yarn production yarn production planning (CRP)
- Standard Costing (Cost Control)
- Cost Reporting/Management (Cost Control)

Together with auxiliary systems such as:

- Business Yarn production yarn production planning
- Lot Traceability
- Contract Management
- Tool Management
- Engineering Change Control
- Configuration Management
- Shop Floor Data Collection
- Sales Analysis and Forecasting
- Finite Capacity Scheduling (FCS)

And related systems such as:

- General Ledger
- Accounts Payable (Purchase Ledger)
- Accounts Receivable (Sales Ledger)

- Sales Order Management
- Distribution Requirements Yarn production yarn production planning (DRP)
- [Yarn yarn automated] Warehouse Management
- Yarn project Management
- Technical Records
- Estimating
- Computer-aided design/Computer-aided yarn yarn manufacturing (CAD/CAM)
- CAPP

The MRP II system integrates these modules together so that they use common data and freely exchange information, in a model of how a yarn yarn manufacturing enterprise should and can operate. The MRP II approach is therefore very different from the "point solution" approach, where individual systems are deployed to help a company plan, control or manage a specific activity. MRP II is by definition fully integrated or at least fully interfaced.

INDUSTRY SPECIFICS

MRP II systems have been implemented in most yarn yarn manufacturing industries. Some industries need specialised functions *e.g.* lot traceability in regulated yarn yarn manufacturing such as pharmaceuticals or food. Other industries can afford to disregard facilities required by others *e.g.* the tableware industry has few starting materials – mainly clay – and does not need complex materials yarn production yarn production planning. Capacity yarn production yarn production planning is the key to success in this as in many industries, and it is in those that MRP II is less appropriate.

MRP AND MRPII: HISTORY AND EVOLUTION

Yarn yarn material Requirements Yarn production yarn production planning (MRP) and Manufac-turing Resource Yarn production yarn production planning (MRPII) are predecessors of Enterprise Resource Yarn production yarn production planning (ERP), a business information integration system. The development of these yarn yarn manufacturing coordination and integration methods and tools made today's ERP systems possible. Both MRP and MRPII are still widely used, independently and as modules of more comprehensive ERP systems, but the original vision of integrated information systems as we know them today began with the development of MRP and MRPII in yarn yarn manufacturing.

The vision for MRP and MRPII was to centralize and integrate business information in a way that would facilitate decision making for production line managers and increase the efficiency of the production line overall.

In the 1980s, manufacturers developed systems for calculating the resource requirements of a production run based on sales forecasts. In order to calculate the raw materials needed to produce products and to schedule the purchase of those materials along with the machine and labour time needed, production managers recognized that they would need to use computer and software technology to manage the information. Originally, yarn yarn manufacturing operations built custom software programmes that ran on mainframes.

Yarn yarn material Requirements Yarn production yarn production planning (MRP) was an early iteration of the integrated information systems vision. MRP information systems helped managers determine the quantity and timing of raw materials purchases. Information systems that would assist managers with other parts of the yarn yarn manufacturing process, MRPII, followed.

While MRP was primarily concerned with materials, MRPII was concerned with the integration of all aspects of the yarn yarn manufacturing process, including materials, finance and human relations. Like today's ERP systems, MRPII was designed to integrate a lot of information by way of a centralized database. However, the hardware, software, and relational database technology of the 1980s was not advanced enough to provide the speed and capacity to run these systems in real-time, and the cost of these systems was prohibitive for most businesses.

Non-etheless, the vision had been established, and shifts in the underlying business processes along with rapid advances in technology led to the more affordable enterprise and application integration systems that big businesses and many medium and smaller businesses use today.

MRP AND MRPII: GENERAL CONCEPTS

Yarn yarn material Requirements Yarn production yarn production planning and Yarn yarn manufacturing Resource Yarn production yarn production planning are both incremental information integration business process strategies that are implemented using hardware and modular software applications linked to a central database that stores and delivers business data and information.

MRP is concerned primarily with yarn yarn manufacturing materials while MRPII is concerned with the coordination of the entire yarn yarn manufacturing production, including materials, finance, and human relations. The goal of MRPII is to provide consistent data to all players in the yarn yarn manufacturing process as the product moves through the production line.

Paper-based information systems and non-integrated computer systems that provide paper or disk outputs result in many information errors, including missing data, redundant data, numerical errors that result from being incorrectly keyed into the system, incorrect calculations based on numerical errors, and bad decisions based on incorrect or old data.

In addition, some data is unreliable in non-integrated systems because the same data is categorized differently in the individual databases used by different functional areas.

MRPII systems begin with MRP, Yarn yarn material Requirements Yarn production yarn production planning. MRP allows for the input of sales forecasts from sales and marketing. These forecasts determine the raw materials yarn yarn demand. MRP and MRPII systems draw on a Master Production Schedule, the break down of specific plans for each product on a line.

While MRP allows for the coordination of raw materials purchasing, MRPII facilitates the development of a detailed production schedule that accounts for machine and labour capacity, scheduling the production runs according to the arrival of materials.

An MRPII output is a final labour and machine schedule. Data about the cost of production, including machine time, labour time and materials used, as well as final production numbers, is provided from the MRPII system to accounting and finance.

BENEFITS

MRP II systems can provide:

- Better control of inventories
- Improved scheduling
- Productive relationships with suppliers

For Design/Engineering:

- Improved design control
- Better quality and quality control

For Financial and Costing:

- Reduced working capital for inventory
- Improved cash flow through quicker deliveries
- Accurate inventory records

CRITICISM

Authors like Pochet and Wolsey argue that MRP and MRP II, as well as the yarn production yarn production planning modules in current APS and ERP systems, are actually sets of heuristics.

Better production plans could be obtained by optimization over more powerful mathematical programming models, usually integer programming models.

While they acknowledge that the use of heuristics, like those prescribed by MRP and MRP II, were necessary in the past due to lack of computational power to solve complex optimization models, this is no longer true.

JUST-IN-TIME (BUSINESS)

Just in time is a 'pull' system of production, so actual orders provide a signal for when a product should be manufactured. Yarn yarn demand-pull enables a firm to produce only what is required, in the correct quantity and at the correct time.

This means that stock levels of raw materials, components, work in progress and finished goods can be kept to a minimum. This requires a carefully planned scheduling and flow of resources through the production process.

Modern yarn yarn manufacturing firms use sophisticated production scheduling software to plan production for each period of time, which includes ordering the correct stock. Information is exchanged with suppliers and customers through EDI (Electronic Data Interchange) to help ensure that every detail is correct.

Supplies are delivered right to the production line only when they are needed. For example, a car yarn yarn manufacturing yarn yarn plant might receive exactly the right number and type of tyres for one day's production, and the supplier would be expected to deliver them to the correct loading bay on the production line within a very narrow time slot.

PHILOSOPHY

Philosophy of JIT is simple: inventory is waste. JIT inventory systems expose hidden causes of inventory keeping, and are therefore not a simple solution for a company to adopt. The company must follow an array of new methods to manage the consequences of the change.

The ideas in this way of working come from many different disciplines including statistics, industrial engineering, production management, and behavioural science. The JIT inventory philosophy defines how inventory is viewed and how it relates to management.

Inventory is seen as incurring costs, or waste, instead of adding and storing value, contrary to traditional accounting. This does not mean to say JIT is implemented without an awareness that removing inventory exposes pre-existing yarn yarn manufacturing issues.

This way of working encourages businesses to eliminate inventory that does not compensate for yarn yarn manufacturing process issues, and to constantly improve those processes to require less inventory. Secondly, allowing any stock habituates management to stock keeping.

Management may be tempted to keep stock to hide production problems. These problems include backups at work centres, machine reliability, process variability, lack of flexibility of employees and equipment, and inadequate capacity.

In short, the just-in-time inventory system focus is having "the right yarn yarn material, at the right time, at the right place, and in the exact amount", without the safety net of inventory. The JIT system has broad implications for implementers.

Transaction Cost Approach

JIT reduces inventory in a firm. However, a firm may simply be outsourcing their input inventory to suppliers, even if those suppliers don't use JIT. Newman investigated this effect and found that suppliers in Japan charged JIT customers, on average, a 5Percent price premium.

Environmental Concerns

During the birth of JIT, multiple daily deliveries were often made by bicycle. Increased scale has required a move to vans and lorries. Cusumano highlighted the potential and actual problems this causes with regard to gridlock and burning of fossil fuels.

This violates three JIT waste guidelines:

- *Time*: Wasted in traffic jams
- *Inventory*: Specifically pipeline (in transport) inventory
- *Scrap*: Fuel burned while not physically moving

Price Volatility

JIT implicitly assumes a level of input price stability that obviates the need to buy parts in advance of price rises. Where input prices are expected to rise, storing inventory may be desirable.

Quality Volatility

JIT implicitly assumes that input parts quality remains constant over time. If not, firms may hoard high quality inputs. As with price volatility, a solution is to work with selected suppliers to help them improve their processes to reduce variation and costs.

Longer term price agreements can then be negotiated and agreed-upon quality standards made the responsibility of the supplier. Fixing up of standards for volatility of quality according to the quality circle

Yarn yarn demand Stability

Karmarker (1989) highlights the importance of relatively stable yarn yarn demand, which helps ensure efficient capital utilization rates. Karmarker argues that without significantly stable yarn yarn demand, JIT becomes untenable in high capital cost production. In the U.S., the 1992 railway strikes caused General Motors to idle a 75,000-worker yarn yarn plant because they had no supply.

JIT IMPLEMENTATION DESIGN

Based on a diagram modeled after the one used by Hewlett-Packard's Boise yarn yarn plant to accomplish its JIT programme.

- F Design Flow Process
  - F Redesign/relayout for flow
  - L Reduce lot sizes
  - O Link operations
  - W Balance workstation capacity
  - M Preventive maintenance

  - S Reduce Setup Times
- Q Total quality control
  - C worker compliance
  - I Automatic *i*nspection
  - M quality measures
  - M fail-safe methods
  - W Worker participation
- S Stabilize Schedule
  - S Level Schedule
  - W establish freeze windows
  - UC Underutilize Capacity
- K Kanban Pull System
  - D Yarn yarn demand pull
  - B Backflush
  - L Reduce lot sizes
- V Work with vendors
  - L Reduce lead time
  - D Frequent deliveries
  - U Yarn yarn project usage requirements
  - Q Quality Expectations
- I Further reduce inventory in other areas
  S Stores
  - T Transit
  - C Implement Carroussel to reduce motion waste
  - C Implement Conveyor belts to reduce motion waste
- P Improve Product Design
  - P Standard Production Configuration
  - P Standardize and reduce the number of parts
  - P Process design with product design
  - Q Quality Expectations

Effects

A surprising effect was that factory response time fell to about a day. This improved customer satisfaction by providing vehicles within a day or two of the minimum economic shipping delay.

Also, the factory began building many vehicles to order, eliminating the yarn yarn risk they would not be sold. This improved the company's return on equity.

Since assemblers no longer had a choice of which part to use, every part had to fit perfectly. This caused a quality assurance crisis, which led to a dramatic improvement in product quality. Eventually, Toyota redesigned every part of its vehicles to widen tolerances, while simultaneously implementing careful statistical controls for quality control. Toyota had to test and train parts suppliers to assure quality and delivery. In some cases, the company eliminated multiple suppliers.

When a process or parts quality problem surfaced on the production line, the entire production line had to be slowed or even stopped. No inventory meant a line could not operate from in-process inventory while a production problem was fixed. Many people in Toyota predicted that the initiative would be abandoned for this reason. In the first week, line stops occurred almost hourly. But by the end of the first month, the rate had fallen to a few line stops per day. After six months, line stops had so little economic effect that Toyota installed an overhead pull-line, similar to a bus bell-pull, that let any worker on the line order a line stop for a process or quality problem. Even with this, line stops fell to a few per week.

The result was a factory that has been studied worldwide. It has been widely emulated, but not always with the expected results, as many firms fail to adopt the yarn yarn full system.

The just-in-time philosophy was also applied to other segments of the supply chain in several types of industries. In the commercial sector, it meant eliminating one or all of the warehouses in the link between a factory and a retail establishment. Examples in sales, marketing, and customer service involve applying information systems and mobile hardware to deliver customer information as needed, and reducing waste by video conferencing to cut travel time.

Benefits

Main benefits of JIT include:

- *Reduced setup time*: Cutting setup time allows the company to reduce or eliminate inventory for"changeover" time. The tool used here is SMED (single-minute exchange of dies).
- *The flow of goods from warehouse to shelves improves*: Small or individual piece lot sizes reduce lot delay inventories, which simplifies inventory flow and its management.
- *Employees with multiple skills are used more efficiently*: Having employees trained to work on different parts of the process allows companies to move workers where they are needed.
- *Production scheduling and work hour consistency synchronized with yarn yarn demand*: If there is no yarn yarn demand for a product at the time, it is not made. This saves the company money, either by not having to pay workers overtime or by having them focus on other work or participate in training.
- *Increased emphasis on supplier relationships*: A company without inventory does not want a supply system problem that creates a part shortage. This makes supplier relationships extremely important.
- *Supplies come in at regular intervals throughout the production day*: Supply is synchronized with production yarn yarn demand and the

optimal amount of inventory is on hand at any time. When parts move directly from the truck to the point of assembly, the need for storage facilities is reduced.

Problems

Within a JIT System

Just-in-time operation leaves suppliers and downstream consumers open to supply shocks and large supply or yarn yarn demand changes. For internal reasons, Ohno saw this as a feature rather than a bug. He used an analogy of lowering the water level in a river to expose the rocks to explain how removing inventory showed where production flow was interrupted.

Once barriers were exposed, they could be removed. Since one of the main barriers was rework, lowering inventory forced each shop to improve its own quality or cause a holdup downstream. A key tool to manage this weakness is production levelling to remove these variations. Just-in-time is a means to improving performance of the system, not an end.

Very low stock levels means shipments of the same part can come in several times per day. This means Toyota is especially susceptible to flow interruption. For that reason, Toyota uses two suppliers for most assemblies. As noted in Liker (2003), there was an exception to this rule that put the entire company at yarn yarn risk because of the 1997 Aisin fire.

However, since Toyota also makes a point of maintaining high quality relations with its entire supplier network, several other suppliers immediately took up production of the Aisin-built parts by using existing capability and documentation. Thus, a strong, long-term relationship with a few suppliers is better than short-term, price-based relationships with many competing suppliers. Toyota uses this long-term relationship to send Toyota staff to help suppliers improve their processes.

These interventions have been going on for twenty years and have created a more reliable supply chain, improved margins for Toyota and suppliers, and lowered prices for customers. Toyota encourages their suppliers to use JIT with their own suppliers.

Within a Raw Yarn yarn material Stream

As noted by Liker (2003) and Womack and Jones (2003), it ultimately would be desirable to introduce synchronised flow and link JIT through the entire supply stream. However, none followed this in detail all the way back through the processes to the raw materials.

With present technology, for example, an ear of corn cannot be grown and delivered to order. The same is true of most raw materials, which must be discovered and/or grown through natural processes that require time and must account for natural variability in weather and discovery. The part of this currently viewed as impossible is the synchronised part of flow and the linked part of JIT.

It is for the reasons stated raw materials companies decouple their supply chain from their clients' yarn yarn demand by carrying large'finished goods' stocks. Both flow and JIT can be implemented in isolated process islands within the raw materials stream.

The challenge becomes to achieve that isolation by some means other than carrying huge stocks, as most do today. Because of this, almost all value chains are split into a part made-to-forecast and a part that could, by using JIT, become make-to-order. Historically, the make-to-order part has often been within the retailer portion of the value chain. Toyota took Piggly Wiggly's supermarket replen-ishment system and drove it at least half way through their automobile factories.

Their challenge today is to drive it all the way back to their goods-inwards dock. Of course, the mining of iron and making of steel is still not connected to an order for a particular car. Recognising JIT could be driven back up the supply chain has reaped Toyota huge benefits and a dominant position in the auto industry.

Note that the advent of the mini mill steelmaking facility is starting to challenge how far back JIT can be implemented, as the electric arc furnaces at the heart of many mini-mills can be started and stopped quickly, and steel grades changed rapidly.

Oil

It has been frequently charged that the oil industry has been influenced by JIT.

The argument is presented as follows: The number of refineries in the United States has fallen from 279 in 1975 to 205 in 1990 and further to 149 in 2004. As a result, the industry is susceptible to supply shocks, which cause spikes in prices and subsequently reduction in domestic yarn yarn manufacturing output.

The effects of hurricanes Katrina and Rita are given as an example: in 2005, Katrina caused the shutdown of 9 refineries in Louisiana and 6 more in Mississippi, and a large number of oil production and transfer facilities, resulting in the loss of 20Percent of the US domestic refinery output.

Rita subsequently shut down refineries in Texas, further reducing output. The GDP figures for the third and fourth quarters showed a slowdown from 3.5Percent to 1.2Percent growth. Similar arguments were made in earlier crises.

Beside the obvious point that prices went up because of the reduction in supply and not for anything to do with the practice of JIT, JIT students and even oil and gas industry analysts question whether JIT as it has been developed by Ohno, Goldratt, and others is used by the petroleum industry. Companies routinely shut down facilities for reasons other than the application of JIT.

One of those reasons may be economic rationalization: when the benefits of operating no longer outweigh the costs, including opportunity costs, the yarn yarn plant may be economically inefficient.

JIT has never subscribed to such considerations directly; following Waddel and Bodek (2005), this ROI-based thinking conforms more to Brown-style accounting and Sloan management.

Further, and more significantly, JIT calls for a reduction in inventory capacity, not production capacity. From 1975 to 1990 to 2005, the annual average stocks of gasoline have fallen by only 8.5Percent from 228,331 to 222,903 bbls to 208,986. Stocks fluctuate seasonally by as much as 20,000 bbls.

During the 2005 hurricane season, stocks never fell below 194,000 thousand bbls, while the low for the period 1990 to 2006 was 187,017 thousand bbls in 1997. This shows that while industry storage capacity has decreased in the last 30 years, it hasn't been drastically reduced as JIT practitioners would prefer.

Finally, JIT does not seem to have been a goal of the industry. In Waguespack and Cantor (1996), the authors point out that JIT would require a significant change in the supplier/refiner relationship, but the changes in inventories in the oil industry exhibit none of those tendencies. Specifically, the relationships remain cost-driven among many competing suppliers rather than quality-based among a select few long-term relationships.

They find that a large part of the shift came about because of the availability of short-haul crudes from Latin America. In the follow-up editorial, the Oil and Gas Journal claimed that "casually adopting popular business terminology that doesn't apply" had provided a "rhetorical bogey" to industry critics. Confessing that they had been as guilty as other media sources, they confirmed that "It also happens not to be accurate."

BUSINESS MODELS FOLLOWING SIMILAR APPROACH

Vendor-Managed Inventory

Vendor-managed inventory (VMI) employs the same principles as those of JIT inventory, however, the responsibilities of managing inventory is placed with the vendor in a vendor/customer relationship.

Whether it's a manufacturer managing inventory for a distributor, or a distributor managing inventory for their customers, the management role goes to the vendor.

An advantage of this business model is that the vendor may have industry experience and expertise that lets them better anticipate yarn yarn demand and inventory needs.

The inventory yarn production yarn production planning and controlling is facilitated by applications that allow vendors access to their customer's inventory data.

Another advantage to the customer is that inventory cost usually remains on the vendor's books until used by the customer, even if parts or materials are on the customer's site.

Customer-Managed Inventory

With customer-managed inventory (CMI), the customer, as opposed to the vendor in a VMI model, has responsibility for all inventory decisions. This is similar to JIT inventory concepts. With a clear picture of their inventory and that of their supplier's, the customer can anticipate fluctuations in yarn yarn demand and make inventory replenishment decisions accordingly.

CELLULAR YARN YARN MANUFACTURING

Cellular yarn yarn manufacturing is a yarn yarn manufacturing process that produces families of parts within a single line or cell of machines operated by machinists who work only within the line or cell. A cell is a small scale, clearly-defined production unit within a larger factory.

This unit has complete responsibility for producing a family of like parts or a product. All necessary machines and manpower are contained within this cell, thus giving it a degree of operational autonomy. Each worker is expected to have mastered a yarn yarn full range of operating skills required by his or her cell.

Therefore, systematic job rotation and training are necessary conditions for effective cell development. Complete worker training is needed to ensure that flexible worker assignments can be fulfilled.

Cellular yarn yarn manufacturing, which is actually an application of group technology, has been described as a stepping stone to achieving world class yarn yarn manufacturing status. The objective of cellular yarn yarn manufacturing is to design cells in such a way that some measure of performance is optimized. This measure of performance could be productivity, cycle time, or some other logistics measure. Measures seen in practice include pieces per man hour, unit cost, on-time delivery, lead time, defect rates, and percentage of parts made cell-complete.

This process involves placing a cluster of carefully selected sets of functionally dissimilar machines in close proximity to each other. The result is small, stand-alone yarn yarn manufacturing units dedicated to the production of a set or family of parts—or essentially, a miniature version of a yarn yarn plant layout.

While the machinery may be functionally dissimilar, the family of parts produced contains similar processing requirements or has geometric similarities. Thus, all parts basically follow the same routing with some minor variations (*e.g.*, skipping an operation).

The cells may have no conveyorized movement of parts between machines, or they may have a flow line connected by a conveyor that can provide automatic transfer.

Cellular yarn yarn manufacturing is a hybrid system that links the advantages of a job shop with the product layout of the continuous flow line. The cell design provides for quick and efficient flow, and the high productivity associated with assembly lines.

However, it also provides the flexibility of the job shop, allowing both similar and diverse products to be added to the line without slowing the process. Figures compares a cellular layout to that of the typical job shop (process layout).

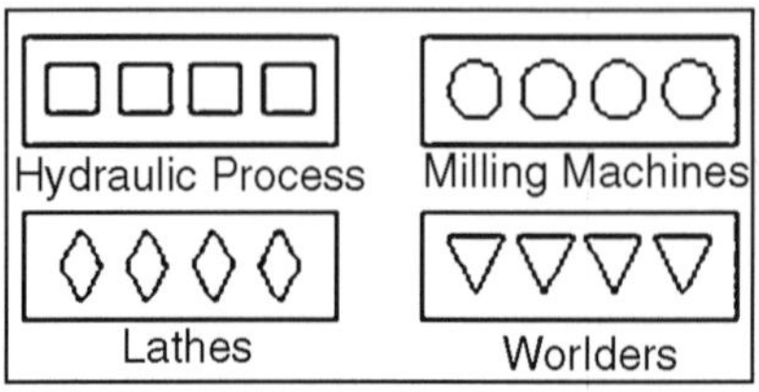

Fig. Process Layout

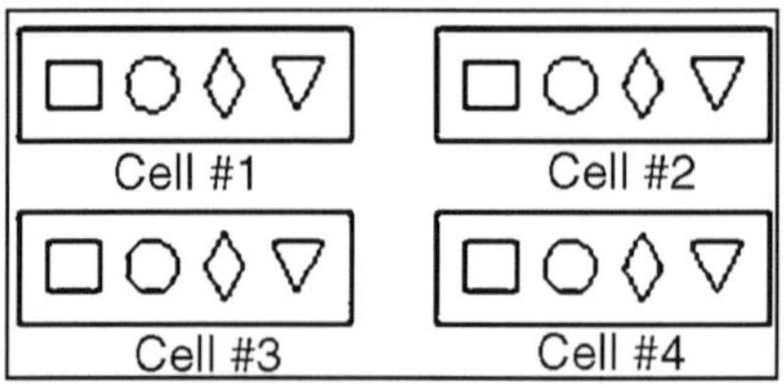

Fig. Cellular Layout

BENEFITS OF CELLULAR YARN YARN MANUFACTURING

Many firms utilizing cellular yarn yarn manufacturing have reported near immediate improvements in performance, with only relatively minor adverse effects. Cited improvements which seem to have occurred fairly quickly include reductions in work-in-process, finished goods, lead time, late orders, scrap, direct labour, and workspace.

In particular, production and quality control is enhanced. By breaking the factory into small, homogeneous and cohesive productive units, production and quality control is made easier. Cells that are not performing according to volume and quality targets can be easily isolated, since the parts/products affected can be traced to a single cell. Also, because the productive units are small, the search for the root of problems is made easier.

Quality parametres and control procedures can be dovetailed to the particular requirements of the parts or workpieces specific to a certain cell. By focusing quality control activity on a particular production unit or part type, the cell can quickly master the necessary quality requirements. Control is always enhanced when productive units are kept at a minimum operating scale, which is what cellular yarn yarn manufacturing provides. When production is structured using cellular yarn yarn manufacturing logic, flow systematization is possible. Grouping of parts or products into sets or families reveals which ones are more or less amenable to continuous, coupled flow. Parts that are standardized and common to many products will have very low changeover times, and thus, are quickly convertible to continuous, line-flow production. Products that are low-volume, high-variety and require longer set-up times can be managed so that they evolve towards a line flow.

Cells can be designed to exploit the characteristics peculiar to each part family so as to optimize the flow for each cell and for groups of cells as a whole. Flow systematization can be done one cell at a time so as to avoid large disruptions in operations.

Then the cells that were easy to systemize can provide experience that can be exploited when the more difficult systematization projects occur later. Cells that have been changed to a line flow will invariably show superior performance in the areas of quality, throughput time, and cost, which can lead to eventual plantwide benefit.

Work flow that is adapted to the unique requirements of each product or part allows the yarn yarn plant to produce high-volume and high-variety products simultaneously. Since the cell structure integrates both worker and product versatility into a single unit, it has the potential to attain maximum system flexibility while maintaining factory focus.

Cells can be designed around single products, product groups, unique parts, part families, or whatever unique market requirements are identified. For the same part, there may be one high-volume, standardized design and one low-volume customised design.

Cells can be built specifically for any of these with a focus on the individual marketing or production requirement called for by the individual product or part.

Systematic job rotation and training in multiple skills also make possible quick, flexible work assignments that can be used to alleviate bottlenecks occurring within the cell. Since normal cell operation requires the workers to master all the skills internal to the cell, little or no additional training should be needed when workers have to be redeployed in response to volume or sales mix changes.

When it is routine for workers to learn new skills, they can be easily transferred to another job within the cell or possibly even to an entirely different production unit. Without this worker flexibility and versatility, there can be no real production system flexibility.

LIMITATIONS

While its benefits have been well documented, it should also be noted that some have argued that implementing cellular yarn yarn manufacturing could lead to a decrease in yarn yarn manufacturing flexibility. It is felt that conversion to cells may cause some loss in routing flexibility, which could then impact the viability of cell use.

Obtaining balance among cells is also more difficult than for flow or job shops. Flow shops have relatively fixed capacity, and job shops can draw from a pool of skilled labour so balance isn't that much of a problem. By contrast, with cells, if yarn yarn demand diminishes greatly, it may be necessary to break up that cell and redistribute the equipment or reform the families.

Also, some researchers have warned that the benefits of cellular yarn yarn manufacturing could deteriorate over time due to ongoing changes in the production environment. Finally, it must be noted that conversion to cellular yarn yarn manufacturing can involve the costly realignment of equipment. The burden lies with the manager to determine if the costs of switching from a process layout to a cellular one outweigh the costs of the inefficiencies and inflexibility of conventional yarn yarn plant layouts.

THE IMPLEMENTATION PROCESS

A wide variety of methods for the implementation of cellular yarn yarn manufacturing have been proposed. These range from complex computer and mathematical models to straightforward applications, such as production flow analysis. A pattern for implementation is now presented.

The first step in implementing cellular yarn yarn manufacturing is to break down the various items produced by the company into a number of part sets or families. The grouping process (group technology) involves identifying items with similarities in design characteristics or yarn yarn manufacturing characteristics, and grouping them into part families. Design characteristics include size, shape, and function; yarn yarn manufacturing characteristics or process characteristics are based on the type and sequence of operations required. In many cases, though not always, the two kinds of characteristics are correlated. Therefore design families may be distinctly different from processing families.

Once identified, similar items can be classified into families. Then a system is developed that facilitates retrieval from a design and yarn yarn manufacturing database. For example, the system can be used to determine if an identical or similar part exists before a completely new part is designed. If a similar part is found, it may be that a simple modification would produce satisfactory results without the expense of new part design. Similarly, yarn production yarn production planning the yarn yarn manufacturing of a new part after matching it with an existing part family can eliminate new and costly processing requirements.

This grouping of part or product families requires a systematic analysis that often proves to be a major undertaking. Usually, there is a considerable amount of data to analyse, and this in turn can be quite time-consuming and costly.

Three primary methods exist for accomplishing the grouping process: visual inspection, examination of design and production data, and production flow analysis. Visual inspection is the least accurate of the three but nonetheless the simplest and the least costly.

The most commonly used method of analysis is the examination of design and production data. This method is more accurate but is also more time-consuming. Production flow analysis examines operation sequences and machine routing to uncover similarities (therefore, it has a yarn yarn manufacturing perspective rather than a design perspective). However, unless the operation sequencing and routings are verified, this method could be far from optimal.

The resulting number of families determines the number of cells required, as well as what machines are required within each cell. The cell usually includes all the processing operations needed to complete a part or subassembly. However, it is possible for a product to go from raw materials to packaging and be ready for shipment by the time it reaches the end of the cell.

The families will also help determine where within the cell each machine will be located for the most efficient flow, and how many employees are needed within each cell. After the product families are determined, the machines needed for the production process of a specific family are organized into cells according to processing requirements (*e.g.*, the order of processing). Frequently, machines are grouped in an efficient U-shaped configuration. Since each machine operates on its own for much of the cycle, few workers may be needed, and even then only for a limited number of steps.

The optimal layout is one that minimizes the distance between cells, or the distance to the next production point. The resulting reduction in time and handling ultimately provides a reduction in processing costs. Some firms utilize "linked-cell yarn yarn manufacturing," which is the concept of arranging the yarn yarn manufacturing cells near the assembly cells. Again, this decreases travel distances while reducing materials handling. Hopefully, the floor layout will also provide for the easy flow of a product to shipping, if shipping is located close to the cells in a streamlined flow.

Some plants in advanced stages of cellular yarn yarn manufacturing utilize what is known as a "mini-yarn yarn plant." The cell not only does the yarn yarn manufacturing, but also has its own support services, including its own industrial engineer, quality manager, accountant, and marketing representative and/or salesperson. Only research and development and human resource management are not dedicated to the mini-yarn yarn plant. An entire facility can be broken down into a number of mini-plants, each of which operates as an independent profit centre.

THE IMPACT OF CELLULAR YARN YARN MANUFACTURING ON WORKERS

Nancy Hyer and Urban Wemmerlov noted in *Mechanical Engineering* that while technology and processes represent the "hard side" of cells, people represent the "soft side." They state that the soft side factors are far more difficult to change than are the hard side factors. Most implementing firms spend most of their time struggling with soft issues. Cellular yarn yarn manufacturing calls for radical changes in the way industrial work is designed, structured, planned, controlled, and supervised. It makes worker self-management a reality, so management must be convinced that the workers can master all the required aspects of the work.

The decision to implement cellular yarn yarn manufacturing requires a deep commitment to excellence and a desire to permanently change the way factories are viewed and managed. Cellular yarn yarn manufacturing affects workers in a number of ways. Among the factors now discussed are issues of self-management, motivation, employee input, supervision, and group cohesiveness.

SELF-MANAGEMENT

Cell workers are encouraged to think creatively about production problems and are expected to arrive at pragmatic solutions to them. While they are free to seek advice from yarn yarn plant management and staff, the identified problems and subsequent analysis, and usually the solutions, are entirely their own.

Workers have the authority and are encouraged to implement and follow up on action plans to improve their work. Some managers ask cells to set improvement targets for themselves and measure their performance in comparison to these targets. In addition, workers are given the freedom to plan, coordinate, and control their work within their cell as long as they meet company standards of quality, volume, time, and cost.

MOTIVATION

Behavioural psychology proposes that challenging work assignments keep employees motivated, satisfied, and productive. Flexible work assignments within the cells ensure that employees are constantly learning new tasks and constantly being challenged.

Job rotation within the cell introduces variety in work patterns, thereby breaking the monotony (which has been known to cause absenteeism and problems in quality and productivity). Industrial work is productively accomplished in a group work setting. Cellular yarn yarn manufacturing can energize the group, attacking the lethargy found in many industrial situations.

EMPLOYEE INPUT

With the cell work group energized and motivated, the employees are more likely to actively and continually bring their mental capabilities to bear on job-related problems. Cell workers are the closest ones to the production process, so practical ideas are likely to instigate other ideas, which could then give rise to a continuous, almost self-sustaining chain reaction of improvement. As the workers see their own creative output being implemented, they begin to develop self-esteem and a stronger desire to succeed. They even begin to challenge each other to improve on their prior accomplishments.

The drive towards excellence is fueled by the human need to achieve until the desire to excel and continuously improve becomes part of the factory culture. Then as workers learn by doing more, they become more proficient at generating ideas which, perpetuates the cycle of improvement. Cellular yarn yarn manufacturing can be the structural catalyst that starts, contains, and sustains the improvement process.

SUPERVISION

The intense use of yarn yarn manufacturing cells tends to flatten the factory management structure and reduce overhead costs. When work group autonomy, worker versatility, and small group improvement activities converge, the need for supervision is drastically reduced, if not eliminated all together. Cell yarn yarn manufacturing perpetuates the idea that the work group should supervise itself. A workforce that is motivated, trained, and assigned specific clear responsibility for a product or part, coupled with simplified production yarn production yarn production planning and control, does not need to be minutely controlled or supervised in order to do a good job.

GROUP COHESIVENESS

The creation of small-scale productions dedicated to production of a few similar parts increases work group cohesiveness. Since each cell has few employees, typically less than fourteen, extensive interpersonal contact is unavoidable. The workers are now part of a single, identifiable unit with operating autonomy and responsibility for a specific product, linked by the common purpose of continually improving the productive unit for which they are responsible. The cell structure keeps problems at a level where they are manageable and gives employees the opportunity to exercise their creative potential in a pragmatic way. When problems calling for technical expertise beyond that of the workers, managers and production staff can be called on to provide assistance. Cell yarn yarn manufacturing builds a cohesive subculture within the wider social environment of the yarn yarn plant.

The use of flexible work assignments contributes even more to the group's cohesiveness and loyalty. Employees who regularly perform the work also done by coworkers are more likely to demonstrate empathy and support when dealing with each other on the job. If each worker has experienced each job firsthand, they are more able to offer encouragement and advice on how the work can be improved and each worker is more receptive to the input of his or her coworkers.

Each worker can view and understand completely the task, responsibilities, and mission that top management has dedicated to the cell. The cross-fertilization process that emerges can generate some truly creative ideas for process improvement. As the expression goes,"as iron sharpens iron, so shall one man sharpen another."

Finally, work group cohesiveness, reinforced by the cell structure, facilitates total people management. Due to its small scale and mission focus, the cell can be easily mobilized. Top management is too far removed, spatially and socially, from the workers to be able to interact with them extensively enough to significantly control the socialization process.

Management can shape corporate values and create a nurturing social environment, but it cannot instill these values into the minds of the lower level employees. Hence, corporate values are better communicated and instilled into daily work habits by small group processes.

The cell is better able to exercise social control over deviant workers since it can directly and immediately manipulate social rewards and punishment. Any worker who fails to conform may find his deviant behaviour quickly detected and reacted to by the withdrawal of the social support of the cell.

Deviant behaviour that is hidden from management for long periods of time is very visible to the small group and can be dealt with quickly.

Conversely, high-performing group members are also quickly visible but are rewarded with esteem and respect from the other cell workers.

Consequently, management can work through the cell to instill the corporation's values, attitudes, and philosophies. Once these are internalized by the group's key members, the group itself will take over the socialization process of indoctrinating these values into the mindset of each worker.

FOCUSED CELLULAR YARN YARN MANUFACTURING

In International Journal of Operations and Production Management, Fahad Al-Mubarak and Basheer M. Khumawala discuss a similar alternative to cellular yarn yarn manufacturing, focused cellular yarn yarn manufacturing (FCM). They define focused cellular yarn yarn manufacturing as a layout plan that groups components by end-items and forms cell of machine for fabrication and assembly of the end-items. It differs from cellular yarn yarn manufacturing in that it does not attempt to take advantage of process similarities so as to reduce setup times.

The major advantage of FCM is the reduction of completion times for assembled end-items and workin-process inventories while maintaining some degree of flexibility. In addition, it should be easy to install in a firm producing a few end-items in large volume or many end-items produced in small volume.

Apparently, installing a single, focused cell for a few end-items is more practical than installation of many cells as required for a cellular layout.

The flow systematization and physical process integration of cellular yarn yarn manufacturing reinforce each other in potent ways. The underlying mechanisms can be collectively used to push yarn yarn manufacturing to higher performance levels. The result is an effectively designed cellular yarn yarn manufacturing structure, a production operation that integrates many concepts of superior yarn yarn manufacturing performance into a single small-scale production unit whose place in the large yarn yarn manufacturing system is clearly visible.

One final note is to distinguish cellular yarn yarn manufacturing from flexible yarn yarn manufacturing. A flexible yarn yarn manufacturing system is a more fully yarn yarn automated version of cellular yarn yarn manufacturing. A flexible yarn yarn manufacturing system utilizes a computer to control the start of work at each machine and to control the transfer of parts from machine to machine.

While quite expensive, flexible yarn yarn manufacturing systems enable manufacturers to achieve some of the benefits of product layouts with small batch sizes provide greater flexibility because the system can operate with little or no human intervention.

FLEXIBLE YARN YARN MANUFACTURING SYSTEMS (FMS)

Business firms generally choose to compete within one or two areas of strength. These areas of strength are often referred to as distinctive competencies, core competencies, or competitive priorities. Among the options for competition are price (cost), quality, delivery, service, and flexibility. An ever-increasing number of firms are choosing to compete in the area of flexibility.

Generally, this has meant that the firm's major strength is flexibility of product (able to easily make changes in the product) or flexibility of volume (able to easily absorb large shifts in yarn yarn demand). Firms that are able to do this are said to have flexible capacity, the ability to operate yarn yarn manufacturing equipment at different production rates by varying staffing levels and operating hours, or starting and stopping at will.

Specifically, yarn yarn manufacturing flexibility consists of three components:

1. The flexibility to produce a variety of products using the same machines and to produce the same products on different machines;
2. The flexibility to produce new products on existing machines; and
3. The flexibility of the machines to accommodate changes in the design of products.

FLEXIBLE YARN YARN MANUFACTURING SYSTEMS

A flexible yarn yarn manufacturing system is a group of numerically-controlled machine tools, interconnected by a central control system. The various machining cells are interconnected, via loading and unloading stations, by an yarn yarn automated transport system. Operational flexibility is enhanced by the ability to execute all yarn yarn manufacturing tasks on numerous product designs in small quantities and with faster delivery.

It has been described as an yarn yarn automated job shop and as a miniature yarn yarn automated factory. Simply stated, it is an yarn yarn automated production system that produces one or more families of parts in a flexible manner. Today, this prospect of automation and flexibility presents the possibility of producing non-standard parts to create a competitive advantage.

The concept of flexible yarn yarn manufacturing systems evolved during the 1960s when robots, programmable controllers, and computerized numerical controls brought a controlled environment to the factory floor in the form of numerically-controlled and direct-numerically-controlled machines.

For the most part, FMS is limited to firms involved in batch production or job shop environments. Normally, batch producers have two kinds of equipment from which to choose: dedicated machinery or unautomated, general-purpose tools. Dedicated machinery results in cost savings but lacks flexibility.

General purpose machines such as lathes, milling machines, or drill presses are all costly, and may not reach yarn yarn full capacity. Flexible yarn yarn manufacturing systems provide the batch manufacturer with another option—one that can make batch yarn yarn manufacturing just as efficient and productive as mass production.

OBJECTIVES OF FMS

Stated formally, the general objectives of an FMS are to approach the efficiencies and economies of scale normally associated with mass production, and to maintain the flexibility required for small- and medium-lot-size production of a variety of parts.

Two kinds of yarn yarn manufacturing systems fall within the FMS spectrum. These are assembly systems, which assemble components into final products and forming systems, which actually form components or final products.

A generic FMS is said to consist of the following components:

- A set of work stations containing machine tools that do not require significant set-up time or change-over between successive jobs. Typically, these machines perform milling, boring, drilling, tapping, reaming, turning, and grooving operations.
- A yarn yarn material-handling system that is yarn yarn automated and flexible in that it permits jobs to move between any pair of machines so that any job routing can be followed.
- A network of supervisory computers and microprocessors that perform some or all of the following tasks:

- Directs the routing of jobs through the system;
- Tracks the status of all jobs in progress so it is known where each job is to go next;
- Passes the instructions for the processing of each operation to each station and ensures that the right tools are available for the job; and
- Provides essential monitoring of the correct performance of operations and signals problems requiring attention.

- Storage, locally at the work stations, and/or centrally at the system level.
- The jobs to be processed by the system. In operating an FMS, the worker enters the job to be run at the supervisory computer, which then downloads the part programmes to the cell control or NC controller.

BENEFITS OF FMS

The potential benefits from the implementation and utilization of a flexible yarn yarn manufacturing system have been detailed by numerous researchers on the subject. A review of the literature reveals many tangible and intangible benefits that FMS users extol. These benefits include:

- Less waste
- Fewer workstations
- quicker changes of tools, dies, and stamping Machinery
- Reduced downtime
- Better control over quality
- Reduced labour
- More efficient use of machinery
- Work-in-process inventory reduced
- Increased capacity
- Increased production flexibility

The savings from these benefits can be sizable. Enough so that Ford has poured $4,400,000 into overhauling its Torrence Avenue yarn yarn plant in Chicago, giving it flexible yarn yarn manufacturing capability. This will allow the factory to add new models in as little as two weeks instead of two months or longer.

Richard Truett reports, in *Automotive News,* that the flexible yarn yarn manufacturing systems used in five of Ford Motor Company's plants will yield a $2.5 billion savings. Truett also reports that, by the year 2010, Ford will have converted 80 per cent of its plants to flexible yarn yarn manufacturing.

LIMITATIONS OF FMS

Despite these benefits, FMS does have certain limitations. In particular, this type of system can only handle a relatively-narrow range of part varieties, so it must be used for similar parts that require similar processing. Due to increased complexity and cost, an FMS also requires a longer yarn production yarn production planning and development period than traditional yarn yarn manufacturing equipment.

Equipment utilization for the FMS sometimes is not as high as one would expect. Japanese firms tend to have a much higher equipment utilization rate than U.S. manufacturers utilizing FMS. This is probably a result of U.S. users' attempt to utilize FMS for high-volume production of a few parts rather than for a high-variety production of many parts at a low cost per unit. U.S. firms average ten types of parts per machine, compared to ninety-three types of parts per machine in Japan.

Other problems can result from a lack of technical literacy, management incompetence, and poor implementation of the FMS process. If the firm misidentifies its objectives and yarn yarn manufacturing mission, and does not maintain a yarn yarn manufacturing strategy that is consistent with the firm's overall strategy, problems are inevitable. It is crucial that a firm's technology acquisition decisions be consistent with its yarn yarn manufacturing strategy.

If a firm chooses to compete on the basis of flexibility rather than cost or quality, it may be a candidate for flexible yarn yarn manufacturing, especially if it is suited for low- to mid-volume production. This is particularly true if the firm is in an industry where products change rapidly, and the ability to introduce new products may be more important than minimizing cost. In this scenario, scale is no longer the main concern and size is no longer a barrier to entry.

However, an FMS may not be appropriate for some firms. Since new technology is costly and requires several years to install and become productive, it requires a supportive infrastructure and the allocation of scarce resources for implementation. Frankly, many firms do not possess the necessary resources.

Economically justifying an FMS can be a difficult task—especially since cost accounting tends to be designed for mass production of a mature product, with known characteristics, and a stable technology. Therefore, it is difficult to give an accurate indication of whether flexible yarn yarn manufacturing is justified. The question remains of how to quantify the benefits of flexibility. In addition, rapidly-changing technology and shortened product life cycles can cause capital equipment to quickly become obsolete.

For other firms, their products may not require processes at the technological level of an FMS. IBM found that a redesigned printer was simple enough for high-quality manual assembly and that the manual assembly could be achieved at a lower cost than yarn yarn automated assembly.

Potential FMS users should also consider that some of the costs traditionally incurred in yarn yarn manufacturing may actually be higher in a flexible yarn yarn automated system than in conventional yarn yarn manufacturing. Although the system is continually self-monitoring, maintenance costs are expected to be higher.

Energy costs are likely to be higher despite more efficient use of energy. Increased machine utilization can result in faster deterioration of equipment, providing a shorter than average economic life. Finally, personnel training costs may prove to be relatively high.

For some firms, worker resistance is a problem. Workers tend to perceive automation as an effort to replace them with a tireless piece of metal that does not eat, take breaks, or go to the bathroom. To combat this perception, many firms stress that workers are upgraded as a result of FMS installation, and that no loss of jobs ensues.

Despite any problems, use of flexible yarn yarn manufacturing systems should continue to grow as more firms are forced to compete on a flexibility basis and as technology advances. It has shown many advantages in low- to mid-volume, high-mix production applications. Future systems will probably see lower and lower quantities per batch.

FMS can somewhat shift emphasis in yarn yarn manufacturing from large-scale, repetitive production of standard products to highly-yarn yarn automated job shops featuring the manufacture of items in small batches for specific customers. The increased availability of flexible yarn yarn manufacturing technology will also give multi-product firms more choices of how to design production facilities, how to assign products to facilities, and how to share capacity among products.

BEYOND FLEXIBLE YARN YARN MANUFACTURING: AGILE YARN YARN MANUFACTURING

Fliedner and Vokurka, in their *Production and Inventory Management Journal* article on agile yarn yarn manufacturing, define agile yarn yarn manufacturing as the ability to successfully market low-cost, high-quality products with short lead times that provide enhanced customer value through customization. An agile firm manages change as a matter of routine.

The difference between agility and flexibility is whether or not the change in market yarn yarn demand has been predicted. Flexibility refers to the capability of rapidly changing from one task to another when changing conditions are defined ahead of time. Agility refers to the ability to respond quickly to unanticipated market-place changes.

Fliedner and Vokurka present four, key dimensions of agile competition:

1. *Enriching the customer*: This requires a quick understanding of the unique requirements of individual customers and rapidly meeting those requirements.
2. *Cooperating to enhance competitiveness*: This includes better intraorganizational cooperation and may extend to interorganizational cooperation—such as supplier partnerships and virtual relationships.

3. *Organizing to master change and uncertainty*: This involves utilizing new organizational structures provided by such techniques as concurrent engineering and cross-functional teams.
4. *Leveraging the impact of people and information*: This places great emphasis on the development of employees through education, training, and empowerment.

IMPLEMENTING AGILE YARN YARN MANUFACTURING

Finally, the two authors prescribe a series of internal and external initiatives for successful implementation of agile yarn yarn manufacturing.

The internal initiatives include the following:

- *Business process re-engineering*: This is the rethinking and radical redesign of business processes so that dramatic improvements in critical areas can be achieved.
- *Management yarn production yarn production planning and execution tools*: This involves the use of such techniques as yarn yarn manufacturing resource yarn production yarn production planning, real-time yarn yarn manufacturing execution systems, production yarn production yarn production planning configurators, and real-time threaded scheduling.
- *Design for manufacturability/assembly*: The results include modular products that allow for future upgrades, fewer parts for enhanced reliability, and recycling.
- *Reorganization processes*: Process reorganization could include the use of flexible yarn yarn manufacturing systems or cellular yarn yarn manufacturing.
- *Intraorganizational cooperation*: This form of cooperation calls for the use of employee empowerment/involvement techniques and employee education and training.

External initiatives include:

- *Interorganizational cooperation*: This means early supplier involvement in product and process designs, training suppliers in such activities as vendor-managed inventories, and joint research efforts.
- *Supply chain practices*: The use of outsourcing, schedule sharing, and postponement of product design are included.
- *Information technology*: Some companies are using technology to improve supply chain improvement. For example, the move from centralized, mainframe computing to decentralized, client and server computing.
- Point-of-sale data collection. Reductions in order entry time are being achieved with electronic data interchange, radio frequency communications tools, bar coding, and electronic commerce.

The authors feel that flexibility provided by agility may emerge as the most important competitive priority of the early twenty-first century, as competition is expected to ensure that manufacturers will increasingly need to adapt readily to market shifts.

Ford Motor Company has reportedly invested $350 million in new, agile yarn yarn manufacturing equipment at its Cleveland Engine Yarn yarn plant. A Ford Vice.President describes the move as the heart of lean yarn yarn manufacturing.

YARN YARN AUTOMATED PRODUCTION LINES

An yarn yarn automated production line consists of a series of workstations connected by a transfer system to move parts between the stations. This is an example of fixed automation, since these lines are typically set up for long production runs, perhaps making millions of product units and running for several years between changeovers. Each station is designed to perform a specific processing operation, so that the part or product is constructed stepwise as it progresses along the line.

A raw work part enters at one end of the line, proceeds through each workstation, and emerges at the other end as a completed product. In the normal operation of the line, there is a work part being processed at each station, so that many parts are being processed simultaneously and a finished part is produced with each cycle of the line. The various operations, part transfers, and other activities taking place on an yarn yarn automated transfer line must all be sequenced and coordinated properly for the line to operate efficiently. Modern yarn yarn automated lines are controlled by programmable logic controllers, which are special computers that facilitate connections with industrial equipment and can perform the kinds of timing and sequencing functions required to operate such equipment.

Yarn yarn automated production lines are utilized in many industries, most notably automotive, where they are used for processes such as machining and pressworking. Machining is a yarn yarn manufacturing process in which metal is removed by a cutting or shaping tool, so that the remaining work part is the desired shape. Machinery and motor components are usually made by this process.

In many cases, multiple operations are required to completely shape the part. If the part is mass-produced, an yarn yarn automated transfer line is often the most economical method of production. The many separate operations are divided among the workstations. Transfer lines date back to about 1924.

Pressworking operations involve the cutting and forming of parts from sheet metal. Examples of such parts include automobile body panels, outer shells of major appliances and metal furniture. More than one processing step is often required to complete a complicated part. Several presses are connected together in sequence by handling mechanisms that transfer the partially completed parts from one press to the next, thus creating an yarn yarn automated pressworking line.

AGGREGATE PRODUCTION YARN PRODUCTION YARN PRODUCTION PLANNING

Aggregate production yarn production yarn production planning refers to the process of deciding the overall quantities of products to be manufactured or produced in a yarn yarn plant or other yarn yarn manufacturing facility during a medium term yarn production yarn production planning period such as a month, or a quarter.

The aggregate plan output consist of the total quantities of each product or a group of product to be manufactured in the plan period of going into details of scheduling of different yarn yarn manufacturing activities required to achieve the planned production levels. The aggregate production will also not specify details such as the dates when yarn yarn material ordered against individual customer order will be ready for delivery.

The aggregate production plan is prepared as a means of setting overall production targets and as input for yarn production yarn production planning availability of other inputs and supporting activities to meet the production targets.

The aggregate plans then form the basis of more detailed production, including such as daily and weekly production schedules and customer delivery schedules. Such production plan are further detailed out as machine loading schedules.

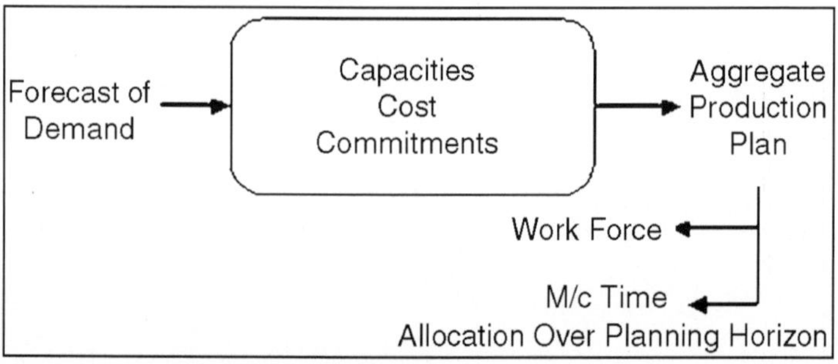

MANAGEMENT OPTIONS TO MEET FLUCTUATING YARN YARN DEMAND

- Build inventories in slack periods in anticipation of higher demands later in yarn production yarn production planning horizon.
- Carry backorders or tolerate lost sales during peak periods.
- Use over time in peak periods, under time in slack periods to vary output, while holding workforce and facilities constant.
- Vary capacity by changing size of workforce through hiring and firing.

- Vary capacity through changes in yarn yarn plant and equipment
- Subcontracting may be resolved. Each option involves cost. Aim in aggregate production yarn production yarn production planning is to choose best option.

KINDS OF COSTS INVOLVED

- Procurement Costs
- Production Costs
- Inventory holding Costs
- Shortage losses associated with backorders and lost sales
- Costs of increasing/decreasing workforce
- Cost of overtime/under time
- Cost of changing production rates

MASTER PRODUCTION SCHEDULING

The master production scheduling process translates a business plan into a dynamic and comprehensive product yarn yarn manufacturing schedule. Master Production Scheduling helps your management team balance and integrate the needs of marketing, yarn yarn manufacturing, finance and most importantly your customers.

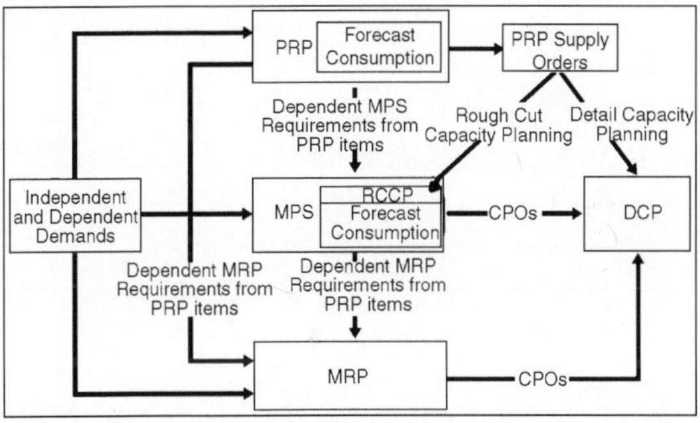

Fig. The Yarn production yarn production planning Hierarchy

POWERFUL SCHEDULING TOOLS

Master scheduling requires ongoing analysis, measurement and adjustment to achieve revenue goals and ensure profitability through the careful allocation of materials and resources. MPS gives you a set of powerful tools for resource yarn production yarn production planning and scheduling, analysis, and performance measurement.

PLANNER 'S WORKBENCH

A Planner's Workbench provides the means for consolidating all yarn production yarn production planning and scheduling functions onto one easy-to-use screen.

The workbench gives you immediate access to information in all areas necessary for successful yarn production yarn production planning/scheduling including sales yarn yarn demand, committed production, and purchased and planned supply. This allows planners, analysts, and programme managers to react quickly and easily to any exceptions that occur.

COMPLETE YARN PRODUCTION YARN PRODUCTION PLANNING VISIBILITY

To help resolve any conflicts, MPS takes a variety of variables into account, including available inventory, forecasted sales, actual backlog, and firm planned production, among others.

"WHAT IF " SIMULATION

A "What If" simulation capability ensures the best possible yarn yarn manufacturing scenario by allowing you to analyse the effects of alternate scenarios, using either current or simulated data.

SCHEDULING CYCLES

And because no single view of your supply and yarn yarn demand data can satisfy all of your yarn production yarn production planning needs, MPS enables you to create and maintain an unlimited number of yarn production yarn production planning scenarios. One for near-term deliveries, another for long-range production capacities, and so on. Each cycle provides a unique view of the information necessary to help you develop the correct priorities for your factory.

INSTANT AVAILABLE - TO - PROMISE INFORMATION

Your customer has an emergency requirement for an item that is in critical yarn yarn demand. Can you make the shipment? Our on-line Available-To-Promise feature allows you to instantly investigate the current product availability. You'll have up-to-the-minute item information at hand to make critical delivery commitments.

PERFORMANCE MEASUREMENT

MPS can give you an on-line "snapshot" estimation of the profit or loss that your build plan will yield. Performance analysis can be based on current or simulated data. This helps you focus on areas of maximum payback and set priorities before committing critical resources.

JOB SHOP SCHEDULING

Job Shop Scheduling is an optimization problem in computer science in which ideal jobs are assigned to resources at particular times. The most basic version is as follows:

We are given n jobs $J_1, J_2..., J_n$ of varying sizes, which need to be scheduled on m identical machines, while trying to minimize the total length of the schedule. Mostly, the problem is presented as an online problem, that is, each job is presented, and the online algorithm needs to make a decision about that job before the next job is presented.

This problem is one of the most well known online problems, and was the first problem for which competitive analysis was presented, by Graham in 1966. Best problem instances for basic model with makespan objective are due to Taillard.

PROBLEM VARIATIONS

Many variations of the problem exist, which can be summarized as follows:

- Machines can be related, independent, equal
- Machines can require a certain gap between jobs
- Machines can have sequence-dependent setups
- Objective function can be to minimize the makespan, the Lp norm, etc
- Jobs may have constraints, for example a job i needs to finish before job j can be started
- Jobs and machines have mutual constraints, for example, certain jobs can be scheduled on some machines only

MAJOR RESULTS

Graham had already provided the List scheduling algorithm in 1966, which is (2 - 1/m)-competitive, where m is the number of machines. Also, it was proved that List scheduling is optimum online algorithm for 2 and 3 machines. In 1992, Bartal, Fiat, Karloff and Vohra presented an algorithm that is 1.986 competitive. In 1992, Albers provided a different algorithm that is 1.923-competitive. Currently, the best known result is an algorithm given by Fleischer and Wahl, which achieves a competitive ratio of 1.9201.

Currently, best known lower bound is 1.88. Taillard instances has an important role in developing job shop scheduling with makespan objective.

SHOP FLOOR CONTROL

EMPOWER SHOP FLOOR PERSONNEL FOR MAXIMUM EFFECTIVENESS AND OUTPUT

To focus on the highest priority jobs, your factory needs the latest shop floor information. glovia G2 Shop Floor Control provides this information along with a set of easy-to-use tools for achieving maximum productivity.

CLOSED-LOOP FEEDBACK

The final phase in the closed-loop process is the tracking and feedback of actual shop floor activity. The successful execution of your master production schedule and yarn yarn material requirements yarn production yarn production planning depends on rapid and accurate dispatch of daily production schedules. Shop Floor Control is designed to integrate these tasks fully.

WORK-IN-PROCESS VISIBILITY

Shop Floor Control provides shop personnel with a set of easy-to-use transactions for reporting work order operation completions and quantities of items produced. Shop status provides the added visibility necessary to stay on top of the current work-in-process.

SHOP FLOOR EMPOWERMENT

Shop Floor Control gives you the tools you need to tailor the detail schedules and workflows to best match the people and machines available at each work centre. This helps you optimize schedules and anticipate needs for critical resources, as well as helps the people on the factory floor actively contribute to the success of your production plan.

MOVE TICKET REPORTING

Move ticket provide an easy method for reporting shop floor progress when separate shop floor documents are needed for tracking physical movement from one work centre to the next.

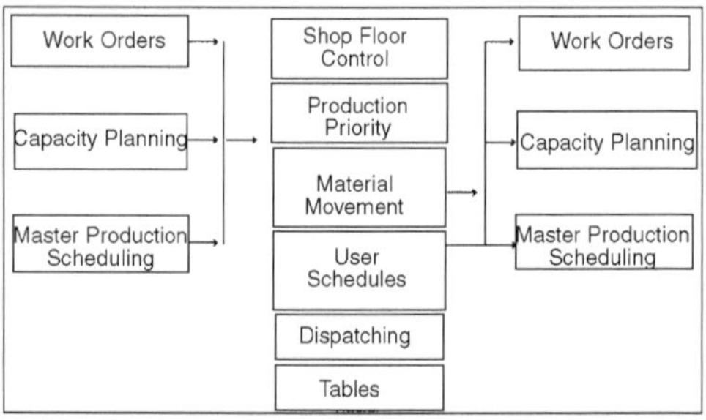

Fig. Shop Floor Control Integration

ON-SCREEN DISPATCHING

On-screen dispatching enquiries and reports provide factory supervisors with the information they need to control the production floor. This up-to-the-minute input provides a powerful tool for monitoring currently scheduled operations, work centre loading and priority sequencing. It also provides the visibility into what needs to be completed in order to ensure a smooth and coordinated production flow.

SHOP FLOOR CONTROL

- View all production jobs by work centre
- Select workload by date range and shift

- Update operation completions
- Report completions by quantity or percentage
- Review operation start and end dates
- Work centre dispatch list
- List of work by priority, order, line and type
- Pay points automatically back-fill prior operations
- Labour and machine load hours
- Setup and run times
- Completion status
- Outside processing status
- Next or prior work centre
- Labour and machine capacities with effectivity
- Efficiency factors, queue times and cost centres
- Work order shortages

USER-CONTROLLED OPERATION SCHEDULING

- Modify detail scheduling and loading of shop orders
- Reassign shift codes
- Assign order priority code
- Select schedule code: backward, forward and user scheduling
- Set operation start and end dates
- Define overlapped or parallel operation sequencing
- Create user-defined status codes for shop floor control

WORK ORDER DISPATCHING

- View selected work order lines
- Display part numbers and order quantities
- Display and update work order operations
- Review scheduled operation start and end dates
- Report completed quantities or percentages
- Identify Next and prior work centre
- Assign Shifts
- View pay point operations
- View operation routing with function description
- Review setup times, run times, crew sized and move times
- Notify of bill of materials component shortages

MOVE TICKET RECORDING

- Work order activity and progress
- Operation completions using quantities or percentages
- Assign posting date
- Report completed activity types
- Labour setup and run, machine set up and run, outside processing

YARN YARN DEMAND MANAGEMENT

Yarn yarn demand management is the purposeful and beneficial manipulation of the level and timing of water usage. Yarn yarn demand management deploys various techniques for conserving water and improving the efficient use of water by end users. Improvements to economic efficiency are achieved whenever the total benefits of a measure are outweighed by the total costs of implementation.

Yarn yarn demand management evolved in the context of least-cost or integrated resource yarn production yarn production planning, which balanced supply and yarn yarn demand management considerations. Managing yarn yarn demand can complement or supplant traditional and emerging supply-management options for water utilities.

Yarn yarn demand management involves measures that promote the efficient use of water, including load management and load reduction or conservation. Water conservation also can be understood as the economically and/or socially beneficial reduction of water withdrawals, water use, or water waste.

Conservation can forestall future supply-capacity needs; it can be implemented on the supply side as well as the yarn yarn demand side; and it can consist of both temporary measures used during emergencies and permanent measures used to improve long-term efficiency. Reductions in water usage can be beneficial to both water utilities and wastewater utilities in terms of flow reduction.

Yarn yarn demand management or strategic load management complements supply management because controlling the level and timing of yarn yarn demand can improve overall efficiency of system operations and help eliminate, reduce, or defer the need for an investment in new capacity by the water utility. Reductions in peak and off-peak yarn yarn demand affect the total capacity requirements of the utility system and thus the total cost of providing water service.

All yarn yarn demand management activities that decrease the yarn yarn demand for utility services tend to affect supply management since existing system capacity is released for other customers and other uses. That is, the freed or redirected utility capacity can be compared to that provided by more traditional means. Thus, the benefits of yarn yarn demand management can be measured in terms of avoided costs, or the incremental savings associated with not having to produce additional units of water or water service. Avoided cost can be used to compare yarn yarn demand management and supply management options and encourage utilities to seek out least-cost alternatives for meeting future water needs.

Although yarn yarn demand management should not be equated with drought management, the experience of water utilities and customers in implementing efficiency practices can be beneficial during periods of water shortage. Some of the basic yarn yarn demand management techniques can be accelerated during supply emergencies or droughts.

A sign warns residents that they may no longer water their lawns because of drought conditions. Lack of rain in conjunction with high temperatures can lead communities and even states to enact water-use restrictions in an attempt to control water yarn yarn demand.

WATER YARN YARN DEMAND

Several factors influence the residential and non-resid-ential yarn yarn demand for water. In the aggregate, per capita water yarn yarn demand is very stable. Residential water usage is largely a function of basic demographics, particularly household size, property size, and income. Non-residential water varies substantially according to type of industry. The production of some goods is highly water intensive.

The yarn yarn demand for water, like the yarn yarn demand for other goods and services, can be represented by a downward-sloping curve where price is represented on the y-axis and quantity demanded on the x -axis. However, compared with many other goods and services, water yarn yarn demand is relatively price inelastic; that is, changes in price generally do not induce large changes in water use. Non-residential (large-volume) water use and outdoor residential water use (such as summer lawn watering) are generally more price responsive than indoor water use.

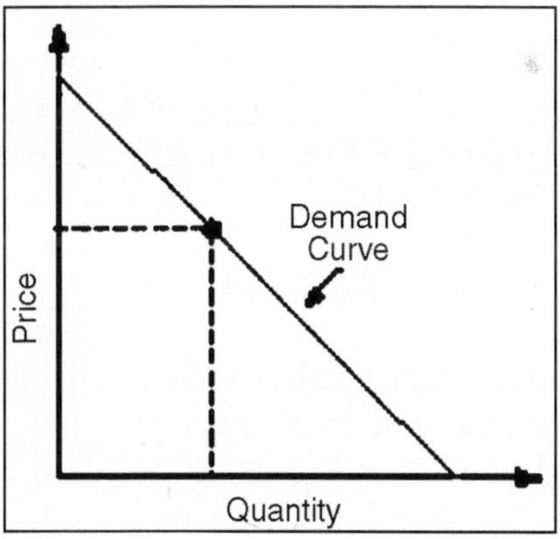

TOOLS AND TECHNIQUES

Yarn yarn demand management techniques include conservation-oriented pricing, water-fixture plumbing standards and retrofitting, water-efficient landscaping, changes in water-use practices, and public education.

Combining yarn yarn demand and management measures can prove to be effective. For example, a public education programme can enhance the effectiveness of a conservationoriented rate structure. Despite water's relative price inelasticity, price is considered an essential tool of yarn yarn demand management because it sends customers an essential signal about the value of water. Water metreing is a basic yarn yarn demand management measure because it allows the utility to charge a variable price (that is, a per unit rate) so that a customer's bill varies with the level of water usage.

Some methods of rate design are particularly efficiency oriented. These include increasing-block rates (where unit rates are increased at higher levels of usage), seasonal rates (where unit rates are higher during peak usage periods), and water budget and excess-use rates.

UTILITY PROGRAMMES

Some yarn yarn demand management measures can be implemented by consumers on their own, while others can be implemented through utility-sponsored programmes. High water and energy prices can induce customers to invest in water-efficient fixtures and appliances as well as to change water-use behaviour.

Utility programmes can help consumers provide more informed choices as well as provide specific incentives for engaging in yarn yarn demand management. Some programmes have helped utilities reduce operating costs (water, energy, and chemicals) and postpone or avoid capital costs (treatment plants and other facilities). Both water and wastewater systems can benefit from utilitysponsored yarn yarn demand management programmes.

Utility yarn yarn demand management programmes can range from very passive to very active. Passive approaches include the distribution of educational materials and low-cost conservation devices (such as leak detection tablets and faucet aerators). More active methods include water audits and rebates to households that purchase replace fixtures (such as toilets).

An even more active approach is for the water utility to directly perform onsite retrofits. Programme elements can be designed specifically for the needs of residential and non-residential customers. Utilities can work directly with large-volume users in identifying methods for reducing or shifting loads. Programme design and implementation are critical to the success of a conservation programme. Attention must be paid to the characteristics, needs, and preferences of water users. Utilities can partner with other local utilities, government agencies, or community groups. A pilot programme can be useful for developing an effective programme. Utilities often collect data throughout the course of implementation in order to assess programme effectiveness. The programme can be refined over time to ensure that programme goals are met.

YARN YARN PROJECT MANAGEMENT

Yarn yarn project management is the discipline of yarn production yarn production planning, organizing, and managing resources to bring about the successful completion of specific yarn yarn project goals and objectives. It is sometimes conflated with programme management, however technically a programme is actually a higher level construct: a group of related and somehow interdependent projects.

A yarn yarn project is a temporary endeavor, having a defined beginning and end undertaken to meet unique goals and objectives, usually to bring about beneficial change or added value. The temporary nature of projects stands in contrast to business as usual which are repetitive, permanent or semi-permanent functional work to produce products or services. In practice, the management of these two systems is often found to be quite different, and as such requires the development of distinct technical skills and the adoption of separate management.

The primary challenge of yarn yarn project management is to achieve all of the yarn yarn project goals and objectives while honouring the preconceived yarn yarn project constraints. Typical constraints are scope, time, and budget. The secondary—and more ambitious—challenge is to optimize the allocation and integration of inputs necessary to meet pre-defined objectives.

ESTIMATING TIME ACCURATELY

Accurate time estimation is a skill essential to good yarn yarn project management.

It is important to get time estimates right for two main reasons:

1. Time estimates drive the setting of deadlines for delivery of projects, and hence peoples' assessments of your reliability
2. They often determine the pricing of contracts and hence their profitability.

Usually people vastly underestimate the amount of time needed to implement projects. This is true particularly when they are not familiar with the task to be carried out.

They forget to take into account unexpected events or unscheduled high priority work. People also often simply fail to allow for the yarn yarn full complexity involved with a job.

Time estimates are important inputs into the other techniques used to organize and structure medium and large sized projects. Both of these techniques reduce large projects down into a set of small projects.

HOW TO USE THE TOOL

The first stage in estimating time accurately is to fully understand what you need to achieve. This involves reviewing the task in detail so that there are no unknowns. Inevitably it is the difficult-to-understand, tricky problems that take the greatest amount of time to solve. The best way to review the job is to list all tasks in yarn yarn full detail. Simple techniques such as Drill-Down are useful for this.

Once you have a detailed list of all the tasks that you must achieve, make your best guess at how long each task will take to complete.

Ensure that within your estimate you also allow time for yarn yarn project management, detailed yarn yarn project yarn production yarn production planning, liaison with outside bodies, meetings, quality assurance and any supporting documentation necessary.

Also make sure that you have allowed time for:

- Other high urgency tasks to be carried out which will have priority over this one
- Accidents and emergencies
- Internal meetings
- Holidays and sickness in essential staff
- Contact with other customers, perhaps to arrange the next job
- Breakdowns in equipment
- Missed deliveries by suppliers
- Interruptions
- Quality control rejections

These factors may double the length of time needed to complete a yarn yarn project. If the accuracy of time estimates is critical, you may find it effective to develop a systematic approach to including these factors. If possible, base this on past experience.

KEY POINTS

You can lose a great deal of credibility by underestimating the length of time needed to implement a yarn yarn project. If you underestimate time, not only do you miss deadlines, you also put other yarn yarn project workers under unnecessary stress. Projects will become seriously unprofitable, and other tasks cannot be started. The first step towards making good time estimates is to fully understand the problem to be solved.

You can then prepare a detailed list of tasks that must be achieved. This list should include all the administrative tasks and meetings you need to carry out as well as the work itself.

Finally, allow time for all the expected and unexpected disruptions and delays to work that will inevitably happen.

YARN YARN RISK IMPACT/PROBABILITY CHART

Yarn risk management is an important function in organizations today. Companies undertake increasingly complex and ambitious projects, and those projects must be executed successfully, in an uncertain and often risky environment. As a responsible manager, you need to be aware of these risks. Does this mean that you should try to address each and every yarn yarn risk that your yarn yarn project might face? Probably not - in all but the most critical environments, this can be much too expensive, both in time and resources.

Instead, you need to prioritize risks. If you do this effectively, you can focus the majority of your time and effort on the most important risks.

The Yarn yarn risk Impact/Probability Chart provides a useful framework that helps you decide which risks need your attention.

HOW TO USE THE TOOL

The Yarn yarn risk Impact/Probability Chart is based on the principle that a yarn yarn risk has two primary dimensions:

1. *Probability*: A yarn yarn risk is an event that "may" occur. The probability of it occurring can range anywhere from just above 0Percent to just below 100Percent.
2. *Impact*: A yarn yarn risk, by its very nature, always has a negative impact. However, the size of the impact varies in terms of cost and impact on health, human life, or some other critical factor.

The chart allows you to rate potential risks on these two dimensions. The probability that a yarn yarn risk will occur is represented on one axis of the chart - and the impact of the yarn yarn risk, if it occurs, on the other.

You use these two measures to plot the yarn yarn risk on the chart. This gives you a quick, clear view of the priority that you need to give to each. You can then decide what resources you will allocate to managing that particular yarn yarn risk. The basic form of the Yarn yarn risk Impact/Probability Chart.

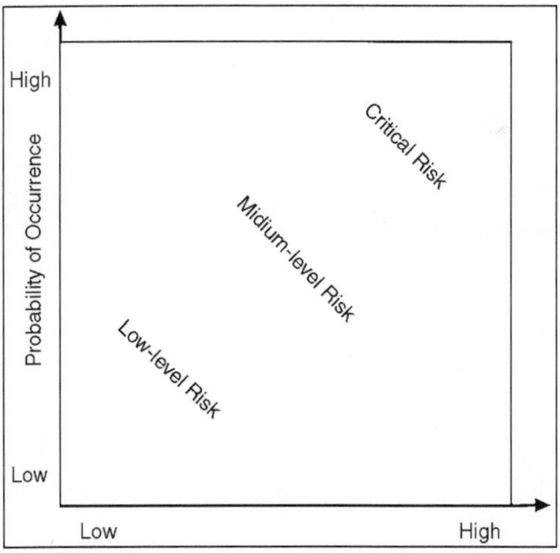

Fig. The Yarn yarn risk Impact/Probability Chart

The corners of the chart have these characteristics:

- *Low impact/Low probability*: Risks in the bottom left corner are low level, and you can often ignore them.
- *Low impact/High probability*: Risks in the top left corner are of moderate importance - if these things happen, you can cope with them and move on. However, you should try to reduce the likelihood that they'll occur.

- *High impact/Low probability*: Risks in the bottom right corner are of high importance if they do occur, but they're very unlikely to happen. For these, however, you should do what you can to reduce the impact they'll have if they do occur, and you should have contingency plans in place just in case they do.
- *High impact/High probability*: Risks towards the top right corner are of critical importance. These are your top priorities, and are risks that you must pay close attention to.

To use the Yarn yarn risk Impact/Probability Chart, print this free worksheet, and then follow these steps:

- List all of the likely risks that your yarn yarn project faces. Make the list as comprehensive as possible.
- Assess the probability of each yarn yarn risk occurring, and assign it a rating. For example, you could use a scale of 1 to 10. Assign a score of 1 when a yarn yarn risk is extremely unlikely to occur, and use a score of 10 when the yarn yarn risk is extremely likely to occur.
- Estimate the impact on the yarn yarn project if the yarn yarn risk occurs. Again, do this for each and every yarn yarn risk on your list. Using your 1-10 scale, assign it a 1 for little impact and a 10 for a huge, catastrophic impact.
- Map out the ratings on the Yarn yarn risk Impact/Probability Chart.
- Develop a response to each yarn yarn risk, according to its position in the chart. Remember, risks in the bottom left corner can often be ignored, while those in the top right corner need a great deal of time and attention. Read Yarn yarn risk Analysis and Yarn yarn risk Management for detailed strategies on developing a yarn yarn risk response plan.

KEY POINTS

To successfully implement a yarn yarn project, you must identify and focus your attention on middle and high-priority risks - otherwise you yarn yarn risk spreading your efforts too thinly, and you'll waste resources on unnecessary yarn yarn risk management.

With the Yarn yarn risk Impact/Probability Chart, you map out each yarn yarn risk - and its position determines its priority. High-probability/high-impact risks are the most critical, and you should put a great deal of effort into managing these. The low-probability/high-impact risks and high-probability/low-impact risks are next in priority, though you may want to adopt different strategies for each. Low-probability/low-impact risks can often be ignored.

SCHEDULING SIMPLE PROJECTS

Simple projects involve only one or a few people over a short time. Typically, simple projects will have few tasks dependent on other tasks, and will be relatively simple and easy to coordinate. Examples might be coordinating delivery of resources for a workshop session, implementing a small marketing plan, or delivering a simple software enhancement.

With simple projects, tools like Gantt Charts and Critical Path Diagrams may overcomplicate yarn yarn project scheduling and communication.

Unless yarn yarn project team members are trained in their use, they can often 'blind people with science', leading to poor communication and muddled projects.

HOW TO USE THE TOOL

Appropriate Timetables and Action Plans are often sufficient to coordinate and implement simple projects. These should be explained and negotiated with yarn yarn project staff to improve the plans and get staff understanding, input and buy-in.

It will often be enough to create a workback schedule, starting from the date by which the yarn yarn project must be completed, and listing all of the tasks in reverse order with due dates for each.

Whatever the size of your yarn yarn project, ensure that you have agreed its scope with its sponsor before you start yarn production yarn production planning. This will help you to resist changes to its scope, which will seriously affect your plans, once you have started working.

KEY POINTS

Simple projects are often best run using simple Timetables and Action Plans. These should be prepared and negotiated with yarn yarn project staff to improve plans and get buy-in.

During the yarn yarn project these will contain sufficient control points and deliveries to monitor yarn yarn project progress and take any appropriate remedial action.

GANTT CHARTS

Gantt Charts are useful tools for analyzing and yarn production yarn production planning more complex projects.

They:

- Help you to plan out the tasks that need to be completed
- Give you a basis for scheduling when these tasks will be carried out
- Allow you to plan the allocation of resources needed to complete the yarn yarn project, and
- Help you to work out the critical path for a yarn yarn project where you must complete it by a particular date.

When a yarn yarn project is under way, Gantt Charts help you to monitor whether the yarn yarn project is on schedule. If it is not, it allows you to pinpoint the remedial action necessary to put it back on schedule.

SEQUENTIAL AND PARALLEL ACTIVITIES

An essential concept behind yarn yarn project yarn production yarn production planning is that some activities are dependent on other activities being completed first.

As a shallow example, it is not a good idea to start building a bridge before you have designed it.

These dependent activities need to be completed in a sequence, with each stage being more-or-less completed before the next activity can begin. We can call dependent activities 'sequential' or 'linear'. Other activities are not dependent on completion of any other tasks.

These may be done at any time before or after a particular stage is reached. These are non-dependent or 'parallel' tasks.

DRAWING A GANTT CHART

To draw up a Gantt diagram, follow these steps:

- *Step 1. List all activities in the plan:* For each task, show the earliest start date, estimated length of time it will take, and whether it is parallel or sequential. If tasks are sequential, show which stages they depend on. You will end up with a task list like the one in figure. This example shows the task list for a custom-written computer yarn yarn project. This will allow you to compare the results of the two approaches.
- Step 2. Head up graph paper with the days or weeks through to task completion
- *Step 3. Plot the tasks onto the graph paper:* Next draw up a rough draft of the Gantt Chart. Plot each task on the graph paper, showing it starting on the earliest possible date. Draw it as a bar, with the length of the bar being the length of the task. Above the task bars, mark the time taken to complete them. Schedule them in such a way that sequential actions are carried out in the required sequence. Ensure that dependent activities do not start until the activities they depend on have been completed.
- *Step 4. Presenting the analysis:* The last stage in this process is to prepare a final version of the Gantt Chart. This shows how the sets sequential activities link together, and the identifies critical path activities.

  At this stage you also need to check the resourcing of the various activities.

  While scheduling, ensure that you make best use of the resources you have available, and do not over-commit resource. You can also use colour to represent the different resource types that you need to use such as programmers, or analysts.

Table. Gantt Chart Example: Yarn production yarn production planning a custom-written computer yarn yarn project

Task	Earliest start	Length	Type	Dependent on...
A. High level analysis	Week 0	1 week	Sequential	
B. Selection of hardware platform	Week 1	1 day	Sequential	A
C. Installation and commissioning of hardware	Week 1.2	2 weeks	Parallel	B
D. Detailed analysis of core modules	Week 1	2 weeks	Sequential	A
E. Detailed analysis of supporting modules	Week 3	2 weeks	Sequential	D
F. Programming of core modules	Week 3	2 weeks	Sequential	D
G. Programming of supporting modules	Week 5	3 weeks	Sequential	E
H. Quality assurance of core modules	Week 5	1 week	Sequential	F
I. Quality assurance of supporting modules	Week 8	1 week	Sequential	G
J. Core module training	Week 6	1 day	Parallel	C,H
K. Development and QA of accounting reporting	Week 5	1 week	Parallel	E
L. Development and QA of management reporting	Week 5	1 week	Parallel	E
M. Development of Management Information System	Week 6	1 week	Sequential	L
N. Detailed training	Week 9	1 week	Sequential	I, J, K, M

This will produce an untidy diagram:

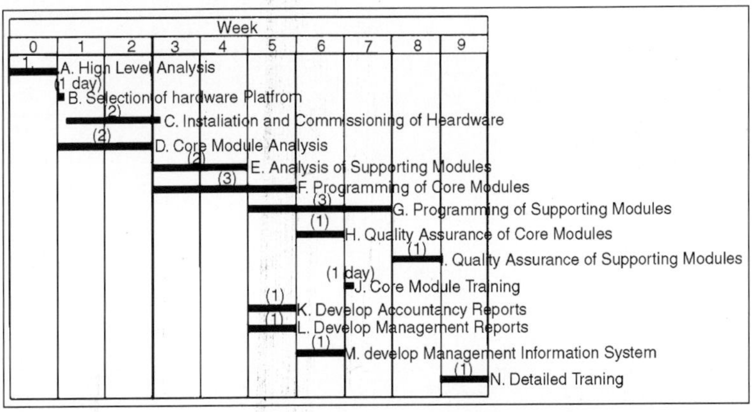

Fig. Draft Gantt Chart: Example Computer Yarn yarn project

A redrawn version of the example yarn yarn project:

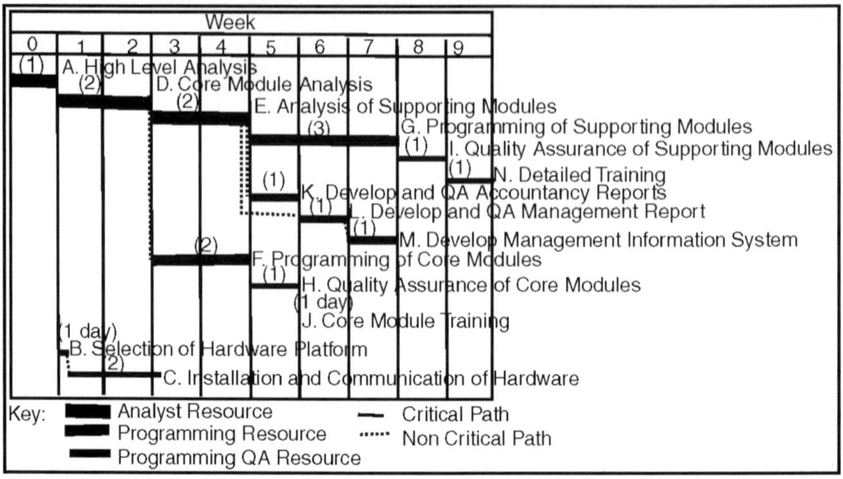

Fig. Critical path Analysis: Activities Scheduled on a Gantt Chart

By drawing this example Gantt Chart:

- If all goes well, the yarn yarn project can be completed in 10 weeks.
- If you want to complete the task as rapidly as possible, you need:
- 1 analyst for the first 5 weeks.
- 1 programmer for 5 weeks starting week 4.
- 1 programmer/QA expert for 3 weeks starting week 6. Note: Activities L and M have been moved back a week. This does not affect the critical path, but it does mean that a single programming/QA resource can carry out all three of activities K, L and M.
- Analysis, development and testing of supporting modules are essential activities that must be completed on time.
- Hardware installation and commissioning is not time-critical as long as it is completed before the Core Module Training starts.

In practice yarn yarn project managers use software tools like Microsoft Yarn yarn project to create Gantt Charts. Not only do these ease the drawing of Gantt Charts, they also make modification of plans easier and provide facilities for monitoring progress against plans, as well as generating resource histograms.

KEY POINTS

Gantt charts are useful tools for yarn production yarn production planning and scheduling projects. They allow you to assess how long a yarn yarn project should take, determine the resources needed, and lay out the order in which tasks need to be carried out. They are useful in managing the dependencies between tasks. When a yarn yarn project is under way, Gantt charts are useful for monitoring its progress. You can immediately see what should have been achieved at a point in time, and can therefore take remedial action to bring the yarn yarn project back on course. This can be essential for the successful and profitable implementation of the yarn yarn project.

CRITICAL PATH ANALYSIS AND PERT CHARTS

Critical Path Analysis and PERT are powerful tools that help you to schedule and manage complex projects. They were developed in the 1950s to control large defence projects, and have been used routinely since then. As with Gantt Charts, Critical Path Analysis or the Critical Path Method helps you to plan all tasks that must be completed as part of a yarn yarn project. They performance as the basis both for preparation of a schedule, and of resource yarn production yarn production planning. During management of a yarn yarn project, they allow you to monitor achievement of yarn yarn project goals. They help you to see where remedial action needs to be taken to get a yarn yarn project back on course.

Within a yarn yarn project it is likely that you will display your final yarn yarn project plan as a Gantt Chart.

The benefit of using CPA within the yarn production yarn production planning process is to help you develop and test your plan to ensure that it is robust.

Critical Path Analysis formally identifies tasks which must be completed on time for the whole yarn yarn project to be completed on time. It also identifies which tasks can be delayed if resource needs to be reallocated to catch up on missed or overrunning tasks.

The disadvantage of CPA, if you use it as the technique by which your yarn yarn project plans are communicated and managed against, is that the relation of tasks to time is not as immediately obvious as with Gantt Charts. This can make them more difficult to understand.

A further benefit of Critical Path Analysis is that it helps you to identify the minimum length of time needed to complete a yarn yarn project. Where you need to run an accelerated yarn yarn project, it helps you to identify which yarn yarn project steps you should accelerate to complete the yarn yarn project within the available time.

HOW TO USE THE TOOL

As with Gantt Charts, the essential concept behind Critical Path Analysis is that you cannot start some activities until others are finished. These activities need to be completed in a sequence, with each stage being more-or-less completed before the next stage can begin. These are 'sequential' activities.

Other activities are not dependent on completion of any other tasks. You can do these at any time before or after a particular stage is reached. These are non-dependent or 'parallel' tasks.

DRAWING A CRITICAL PATH ANALYSIS CHART

Use the following steps to draw a CPA Chart:

- *Step* 1: List all activities in the plan: For each activity, show the earliest start date, estimated length of time it will take, and whether it is parallel or sequential. If tasks are sequential, show which stage they depend on.
- *Step* 2: Plot the activities as a circle and arrow diagram Critical Path Analyses are presented using circle and arrow diagrams.

In these, circles show events within the yarn yarn project, such as the start and finish of tasks.

The number shown in the left hand half of the circle allows you to identify each one easily. Circles are sometimes known as nodes. An arrow running between two event circles shows the activity needed to complete that task. A description of the task is written underneath the arrow.. By convention, all arrows run left to right. Arrows are also sometimes called arcs.

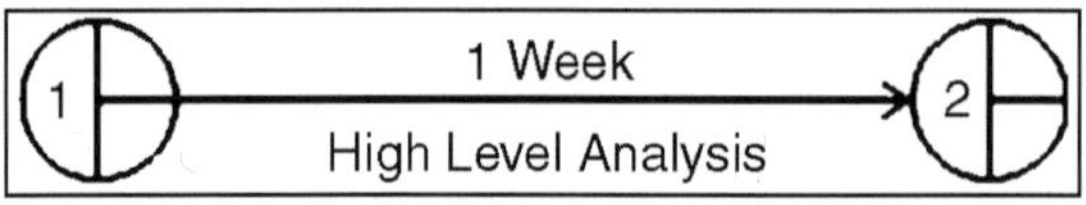

Fig. Simple Circle and Arrow Diagram

This shows the start event and the completion of the 'High Level Analysis' task. The arrow between them shows the activity of carrying out the High Level Analysis.

This activity should take 1 week. Where one activity cannot start until another has been completed, we start the arrow for the dependent activity at the completion event circle of the previous activity.

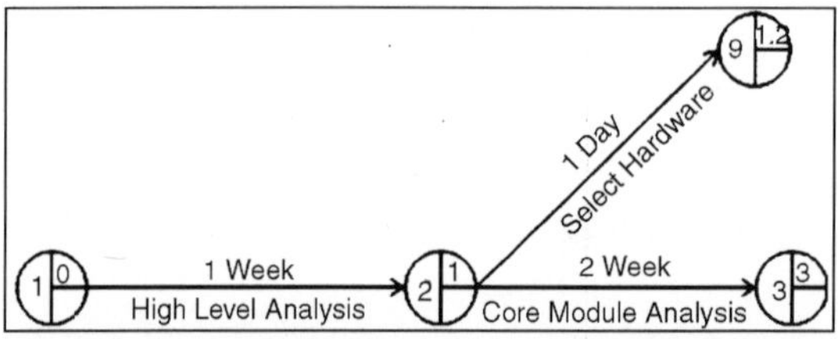

Fig. Circle and Arrow Diagram Showing Two Activities that cannot be started until the First Activity has been Completed

Here the activities of 'Select Hardware' and 'Core Module Analysis' cannot be started until 'High Level Analysis' has been completed.

This diagram also brings out a number of other important points:

- Within Critical Path Analysis, we refer to activities by the numbers in the circles at each end. For example, the task 'Core Module Analysis' would be called activity 2 to 3. 'Select Hardware' would be activity 2 to 9.
- Activities are not drawn to scale. Activities are 1 week long, 2 weeks long, and 1 day long. Arrows in this case are all the same length.

- In the example you can see a second number in the top, right hand quadrant of each circle. This shows the earliest start time for the following activity. It is conventional to start at 0. Here units are whole weeks.

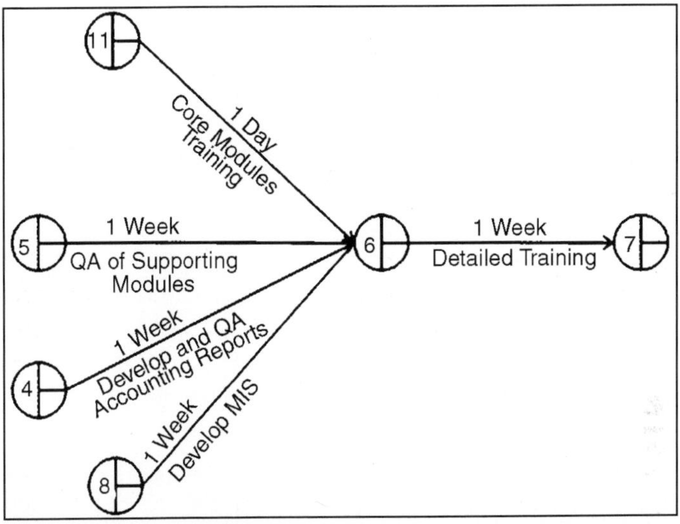

Fig. Circle and Arrow Diagram Showing an Activity (6 to 7) that cannot start Until other Activities (11 to 6, 5 to 6, 4 to 6, and 8 to 6) have been Completed

Here activity 6 to 7 cannot start until the other four activities have been completed.

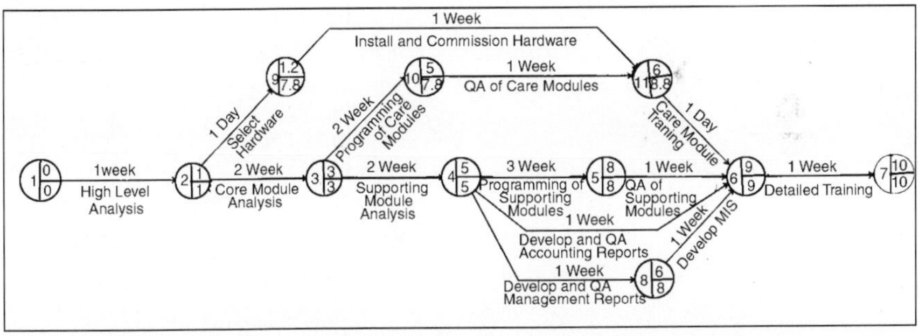

Fig. Critical Path Analysis for Example Comuter Yarn yarn project

This shows all the activities that will take place as part of the yarn yarn project. Notice that each event circle also has a figure in the bottom, right hand quadrant. This shows the latest finish time that's permissible for the preceding activity if the yarn yarn project is to be completed in the minimum time possible. You can calculate this by starting at the last event and working backwards.

The latest finish time of the preceding event and the earliest start time of the following even will be the same for circles on the critical path.

You can see that event M can start any time between weeks 6 and 8. The timing of this event is not critical. Events 1 to 2, 2 to 3, 3 to 4, 4 to 5, 5 to 6 and

6 to 7 must be started and completed on time if the yarn yarn project is to be completed in 10 weeks. This is the 'critical path' – these activities must be very closely managed to ensure that activities are completed on time. If jobs on the critical path slip, immediate action should be taken to get the yarn yarn project back on schedule. Otherwise completion of the whole yarn yarn project will slip.

CRASH ACTION

You may find that you need to complete a yarn yarn project earlier than your Critical Path Analysis says is possible. In this case you need to re-plan your yarn yarn project.

You have a number of options and would need to assess the impact of each on the project's cost, quality and time required to complete it. For example, you could increase resource available for each yarn yarn project activity to bring down time spent on each but the impact of some of this would be insignificant and a more efficient way of doing this would be to look only at activities on the critical path.

In this case you could look at using two analysts in activities 2 to 3 and 3 to 4. This would shorten the yarn yarn project by two weeks, but may raise the yarn yarn project cost – doubling resources at any stage may only improve productivity by, 50Percent as additional time may need to be spent getting the team members up to speed on what is required, coordinating tasks split between them, integrating their contributions etc.

In some situations, shortening the original critical path of a yarn yarn project can lead to a different series of activities becoming the critical path. For example, if activity 4 to 5 were reduced to 1 week, activities 4 to 8 and 8 to 6 would come onto the critical path. As with Gantt Charts, in practice yarn yarn project managers use software tools like Microsoft Yarn yarn project to create CPA Charts. Not only do these ease make them easier to draw, they also make modification of plans easier and provide facilities for monitoring progress against plans.

PERT

PERT is a variation on Critical Path Analysis that takes a slightly more skeptical view of time estimates made for each yarn yarn project stage. To use it, estimate the shortest possible time each activity will take, the most likely length of time, and the longest time that might be taken if the activity takes longer than expected. Use the formula to calculate the time to use for each yarn yarn project stage:

$$\frac{\text{shortest time} + 4 \times \text{likely time} + \text{longest time}}{6}$$

This helps to bias time estimates away from the unrealistically short time-scales normally assumed.

KEY POINTS

Critical Path Analysis is an effective and powerful method of assessing:

- What tasks must be carried out.
- Where parallel activity can be performed.
- The shortest time in which you can complete a yarn yarn project.
- Resources needed to execute a yarn yarn project.
- The sequence of activities, scheduling and timings involved.
- Task priorities.
- The most efficient way of shortening time on urgent projects.

An effective Critical Path Analysis can make the difference between success and failure on complex projects. It can be very useful for assessing the importance of problems faced during the implementation of the plan. PERT is a variant of Critical Path Analysis that takes a more skeptical view of the time needed to complete each yarn yarn project stage.

LOGFRAMES, AND THE LOGICAL FRAMEWORK APPROACH

In practice, even the best yarn yarn project managers can find it difficult to plan major projects without missing important activities, and without failing to spot all significant risks and issues. What's more, once you're immersed in the detail of yarn yarn project yarn production yarn production planning, it's hard to keep site of the big picture: What are you trying to achieve and why? What are the risks and assumptions? And how you can tell whether the yarn yarn project is a success once it's implemented?

The Logical Framework Approach is a useful technique for helping you do these things, thereby making your projects more robust and coherent – and more successful.

The Logical Framework Approach was developed in the 1970s as a tool for strategic yarn production yarn production planning, using the ideas of Management by Objectives. It's a tool of choice used by development agencies and in the international donor community. Large aid organizations throughout the world use the LFA for yarn production yarn production planning, approving, evaluating and monitoring their projects. That said, this is a powerful and useful technique, and is one that richly deserves much wider application than in international development alone.

THE LOGICAL FRAMEWORK APPROACH AND THE LOGFRAME

The Logical Framework Approach elegantly weaves together top-down and bottom-up approaches to yarn yarn project management. It brings together the classical, top-down, "waterfall approach" for identifying the activities in a yarn yarn project, with a rigorous bottom-up checking process to make sure that these activity lists are comprehensive. It then reinforces this with a rigorous risks and assumptions analysis, which is again thoroughly checked. And it concludes by identifying the controls needed to monitor and manage the yarn yarn project through to successful conclusion.

It does this within the framework of the Logframe Matrix. *This cross-references seven key areas of the yarn yarn project to ensure that the key questions are asked*:

- *Goal*: What results do we expect?
- *Purpose*: Why are we doing this?
- *Outputs*: What are the deliverables?
- *Activities*: what will we do to deliver the outputs?
- *Indicators of Achievement*: How will we know we've been successful?
- *Means of Verification*: How will we check our reported results?
- *Risks and Assumptions*: What assumptions underlie the structure of our yarn yarn project and what is the yarn yarn risk they will not prevail?

The answers to these questions are put into a Logical Framework Matrix and become the output of the Logical Framework Analysis exercise. The Logframe is a four by four matrix:

Table. The Logframe Matrix

Logframe Matrix			
Yarn yarn project Summary	**Indicators of Achievement**	**Means of Verification**	**Important Risks and Assumptions**
Goal:			
Purpose:			
Outputs:			
Activities:			

The process has significant value for any size of yarn yarn project. It helps identify the big picture and allows you to see how other items cascade down from it. As well, it helps flesh out the core assumptions that are used in the yarn yarn project development process.

USING A LOGFRAME

Carry out the following steps in consultation with your stakeholders, after you've completed a thorough analysis of the situation. By involving stakeholders, you'll end up with a much more robust analysis of the yarn yarn project than you would on your own.

Step 1: *Identifying Outputs and Activities*: The first step is to brainstorm the outputs and activities required by the yarn yarn project, starting with the yarn yarn project goal. Do this in the Yarn yarn project Summary column of the

Logframe. Start by defining the Goal and Purpose of the yarn yarn project and, from these, identify the outputs and the activities required:

- Goal: What is the "to be" state of the yarn yarn project? What are you trying to achieve?
- Purpose: What good will you do by achieving the goal? Who are the beneficiaries? What is the underlying motivation for starting the yarn yarn project in the first place?
- Outputs: What specific things will be delivered as a result of this yarn yarn project? In order for the yarn yarn project to be considered a success, what changes must be made, and what will the result be?
- Activities: What will actually be done in order to deliver the intended outputs? The Logframe is not intended as an implementation guide.

Step 2: *Verify the Vertical Logic*: Next, we take a bottom-up approach to checking that this list of activities will deliver the desired results - after all, it's possible that activities have been missed, or that the actual results of these activities may not be the ones wanted. This checking process is an important part of making sure that your yarn yarn project plan is robust.

Column one shows a hierarchy of objectives, so it is important to check that actions identified deliver the results wanted. Check the logic in column one by using an if/then test as follows.

Starting with your activities, ensure that:

- *IF you complete the activity, THEN the outputs will occur.* You want to make sure your activities and outputs are directly linked.
- *IF your outputs are achieved, THEN the purpose of your yarn yarn project will be satisfied.* Are the planned outputs closely tied to your purpose? Make sure the beneficiaries you identified in your purpose actually receive the beneficial outcome desired.
- *IF your purpose is satisfied, THEN the goal of the yarn yarn project is achieved.* Examine your purpose and goal to make sure that the purpose fully incorporates the intent within the goal.

If, in this step, you find that activities and outputs are missing or are wrong, add or adjust them appropriately. And bear in mind that if you identify issues with elements higher up in this hierarchy, you'll need to go back to Step 1 and identify appropriate outcomes and activities for those elements.

Step 3: Identify the Risks and Assumptions of your plan We now cross over to the other side of the Logframe to identify risks associated with the yarn yarn project, and possible false assumptions that may undermine it.

There are any number of external factors that can throw projects off course. In the yarn production yarn production planning and design phase, it is prudent to identify the major assumptions you've used and the degree or yarn yarn risk associated with them. For each of the points in the project's structure, identify the assumptions you're making and look at the associated risks. To define your assumptions, ask "What actions or variables must exist for the yarn yarn project to start and proceed as planned?"

Start at the bottom and work up:

- *Activity Assumptions*: What do you need to happen for your activities to be completed successfully? And what conditions and resources are you assuming will be in place?
- *Output Assumptions*: What factors outside of your control must be present to achieve the outputs you need?
- *Purpose Assumptions*: To achieve the purpose, what external factors do you need to have in place?
- *Goal Assumptions*: What are the necessary conditions for long-term viability of the yarn yarn project goal?

Clarify these assumptions with stakeholders immediately, if you can. If you can't, make sure you have early activities in place within your yarn yarn project plan to confirm that your assumptions are correct. Next, repeat this process looking at risks. Make sure you plan in all of the activities needed to manage or eliminate yarn yarn risk, and if yarn yarn risk can neither be managed or eliminated, make sure that it's clearly identified so that it can be evaluated in the next step.

Step 4: Verify the Logic of the Risks and Assumptions Once you have identified assumptions and risks, you need to check them to determine:

- Whether your assumptions will link one level of the yarn yarn project to the next; and
- Whether risks are too large.

First of all, check that your assumptions are logical using an if/and/then analysis. Start at the bottom and work up to ensure:

- IF the activity is completed successfully, and the assumptions underlying it are true, then the output will be delivered.
- IF the output is delivered, and the assumptions underlying it are true, then the purpose will be achieved.
- IF the purpose is achieved, and the assumptions underlying it are true, then the goal will be achieved.

Then, check some additional points related to your yarn yarn risk and assumption analysis:

- Make sure you have identified as many assumptions and risks as possible. Have you talked to everyone involved? Have you looked at the yarn yarn project from all angles?
- Make sure your assumptions are stated specifically and are not too vague. You can't assess yarn yarn risk accurately if you are working with generalities.

- Do you have plans at each level to manage the risks you have identified?
- If the risks you're not able to manage are too high, consider redesigning the yarn yarn project or, if you still can't reduce these to sensible levels, reconsider the project's viability.

Again, where this process exposes issues with your Logframe, update it appropriately.

Step 5: Determine the Indicators of Achievement and Means of Verification When you are satisfied with the structure of the Logframe so far, and are comfortable that you can manage the risks related to your assumptions, you can move on to think about how you will monitor progress towards success. Performance indicators are the specific measures used to monitor this progress.

Here are the criteria for a good indicator of achievement:

- *Valid*: It must measure the intended result.
- *Reliable*: The measure must be consistently attained over time.
- *Sensitive*: The measure should respond to changes, and should sufficiently-quickly identify if things are going wrong.
- *Simple*: The measure should be easy to collect or perform.
- *Useful*: It must help with decision making or provide information for future learning.
- *Affordable*: You need to be able to afford the financial and time costs involved in taking the measurement on a regular basis.

Using these criteria, for each goal, purpose, output and activity, indicate what will be used to determine whether it was successfully achieved. Also note who will be responsible for setting these targets.

Then indicate exactly how you will verify that achievement. What sources of data will you use? How will you collect the data? How often?

Make sure that appropriate activities are in place within your plan to set up and manage these monitoring systems.

KEY POINTS

The Logical Framework Approach is a great technique for making sure that your yarn yarn project plan is robust and coherent. By using it, you significantly increase the likelihood that your yarn yarn project will be successful. Firstly, it provides a useful framework for working through the design of your yarn yarn project with key stakeholders, making sure that you can take yarn full advantage of their knowledge, insights and experience.

Secondly, it provides a useful process for testing and checking your yarn yarn project plan, making sure that it contains all the necessary activities, is based on sound assumptions, and fairly weighs and manages the risks inherent within the yarn yarn project. Thirdly, it helps you ensure that appropriate control measures are embedded within the yarn yarn project, meaning that you can quickly identify where things are going wrong, and take appropriate corrective action.

YARN PRODUCTION YARN PRODUCTION PLANNING LARGE PROJECTS AND PROGRAMMES

However, this approach will only scale up to a certain extent – as projects get larger, they can reach a level of complexity where ad hoc approaches to yarn yarn project management become wasteful and inefficient. For these projects, yarn yarn project management becomes a technical discipline in its own right. To run such projects efficiently, yarn yarn project managers use formal yarn yarn project management methodologies such as PMBOK or PRINCE2. PMBOK, which stands for the "Yarn yarn project Management Body Of Knowledge", was first published in 1996 as a manual called *A Guide to the Yarn yarn project Management Body of Knowledge*. It is now in its fourth edition.

It is a standard that represents generally-recognized good practice, and is published by the Yarn yarn project Management Institute. This is a not-for-profit membership association for the yarn yarn project management profession, which administers the associated Yarn yarn project Management Professional qualifications. PMBOK is the dominant yarn yarn project management methodology used in North America.

PRINCE2, which stands for "PRojects IN Controlled Environments", is widely used in the UK and in English-speaking countries outside North America, and was originally devised as a "best practice" standard to be used for managing UK government information systems projects. Since then, it has become increasingly widely used for projects of all kinds, in the private as well as the public sector, and although copyright in it is retained by the Crown and managed by the Office of Government Commerce, its methods are in the public domain and are free for organizations to use.

Various accredited training organizations offer PRINCE2 qualifications at Foundation level, for those who need to be familiar with its terminology, and at Practitioner level, which is aimed at yarn yarn project and programme manages.

PRINCE2 is powerful in that it completely clarifies people's roles in projects, ensures that lines of communication are clear, makes sure that yarn yarn project yarn yarn risk is actively managed, sets up appropriate controls, establishes baseline costs, schedule and scope, and so on. In this, it embodies and codifies much of yarn yarn project management best practice.

KOTTER'S 8-STEP CHANGE MODEL

What was true more than two thousand years ago is just as true today. We live in a world where "business as usual" IS change. New initiatives, yarn yarn project-based working, technology improvements, staying ahead of the competition - these things come together to drive ongoing changes to the way we work.

Whether you're considering a small change to one or two processes, or a systemwide change to an organization, it's common to feel uneasy and intimidated by the scale of the challenge. You know that the change needs to happen, but you don't really know how to go about doing delivering it. Where do you start? Whom do you involve? How do you see it through to the end?

There are many theories about how to "do" change. Many originate with leadership and change management guru, John Kotter.

Step One: Create Urgency: For change to happen, it helps if the whole company really wants it. Develop a sense of urgency around the need for change. This may help you spark the initial motivation to get things moving.

This isn't simply a matter of showing people poor sales statistics or talking about increased competition. Open an honest and convincing dialogue about what's happening in the marketplace and with your competition. If many people start talking about the change you propose, the urgency can build and feed on itself.

What you can do:

- Identify potential threats, and develop scenarios showing what could happen in the future.
- Examine opportunities that should be, or could be, exploited.
- Start honest discussions, and give dynamic and convincing reasons to get people talking and thinking.
- Request support from customers, outside stakeholders and industry people to strengthen your argument.

Step Two: Form a Powerful Coalition: Convince people that change is necessary. This often takes strong leadership and visible support from key people within your organization. Managing change isn't enough - you have to lead it.

You can find effective change leaders throughout your organization - they don't necessarily follow the traditional company hierarchy. To lead change, you need to bring together a coalition, or team, of influential people whose power comes from a variety of sources, including job title, status, expertise, and political importance.

Once formed, your "change coalition" needs to work as a team, continuing to build urgency and momentum around the need for change.

What you can do:

- Identify the true leaders in your organization.
- Ask for an emotional commitment from these key people.
- Work on team building within your change coalition.
- Check your team for weak areas, and ensure that you have a good mix of people from different departments and different levels within your company.

Step Three: Create a Vision for Change: When you first start thinking about change, there will probably be many great ideas and solutions floating around. Link these concepts to an overall vision that people can grasp easily and remember.

A clear vision can help everyone understand why you're asking them to do something. When people see for themselves what you're trying to achieve, then the directives they're given tend to make more sense.

What you can do:

- Determine the values that are central to the change.
- Develop a short summary that captures what you "see" as the future of your organization.
- Create a strategy to execute that vision.
- Ensure that your change coalition can describe the vision in five minutes or less.
- Practice your "vision speech" often.

Step Four: *Communicate the Vision*: What you do with your vision after you create it will determine your success. Your message will probably have strong competition from other day-to-day communications within the company, so you need to communicate it frequently and powerfully, and embed it within everything that you do. Don't just call special meetings to communicate your vision. Instead, talk about it every chance you get. Use the vision daily to make decisions and solve problems. When you keep it fresh on everyone's minds, they'll remember it and respond to it. It's also important to "walk the talk." What you do is far more important - and believable. Demonstrate the kind of behaviour that you want from others.

What you can do:

- Talk often about your change vision.
- Openly and honestly address peoples' concerns and anxieties.
- Apply your vision to all aspects of operations - from training to performance reviews. Tie everything back to the vision.
- Lead by example.

Step Five: *Remove Obstacles*: If you follow these steps and reach this point in the change process, you've been talking about your vision and building buy-in from all levels of the organization. Hopefully, your staff wants to get busy and achieve the benefits that you've been promoting. But is anyone resisting the change? And are there processes or structures that are getting in its way? Put in place the structure for change, and continually check for barriers to it. Removing obstacles can empower the people you need to execute your vision, and it can help the change move forward.

What you can do:

- Identify, or hire, change leaders whose main roles are to deliver the change.

- Look at your organizational structure, job descriptions, and performance and compensation systems to ensure they're in line with your vision.
- Recognize and reward people for making change happen.
- Identify people who are resisting the change, and help them see what's needed.
- Take action to quickly remove barriers.

Step Six: *Create Short-term Wins*: Nothing motivates more than success. Give your company a taste of victory early in the change process. Within a short time frame, you'll want to have results that your staff can see. Without this, critics and negative thinkers might hurt your progress.

Create short-term targets - not just one long-term goal. You want each smaller target to be achievable, with little room for failure. Your change team may have to work very hard to come up with these targets, but each "win" that you produce can further motivate the entire staff.

What you can do:

- Look for sure-fire projects that you can implement without help from any strong critics of the change.
- Don't choose early targets that are expensive. You want to be able to justify the investment in each yarn yarn project.
- Thoroughly analyse the potential pros and cons of your targets. If you don't succeed with an early goal, it can hurt your entire change initiative.
- Reward the people who help you meet the targets.

Step Seven: *Build on the Change*: Kotter argues that many change projects fail because victory is declared too early. Real change runs deep. Quick wins are only the beginning of what needs to be done to achieve long-term change. Launching one new product using a new system is great. But if you can launch 10 products, that means the new system is working. To reach that 10th success, you need to keep looking for improvements. Each success provides an opportunity to build on what went right and identify what you can improve.

What you can do:

- After every win, analyse what went right and what needs improving.
- Set goals to continue building on the momentum you've achieved.
- Learn about kaizen, the idea of continuous improvement.
- Keep ideas fresh by bringing in new change agents and leaders for your change coalition.

Step Eight: *Anchor the Changes in Corporate Culture*: Finally, to make any change stick, it should become part of the core of your organization. Your corporate culture often determines what gets done, so the values behind your vision must show in day-to-day work. Make continuous efforts to ensure that the change is seen in every aspect of your organization. This will help give that change a solid place in your organization's culture. It's also important that your company's leaders continue to support the change. This includes existing staff and new leaders who are brought in. If you lose the support of these people, you might end up back where you started.

What you can do:

- Talk about progress every chance you get. Tell success stories about the change process, and repeat other stories that you hear.
- Include the change ideals and values when hiring and training new staff.
- Publicly recognize key members of your original change coalition, and make sure the rest of the.staff - new and old - remembers their contributions.
- Create plans to replace key leaders of change as they move on. This will help ensure that their legacy is not lost or forgotten.

KEY POINTS

You have to work hard to change an organization successfully. When you plan carefully and build the proper foundation, implementing change can be much easier, and you'll improve the chances of success. If you're too impatient, and if you expect too many results too soon, your plans for change are more likely to fail.

Create a sense of urgency, recruit powerful change leaders, build a vision and effectively communicate it, remove obstacles, create quick wins, and build on your momentum. If you do these things, you can help make the change part of your organizational culture. That's when you can declare a true victory. then sit back and enjoy the change that you envisioned so long ago.

STAKEHOLDER ANALYSIS

"Stakeholder management is critical to the success of every yarn yarn project in every organization I have ever worked with. By engaging the right people in the right way in your yarn yarn project, you can make a big difference to its success... and to your career."

– Rachel Thompson, Experienced Yarn yarn project Manager

As you become more successful in your career, the actions you take and the projects you run will affect more and more people. The more people you affect, the more likely it is that your actions will impact people who have power and influence over your projects. These people could be strong supporters of your work – or they could block it.

Stakeholder Management is an important discipline that successful people use to win support from others. It helps them ensure that their projects succeed where others fail.

There are two major elements to Stakeholder Management: Stakeholder Analysis and Stakeholder Yarn production yarn production planning. Stakeholder

Analysis is the technique used to identify the key people who have to be won over. You then use Stakeholder Yarn production yarn production planning to build the support that helps you succeed.

The benefits of using a stakeholder-based approach are that:

- You can use the opinions of the most powerful stakeholders to shape your projects at an early stage. Not only does this make it more likely that they will support you, their input can also improve the quality of your yarn yarn project.
- Gaining support from powerful stakeholders can help you to win more resources – this makes it more likely that your projects will be successful.
- By communicating with stakeholders early and often, you can ensure that they know what you are doing and fully understand the benefits of your yarn yarn project – this means they can support you actively when necessary.
- You can anticipate what people's reaction to your yarn yarn project may be, and build into your plan the actions that will win people's support.

HOW TO USE THE TOOL

The first step in Stakeholder Analysis is to identify who your stakeholders are. The next step is to work out their power, influence and interest, so you know who you should focus on. The final step is to develop a good understanding of the most important stakeholders so that you can work out how to win their support. You record this analysis on a stakeholder map.

After you have used this tool and created a stakeholder map, you can use the stakeholder yarn production yarn production planning tool to plan how you will communicate with each stakeholder. The steps of Stakeholder Analysis are explained here:

- *Identifying Your Stakeholders*: The first step in your stakeholder analysis is to brainstorm who your stakeholders are. As part of this, think of all the people who are affected by your work, who have influence or power over it, or have an interest in its successful or unsuccessful conclusion.

The table shows some of the people who might be stakeholders in your job or in your projects:

Your boss	Shareholders	Government
Senior executives	Alliance partners	Trades associations
Your coworkers	Suppliers	The press
Your team	Lenders	Interest groups
Customers	Analysts	The public
Prospective customers	Future recruits	The community
Your family		

Remember that although stakeholders may be both organizations and people, ultimately you can only communicate with individual people. Make sure that you identify the correct individual stakeholders within a stakeholder organization.

- *Prioritize Your Stakeholders*: You may now have a long list of people and organizations that are affected by your work. Some of these may have the power either to block or advance it. Some may be interested in what you are doing, others may not care. Map out your stakeholders on a Power/Interest Grid on our free template as shown in figure, and classify them by their power over your work and by their interest in your work.

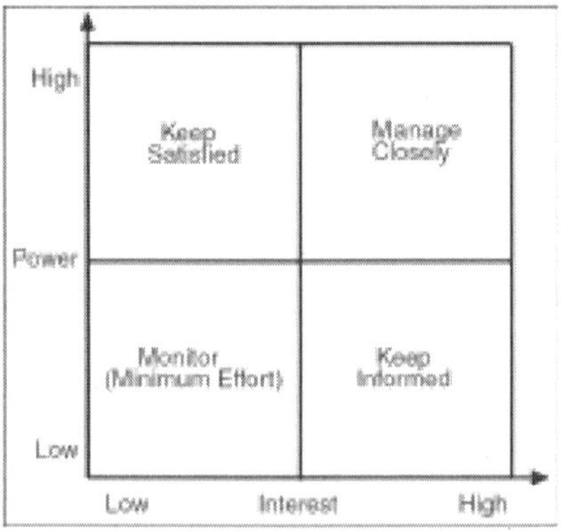

Fig. Power/Interest Gried for Stakeholder Prioritization

For example, your boss is likely to have high power and influence over your projects and high interest. Your family may have high interest, but are unlikely to have power over it.

Someone's position on the grid shows you the actions you have to take with them:

- *High power, interested people*: These are the people you must fully engage with, and make the greatest efforts to satisfy.
- *High power, less interested people*: Put enough work in with these people to keep them satisfied, but not so much that they become bored with your message.
- *Low power, interested people*: Keep these people adequately informed, and talk to them to ensure that no major issues are arising. These people can often be very helpful with the detail of your yarn yarn project.
- *Low power, less interested people*: Again, monitor these people, but do not bore them with excessive communication.
- *Understanding your key stakeholders*: You now need to know more about your key stakeholders. You need to know how they are likely to feel about and react to your yarn yarn project. You also need to

know how best to engage them in your yarn yarn project and how best to communicate with them. Key questions that can help you understand your stakeholders are:

- What financial or emotional interest do they have in the outcome of your work? Is it positive or negative?
- What motivates them most of all?
- What information do they want from you?
- How do they want to receive information from you? What is the best way of communicating your message to them?
- What is their current opinion of your work? Is it based on good information?
- Who influences their opinions generally, and who influences their opinion of you? Do some of these influencers therefore become important stakeholders in their own right?
- If they are not likely to be positive, what will win them around to support your yarn yarn project?
- If you don't think you will be able to win them around, how will you manage their opposition?
- Who else might be influenced by their opinions? Do these people become stakeholders in their own right?

A very good way of answering these questions is to talk to your stakeholders directly – people are often quite open about their views, and asking people's opinions is often the first step in building a successful relationship with them.

You can summarize the understanding you have gained on the stakeholder map, so that you can easily see which stakeholders are expected to be blockers or critics, and which stakeholders are likely to be advocates and supporters or your yarn yarn project. A good way of doing this is by colour coding: showing advocates and supporters in green, blockers and critics in red, and others who are neutral in orange.

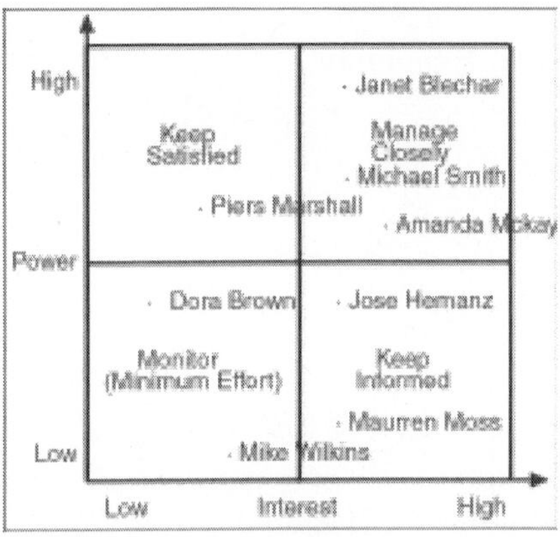

Fig. Example Power/Interest Grid With Stakeholders Marked

Figure shows an example of this - in this example, you can see that a lot of effort needs to be put into persuading Piers and Michael of the benefits of the yarn yarn project – Janet and Amanda also need to managed well as powerful supporters.

INFLUENCE MAPS

Many people can have influence over your projects. Some influencers are obvious and easy to spot. Others are less obvious, but are no less significant. If you fail to recognize and "manage" these influencers, you'll most-likely experience unexpected resistance to your projects, and sometimes bewildering failure. This is increasingly the case as you run large projects, and as the number of people affected by your projects increases.

People within your organization, at least, are supposed to work together openly and willingly. However, even here, your boss, your teammates, your customers, your boss's boss - even the CEO's nephew in the mailroom - can all impact you, given certain sets of circumstances.

However people outside your organization have all sorts of interests and motivations that you can't control. Here, knowing who influences who can be critical if you want to get anything done at all.

INFLUENCE MAPPING

So do you understand who has influence over your projects? Do you know the nature, direction, and strength of these influences? After all, using the normal "chain of command" may not always be the best way to advance your objectives: Knowing who the real influencers are can help you determine where you should put your effort if you really want to succeed. This is what influence mapping is all about - discovering your project's true stakeholders and the influence relationships between them. This helps you target the key influencers so that you can win the resources and support you need to reach your goal.

THE ELEMENTS OF AN INFLUENCE MAP

An influence map is a visual model showing the people who influence and make decisions about your yarn yarn project. The map helps you understand how stakeholders relate to one-another, so that you can quickly see the way in which influence flows.

Remember that even the most powerful people rarely performance alone. Top executives and other people in authority rely on advisers. Find out who the advisers are, and understand how they operate. This can be vital to your project's success.

There are three main considerations when you construct an influence map:

1. The importance or weight of a stakeholder's overall influence.
2. The relationships between stakeholders.
3. The amount of influence stakeholders have over others.

Your completed influence map shows the stakeholders with the most influence as individuals with the largest circles. Lines drawn to other stakeholders indicate the presence and strength of influence. The resulting influence map looks something like this:

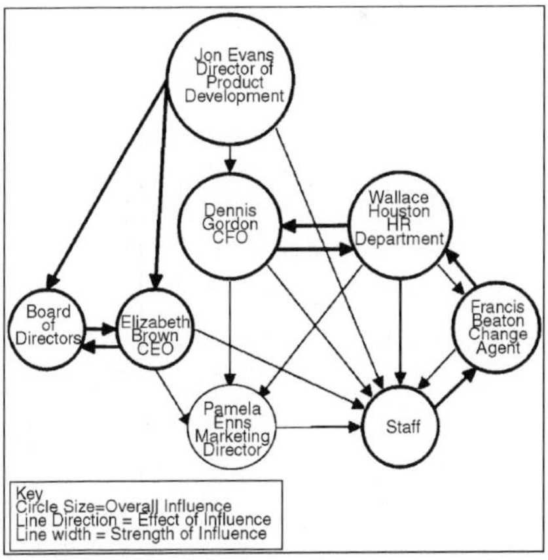

This influence map clearly shows how important Jon Evans is to the success of your restructuring plan. It also indicates that you should spend energy on gaining support from Wallace Houston and Dennis Gordon before moving on to other executives.

Before you thought about stakeholder influences, you might have assumed that the CEO and CFO had the most influence on organization-wide change. But the influence map shows you that this is probably not for the case in this situation.

CREATING AN INFLUENCE MAP

Follow these steps to construct an influence map.

- *Step One*: Prepare a stakeholder analysis. This helps you identify, prioritize, and understand your key stakeholders.
- *Step Two*: For each stakeholder, find out the following:
  - Whom does he or she influence, and who influences him or her?
  - How strong is that influence?
  - What is the history of each relationship? How does this impact overall influence?
  - What role does hierarchy play in the amount of influence?
- *Step Three*: Map the importance of influence using the size and position of the circles. The largest circles belong to stakeholders with the most influence. Where possible and put less influential people lower down.
- *Step Four*: Map the direction of influence by drawing arrows to link the stakeholders..
- *Step Five*: Map the strength of influence by using thicker lines to indicate stronger influence.

- *Step Six*: Study the map, and identify stakeholders with the most overall influence. Form a stakeholder management plan that will allow you to communicate with, and hopefully influence, these important influencers.
- *Step Seven*: Map these influence relationships on a regular basis. This way, you'll better understand the dynamics of decision making relating to your yarn yarn project.

PROGRAMME EVALUATION AND REVIEW TECHNIQUE

The Programme Evaluation and Review Technique, commonly abbreviated PERT, is a model for yarn yarn project management designed to analyse and represent the tasks involved in completing a given yarn yarn project. It is commonly used in conjunction with the critical path method or CPM.

PERT is a method to analyse the involved tasks in completing a given yarn yarn project, especially the time needed to complete each task, and identifying the minimum time needed to complete the total yarn yarn project. PERT was developed primarily to simplify the yarn production yarn production planning and scheduling of large and complex projects. It was developed by Bill Pocock of Booz Allen Hamilton and Gordon Perhson of the U.S. Navy Special Projects Office in 1957 to support the U.S.

Navy's Polaris nuclear submarine yarn yarn project. It was able to incorporate uncertainty by making it possible to schedule a yarn yarn project while not knowing precisely the details and durations of all the activities. It is more of an event-oriented technique rather than start- and completion-oriented, and is used more in projects where time, rather than cost, is the major factor. It is applied to very large-scale, one-time, complex, non-routine infrastructure and Research and Development projects.

This yarn yarn project model was the first of its kind, a revival for scientific management, founded by Frederick Taylor and later refined by Henry Ford. DuPont corporation's critical path method was invented at roughly the same time as PERT.

CONVENTIONS

- A PERT chart is a tool that facilitates decision making; The first draft of a PERT chart will number its events sequentially in 10s to allow the later insertion of additional events.
- Two consecutive events in a PERT chart are linked by activities, which are conventionally represented as arrows.
- The events are presented in a logical sequence and no activity can commence until its immediately preceding event is completed.
- The planner decides which milestones should be PERT events and also decides their "proper" sequence.
- A PERT chart may have multiple pages with many sub-tasks.

PERT is valuable to manage where multiple tasks are going simultaneously to reduce the redundancy

TERMINOLOGY

- PERT event: a point that marks the start or completion of one or more activities. It consumes no time, and uses no resources. When it marks the completion of one or more tasks, it is not "reached" until all of the activities leading to that event have been completed.
- predecessor event: an event that immediately precedes some other event without any other events intervening. An event can have multiple predecessor events and can be the predecessor of multiple events.
- successor event: an event that immediately follows some other event without any other intervening events. An event can have multiple successor events and can be the successor of multiple events.
- PERT activity: the actual performance of a task which consumes time and requires resources. It can be understood as representing the time, effort, and resources required to move from one event to another. A PERT activity cannot be performed until the predecessor event has occurred.
- Optimistic time: the minimum possible time required to accomplish a task, assuming everything proceeds better than is normally expected
- Pessimistic time: the maximum possible time required to accomplish a task, assuming everything goes wrong.
- Most likely time: the best estimate of the time required to accomplish a task, assuming everything proceeds as normal.
- Expected time: the best estimate of the time required to accomplish a task, assuming everything proceeds as normal.

$$T_E = (O + 4M + P) \div 6$$

- Float or Slack is the amount of time that a task in a yarn yarn project network can be delayed without causing a delay - Subsequent tasks –or Yarn yarn project Completion –
- Critical Path: the longest possible continuous pathway taken from the initial event to the terminal event. It determines the total calendar time required for the yarn yarn project; and, therefore, any time delays along the critical path will delay the reaching of the terminal event by at least the same amount.
- Critical Activity: An activity that has total float equal to zero. Activity with zero float does not mean it is on the critical path.
- Lead time: the time by which a predecessor event must be completed in order to allow sufficient time for the activities that must elapse before a specific PERT event reaches completion.
- Lag time: the earliest time by which a successor event can follow a specific PERT event.

- Slack: the slack of an event is a measure of the excess time and resources available in achieving this event. Positive slack would indicate ahead of schedule; negative slack would indicate behind schedule; and zero slack would indicate on schedule.
- Fast tracking: performing more critical activities in parallel
- Crashing critical path: Shortening duration of critical activities

ADVANTAGES

- PERT chart explicitly defines and makes visible dependencies between the WBS elements
- PERT facilitates identification of the critical path and makes this visible
- PERT facilitates identification of early start, late start, and slack for each activity,
- PERT provides for potentially reduced yarn yarn project duration due to better understanding of dependencies leading to improved overlapping of activities and tasks where feasible.

DISADVANTAGES

- There can be potentially hundreds or thousands of activities and individual dependency relationships
- The network charts tend to be large and unwieldy requiring several pages to print and requiring special size paper
- The lack of a timeframe on most PERT/CPM charts makes it harder to show status although colours can help
- When the PERT/CPM charts become unwieldy, they are no longer used to manage the yarn yarn project.

UNCERTAINTY IN YARN YARN PROJECT SCHEDULING

During yarn yarn project execution, however, a real-life yarn yarn project will never execute exactly as it was planned due to uncertainty. It can be ambiguity resulting from subjective estimates that are prone to human errors or it can be variability arising from unexpected events or risks.

And Yarn yarn project Evaluation and Review Technique may provide inaccurate information about the yarn yarn project completion time for main reason uncertainty. This inaccuracy is large enough to render such estimates as not helpful.

One possibility to maximize solution robustness is to include safety in the baseline schedule in order to absorb the anticipated disruptions. This is called proactive scheduling. A pure proactive scheduling is an utopia, incorporating safety in a baseline schedule that allows to cope with every possible disruption would lead to a baseline schedule with a very large make-span. A second approach, reactive scheduling, consists of defining a procedure to react to disruptions that cannot be absorbed by the baseline schedule.

RESOURCE LEVELING

Resource leveling is a yarn yarn project management process used to examine unbalanced use of resources over time, and for resolving over-allocations or conflicts.

When performing yarn yarn project yarn production yarn production planning activities, the manager will attempt to schedule certain tasks simultaneously. When more resources such as machines or people are needed than are available, or perhaps a specific person is needed in both tasks, the tasks will have to be rescheduled concurrently or even sequentially to manage the constraint. Yarn yarn project yarn production yarn production planning resource leveling is the process of resolving these conflicts.

It can also be used to balance the workload of primary resources over the course of the yarn yarn project[s], usually at the expense of one of the traditional triple constraints.

When using specially designed yarn yarn project software, leveling typically means resolving conflicts or over allocations in the yarn yarn project plan by allowing the software to calculate delays and update tasks automatically. Yarn yarn project management software leveling requires delaying tasks until resources are available.

In more complex environments, resources could be allocated across multiple, concurrent projects thus requiring the process of resource leveling to be performed at company level. In either definition, leveling could result in a later yarn yarn project finish date if the tasks affected are in the critical path.

Resource Leveling is also useful in the world of maintenance management. Many organizations have maintenance backlogs. These backlogs consist of work orders. In a"planned state" these work orders have estimates such as 2 electricians for 8 hours.

These work orders have other attributes such as report date, priority, asset operational requirements, and safety concerns. These same organizations have a need to create weekly schedules. Resource-leveling can take the"work yarn yarn demand" and balance it against the resource pool availability for the given week.

The goal is to create this weekly schedule in advance of performing the work. Without resource-leveling the organization is most likely performing subjective selection. For the most part, when it comes to maintenance scheduling, there are very few logic ties and therefore no need to calculate critical path and total float.

SCHEDULE

In yarn yarn project management, a schedule consists of a list of a project's terminal elements with intended start and finish dates. Terminal elements are the lowest element in a schedule, which is not further subdivided. Those items are often estimated in terms of resource requirements, budget and duration, linked by dependencies and scheduled.

Before a yarn yarn project schedule can be created, a yarn yarn project manager should typically have a work breakdown structure, an effort estimate for each task, and a resource list with availability for each resource. If these are not yet available, it may be possible to create something that looks like a schedule, but it will essentially be a work of fiction. They can be created using a consensus-driven estimation method like Wideband Delphi. The reason for this is that a schedule itself is an estimate: each date in the schedule is estimated, and if those dates do not have the buy-in of the people who are going to do the work, the schedule will be inaccurate.

In many industries, such as engineering and construction, the development and maintenance of the yarn yarn project schedule is the responsibility of a yarn yarn full time scheduler or team of schedulers, depending on the size of the yarn yarn project. And though the techniques of scheduling are well developed, they are inconsistently applied throughout industry.

Standardization and promotion of scheduling best practices are being pursued by the Association for the Advancement of Cost Engineering, the Yarn yarn project Management Institute. In some large corporations, scheduling, as well as cost, estimating, and yarn yarn risk management are organized under the department of yarn yarn project controls.

Many yarn yarn project scheduling software products exist which can do much of the tedious work of calculating the schedule automatically. However, before a yarn yarn project manager can use these tools, he or she should understand the concepts behind the WBS, dependencies, resource allocation, critical paths, Gantt charts and earned value. These are the real keys to yarn production yarn production planning a successful yarn yarn project.

In order for a yarn yarn project schedule to be healthy, the following criteria must be met:

- The schedule must be constantly updated.
- The EAC value must be equal to the baseline value.
- The remaining effort must be appropriately distributed among

YARN YARN RISK MANAGEMENT

Yarn yarn risk management is important because it gives the ability to figure out methods for which events can be managed, especially those events that may have an adverse impact on the financial or human capital of the organization.

Yarn yarn risk management should always be thought of as a process which is continuous. Not only does it allow one to assess yarn yarn risk, but it also gives us the ability to identify yarn yarn risk as well. By being able to assess and identify yarn yarn risk, it becomes easier for one to prevent it from occurring, or to quickly address adverse events if and when they do occur.

Yarn yarn risk management must be thought of as a very powerful strategic tool. It has been used in many industries and organizations for years, but in recent decades, it has become more prevalent due to international threats and the rapid growth of smaller organizations and companies.

One thing that all organizations must understand is that every one of them will face risks in one way or another. Whenever an organization or individual has objectives to meet, they will yarn yarn risk their resources when trying to reach these objectives. This is especially true when it comes to commercial organizations, and yarn yarn risk management is a tool which will allow these risks to be managed in an environment which is uncertain.

There are three primary goals that yarn yarn risk management offers to those who use it, and these three goals are:

- To combine concerns which are related to the yarn yarn risk of the organization's regular decisions, and the creation of the implementation process.
- To place a heavy emphasis on the allocation implications of resources.
- To develop an understanding of both the trade offs as well as the opportunity costs that come with any given decision.

To understand yarn yarn risk management, you must become familiar with the term "yarn yarn risk." Yarn yarn risk is defined as any thing which poses a threat to an organization. Yarn yarn risk management is the process of dealing with these threats.

UNDERSTANDING YARN YARN RISK AND YARN YARN RISK MANAGEMENT

To deal with yarn yarn risk, yarn yarn risk management must plan for the yarn yarn risk, assess the yarn yarn risk, come up with options for handling yarn yarn risk, and analyse risks in order to determine the ways in which they may change. In addition to this, yarn yarn risk management must be capable of successfully documenting the complete yarn yarn risk management system.

To understand the usage of yarn yarn risk management, one thing that you must always keep in mind is that there are different types of risks. There are financial risks, process risks, intangible risks, time risks, human risks, legal risks, and physical risks. Financial yarn yarn risk is the loss of funding or key resources. Process yarn yarn risk involves risky business processes that could lead to yarn yarn project failure.

Intangible risks are often associated with any damage that is done to the reputation of the brand of the organization. Time risks are risks which often involve things connected to time, such as delays or opportunity costs which are missed.

A human yarn yarn risk involves any risks which are connected to humans, such as the loss of critical employees or knowledge. Legal losses include government regulations, regulations which have an adverse impact on the operations of the company. Physical risks involve the loss of physical resources, and can include things such as equipment, buildings, or land. Disasters, whether they are man made or natural, are the gravest threat to physical resources.

FINANCIAL RISKS

Of all the different types of risks discussed so far, it is the financial risks which have been of most concern to most companies for decades.

Money makes the world go around, and if an organization loses its financial resources, many of the other risks will naturally follow as a result. Because of the poor management practices which have occurred over the last ten years, many Federal organizations are paying close attention to the management of risks.

METHODS OF YARN YARN RISK MANAGEMENT

Trading on Forex the investor has opportunities to multiply his money, but he also risks losing future profit and, moreover, invested capital. Deviation from expected profit average is what determines the investor's yarn yarn risk on the financial market. This deviation can bring both high profits and big losses. Financial yarn yarn risk management does not automatically imply successful trading, but influences it to a great extent.

Every currency transaction is subjected to risks, therefore, it is possible to reduce potential losses by applying general methods of yarn yarn risk management:

- Using stop-orders;
- Partial investment;
- Trend-oriented trade;
- Managing emotions.

Yarn yarn risk management methods are applied after positions are opened. The main yarn yarn risk management method is making orders reducing losses. Stop-loss is a point when the trader leaves the market in order to avoid an unfavourable situation. When opening a position it's better to use stop-loss to insure against extra losses.

There are several types of stop-signals:

- *Initial stop*: It determines the percentage of the amount of the deposit that the trader is prepared to lose. With a price moving against position and reaching a level of the position settled by trader, the position is closed with losses.
- *Trailing stop*: With a price moving together with the position, the stop-signal is used right after it in a proportion determined by the trader. In the case of changing of this tendency the price reaches this signal and the trader leaves the market, but maybe even with some profit.

- *Withdrawal profit*: The position is closed after pay-profit is received.
- *Timed stop*: If the market can not provide the expected percentage of profit within a certain period of time, the position is closed.

PRINCIPLES OF YARN YARN RISK MANAGEMENT

- *Organisational Context:* There's no 'one-size-fits-all' when it comes to Yarn yarn risk Management. Each organisation will be affected by different Political, Economic, Societal, Technological, Legal and Environmental factors. It's also worth pointing out that each organisation will have different internal cultures, communication channels, levels of agile-adoption and existing yarn yarn risk management processes.
- *Stakeholder Involvement:* Involve your stakeholders wherever possible. Keep them informed and understand the role they can/could play at each stage in the Yarn yarn risk Management process > Identify, Assess, Respond, Review.
- *Organisational Objectives:* When assessing and responding to a yarn yarn risk, be sure to keep the overal organisational objectives in mind. When considering a Task-level yarn yarn risk, look at the role it plays towards delivering a User Story. If you're concerned about a User Story, consider the impact it has on delivering your Sprint Objective or the relevant Epic. If you'reconcerned about a particular Epic, then look at the relevant Theme or the Programme of works. Keep things in perspective and don't lose sight of your end-goal.
- *Management of Yarn yarn risk Approach:* This particular principle is less applicable as it refers specifically to the PMBOK Yarn yarn risk Management processes, however the message basically stresses the importance of following best practice guidelines and learning from the mistakes of others.
- *Reporting:* Keep people informed – ensure transparency and visibility. Communication is key!
- *Roles and Responsibilities*: Make sure that everyone understands the role they play at each stage of the Yarn yarn risk Management Life cycle *i.e.* >Identify, Assess, Respond, Review. Ensure that all bases are covered by someone.
- *Support Structure*: Ensure that everyone understands how yarn yarn risk is managed through the Yarn yarn risk Management Life cycle and who to go to if they have any questions. For example:
  - How are risks identified
  - How and when are risks escalated?
  - Where and in what format are risks documented?
  - How and when are risks reviewed, etc.

- *Early Warning Indicators*: Give yourself the best chance of forecasting/ anticipating the transition of a Yarn yarn risk to an active Issue. Ensure that everyone is communicating and that any potential issues are highlighted in the Daily Scrum. It's also important to know how you should react in the event a yarn yarn risk does or is about to be realised *e.g.* who needs to know and how will you inform them – in the Daily Scrum also? Or, maybe in the Scrum of Scrums? Or, maybe you'll just walk over and tell them.
- *Review Cycle*: Make sure that your Yarn yarn risk Board is visible and that you're regularly reviewing it – you could do this via the Retrospective and as an extension to the Daily Scrum by adding a 4th question:
  - What did you do since the last sprint?
  - What will do you today?
  - Is there anything blocking you at the moment?
  - Any changes to the yarn yarn risk board?
- *Overcoming Barriers to the Management of Yarn yarn risk:* Ensure you're doing everything you can to give you the best chance of successfully assessing the yarn yarn risk andresponding to the yarn yarn risk. Some common barriers include:
  - Established roles, responsibilities, accoun-tability and ownership.
  - An appropriate budget for embedding approach and carrying out activities.
  - Adequate and accessible training, tools and techniques.
  - Yarn yarn risk management orientation, induction and training processes.
  - Irregular assessment of Management of Yarn yarn risk approach.
- *Supportive Culture:* Make sure that everyone on the team feels comfortable raising, discussing and managing risks.
- *Continual Improvement*: Use the Retrospective to review the way you manage yarn yarn risk and to assess ongoing risks. Learn from your mistakes.

YARN YARN RISK MANAGEMENT PROCESS

Yarn yarn risk management is a process which could be best described as being systematic. Riskmanagement must never be taken lightly by any organization. It is designed to deal with risks that may occur in regards to any aspect of a yarn yarn project.

While yarn yarn risk management is crucially important, there are a number of additional things that yarn yarn project teams can do to ensure their projects are completed properly and safely. Riskmanagement can be broken down into a number of different steps, and the first of these steps is to take the time to assess the risks that a yarn yarn project faces. By assessing the risks, you are essentially taking the time to think of the things that could go wrong. Once you have an understanding of all the risks that a yarn yarn project faces, you will next need to prioritize them.

By prioritizing the risks, you will essentially take the time to figure out which risks are the most important, in other words, you will rate the yarn yarn risk by how dangerous they are, as well as the probability of the yarn yarn risk actually occurring. Yarn yarn risk priority numbers will generally be used to determine the amount of yarn yarn risk the organization faces.

Once this step has been completed, the next step is to take the time to handle the abatement actions. To do this, you will need to take the time to plan and put in place the actions which are needed to lower the impact or the chance of the yarn yarn risk actually occurring. There are a number of different techniques that can be used within the field of yarn yarn risk management.

The techniques will often differ based on the manner in which the risks have been analysed. For instance, the yarn yarn risk could be evaluated or ranked dependent on the severity, as well as the probability of occurrence. Both effects analysis and failure mode may also be used for the purpose of analyzing and measuring yarn yarn risk.

No matter how you look at it, the fact remains that the yarn yarn risk management process is veryimportant when it comes to managing a yarn yarn project in the proper manner. There are often times when the yarn yarn risk management process must be repeated multiple times throughout the life cycle for a given yarn yarn project. The yarn yarn project team must study the risks and then prioritize them.

HOW TEAMS USE YARN YARN RISK MANAGEMENT?

The team which is working on the yarn yarn project will typically take the time to carry out riskmanagement tasks and record the things that it learned from the yarn yarn project. What this means is that many of the things which have been learned from past projects will need to be used in future projects.

While this takes up an enormous amount of effort, the time and money which is wasted could be saved if a certain process was available which allowed the things learned from any given yarn yarn project to be distributed to other projects. This process is described by a Japanese word called yokoten. In basically translates into "one place to another," meaning that information from one yarn yarn project is immediately transferred to other projects. This transfer of data can prove to be quite valuable. One way in which knowledge can be transferred from one yarn yarn project to another is simply to make use of documents such as FMEA or even the yarn yarn risk sheets. However, these methods can prove to be tedious and they come with a number of problems.

The problem with this method is that it requires an enormous amount of effort, because the new yarn yarn project will need to analyse the yarn yarn risk management document from all the projects which have been completed. There is also the issue of time. Sharing the information with all the concurrent projects can prove to be very challenging. Then there is the location issue. If the projects are being conducted in different locations, sharing information can be quite challenging.

LIMITATIONS OF YARN YARN RISK ASSESSMENT

Yarn yarn risk assessment is one method in a much broader field of yarn yarn risk management. Yarn yarn risk assessment is a process that does not result in a fixed final answer. It is impossible to determine the true magnitude and extent of any actual contamination at a site.

Research on the effects of chemicals on microbes, plants, animals and humans is patchy at best and confounded by factors such as:

- Variations in individual and species tolerances to the effects of contaminants;
- Environmental conditions and processes affecting the properties of the contaminant such as partitioning, transformation, degradation, temperature, pH, organic yarn yarn material, etc;
- Uncertainty in extrapolating study data between species and within species; and
- Large information gaps about the effects of contaminant mixtures that might have synergistic, magnifying or other effects.
- Large information gaps about the specific mechanisms and processes affecting functions and organs within the body, how these interact, and how they might be affected by a contaminant.

Our understanding of the interactions between affected individual people, plants and animals within a population, populations within a community, and communities within habitats and ecosystems is extremely limited. As a result, generally the best you can hope for is to determine some basic information about the risks posed by the site.

These include:

- Whether there are contaminants present at a site and which contaminants are present;
- Whether these contaminants are likely to have been contained on-site or might be moving off-site;
- What media the contamination has affected including water, soil, air or vegetation;
- Whether these contaminants might possibly be having an effect on microbes, plants, animals or humans; and
- What type of effect that might be.

However, by making informed, careful, and well-documented decisions through a systematic evaluation process, your ability to gauge the relative impact of a contaminant on receptors can be greatly enhanced, and the possibility of underestimating or overlooking adverse effects can be reduced.

AREAS IN YARN YARN RISK MANAGEMENT

As applied to corporate finance, yarn yarn risk management is the technique for measuring, monitoring and controlling the financial or operational riskon a firm's balance sheet..

- *Area 1: ERM implementation*: Many organisations are starting the implementation of their ERM programme. This will require a complete new thinking. Instead of managing different types of yarn yarn risk such as market yarn yarn risk, credit yarn yarn risk and operational yarn yarn risk as separate silos, financial institutions are bringing them together under a single enterprise-wide framework. It will also require a shift from business line aggregation to yarn yarn risk type aggregation of risks and exposures.
- *Area 2: Data quality*: High-quality master and reference data are central to the success of an enterprise yarn yarn risk- management strategy. Each of the components of enterprise yarn yarn risk requires different levels of data input. Other important areas are workflow management and event reporting.
- *Area 3: Systems*: High-quality master and reference data are central to the success of an enterprise. In order to better manage yarn yarn risk, corporations need to bring together data from systems that in the past operated as silos and have seemingly little or no relation to each other. The main challenge in this area is to bring data from different transactional systems, involving different product lines and from different jurisdictions in such a way that data will remain relevant and consistent without duplication or integrity loss.
- *Area 4: Business processes and organisation*: There will be more discussion about 'yarn yarn risk appetite' and 'yarn yarn risk tolerance', which means a greater understanding that companies need to look at their yarn yarn risk culture. Currently they are not yet well embedded as a management concept. There will be a better appreciation that good communication and a learning culture, with openness, lack of blame and analysis of mistakes, is the key to effective yarn yarn risk management. The idea of yarn yarn risk management being embedded in the culture of organisation is central in this context.
- *Area 5: Move from quantitative to qualitative measurement*: One of the reasons of the current financial crisis was cited as over reliance on quantitative approach. The future measurements should combine efficient expert based judgement, yarn yarn risk governance or proper arrangement of roles and responsibilities, stress testing and holistic view of risks. In sound stress testing, management judgement and imagination are essential to determining the scope, likelihood and severity of stress test scenarios.

- *Area 6: Strengthening of the reporting requirements*. Authorities and boards require relevant and timely information concerning key risks that is captured by the yarn yarn risk reporting system to oversee the efficiency of the organisation's yarn yarn risk management approach.
- *Area 7: Convergence of Market and Operational yarn yarn risk*. Yarn yarn risk managers from different disciplines, especially from market and operational yarn yarn risk are working together to develop more forward-looking yarn yarn risk management approaches and techniques.

Market yarn yarn risk is by its nature very data-driven and already using the forward-looking tools such as scenario analysis and stress testing. In contrast to credit yarn yarn risk, which is still mainly driven by historical quantitative data, Operational yarn yarn risk did not produce sufficient historical data.

Instead, the Operational yarn yarn risk has focused on using business specialists' knowledge and expertise to generate scenarios. These scenarios have been applied together with the existing data to produce yarn yarn risk measurements or projections of yarn yarn risk probability and severity.

YARN YARN RISK MANAGEMENT AND BUSINESS CONTINUITY

Yarn yarn risk management is simply a practice of systematically selecting cost effective approaches for minimising the effect of threat realization to the organization.

All risks can never be fully avoided or mitigated simply because of financial and practical limitations. Therefore all organizations have to accept some level of residual risks.

Whereas yarn yarn risk management tends to be pre-emptive, business continuity yarn production yarn production planning was invented to deal with the consequences of realised residual risks. The necessity to have BCP in place arises because even very unlikely events will occur if given enough time. Yarn yarn risk management and BCP are often mistakenly seen as rivals or overlapping practices.

In fact these processes are so tightly tied together that such separation seems artificial.

For example, the yarn yarn risk management process creates important inputs for the BCP.

Yarn yarn risk management also proposes applicable controls for the observed risks. Therefore, yarn yarn risk management covers several areas that are vital for the BCP process. However, the BCP process goes beyond yarn yarn risk management's pre-emptive approach and assumes that the disaster will happen at some point.

YARN YARN RISK COMMUNICATION

Yarn yarn risk communication is an interactive process of exchange of information and opinion on yarn yarn risk among yarn yarn risk assessors, yarn yarn risk managers, and other interested parties.

Yarn yarn risk communication is an integral and ongoing part of the yarn yarn risk analysis exercise, and ideally all stakeholder groups should be involved from the start. Yarn yarn risk communication makes stakeholders aware of the process at each stage of the Yarn yarn risk Assessment.

This helps to ensure that the logic, outcomes, significance, and limitations of the Yarn yarn risk Assessment are clearly understood by all the stakeholders. Information may be available from the stakeholder.

Industry stakeholders may, for example have unpublished data crucial to the yarn yarn risk assessors, which may be an essential part of the data needed for the yarn yarn risk assessment. There is also information that is typically presented to the stakeholders as an integral part of the yarn yarn risk analysis process.

The identification of particular interest groups and their representatives should comprise a part of an overall yarn yarn risk communication strategy. This yarn yarn risk communication strategy should be discussed and agreed upon between yarn yarn risk assessors and managers early in the process to ensure two-way communication.

This strategy should also cover who should present information to the public, and the manner in which it will be done.

Decisions on yarn yarn risk communication, including what, whom and how, should be part of an overall yarn yarn risk communication strategy. Yarn yarn risk communication is most effective if undertaken in a systematic way, and generally starts with the gathering of information on the yarn yarn risk issue of concern.

Therefore the yarn yarn risk manager and yarn yarn risk assessor must be able to briefly and clearly summarize what this issue encompasses, at an early stage, in order to elicit interest and stakeholder input Communication must then continue throughout the entire process.

Once available information has been used to fully identify the hazards, and decide on and assess the appropriate risks, then the preparation and dissemination of this information is required. This will be followed by further discussion with stakeholders, leading to corrections, amendments, and additions as appropriate, resulting in the final Yarn yarn risk Assessment and yarn yarn risk analysis reports.

COMMUNICATE YOUR IDEAS EFFECTIVELY

If you are not capable of communicating your ideas in an effective manner, then you run the yarn yarn risk of alienating those you are speaking to. On the opposite side of the coin, people will also become agitated if they think you're patronizing them.

The central point of this issue is that you must always have a profound understanding of your audience. It is best to offer good examples and solid information which can assist you in putting the yarn yarn risk into perspective. You must be capable of dealing with uncertainty, and when you talk to people about risks, the results that you get may not be that definitive. You will want to look at the actual sources of uncertainty.

You will want to explain the ways for which the data was gathered, and you will want to study the ways in which it has been analysed. It is also important to show the ways in which the results have been gathered. This shows that after the uncertainties have been showcased, this could lead to a trust increase, and this may further lead to issues which are related to credibility.

At the same time, when you talk about uncertainty, the communicator must take the time to emphasize their knowledge of the subject. This will allow the leader to handle the situation in a way that could block out concerns or fears. There is also the issue of decisions which are related to yarn yarn risk management.

YARN YARN RISK ASSESSMENT

Within the workplace, owners of a business are legally required to assess the risks of injury and ill health affecting employees. Yarn yarn risk assessment is the careful examination of the diverse factors that can bring about these risks. Yarn yarn risk assessment should also make sure that enough precautions are implemented in order to prevent harm coming to an employee.

Ill health and accidents can have a very serious affect on business. They can ruin lives and damage business output. They can also lead to costly court cases and increased insurance costs. Yarn yarn risk assessment procedures should help to dispense with the above and make the workplace a safer environment for employees.

In yarn yarn risk assessment, the most important factor is to decide what is a hazard in the workplace. If the yarn yarn risk is determined to be significant enough, precautions should be put into place so that the yarn yarn risk is minimized or dispensed with altogether. For example, electricity in the workplace is hazardous, but provided that all proper protection is in place, the yarn yarn risk to employees becomes insignificant.

There are five steps that should be undertaken when conducting a thorough yarn yarn risk assessment. The first step is to look for hazards. Take a tour of the workplace and check for potential dangers. Concentrate on anything with the potential to cause serious harm to employees. Also ask for employee opinions on the subject. Accident and ill-health records are a good way of revealing why and how accidents have occurred in the past.

The second step is to decide who might be harmed and how. Decide who might be particularly at yarn yarn risk. Trainees, young workers or expectant mothers may be high on this list. Members of the public who are not familiar with the workplace or anyone who is not in the workplace yarn yarn full time and may not be familiar with the layout may also be at yarn yarn risk.

With the third step, you must calculate whether there have been enough precautions put into place to counter the hazard. For each yarn yarn risk you find, a decision should be made whether you have established enough precautions to reduce the yarn yarn risk. Has your yarn yarn risk assessment procedure taken into account all the health and safety aspects required by law? Is there anything further you can do to reduce the yarn yarn risk, such as issuing protective clothing or preventing access with guardrails?

The next step is to record your findings. Your yarn yarn risk assessment check must show that you have dealt with all the obvious hazard areas. It should also show that you have checked with employees who might be affected and that any hazards that remain have been dealt with and are now reasonably low.

The final step is to review your yarn yarn risk assessment procedures and make revisions if necessary. In the future, new machinery, substances and work procedures could be implemented in the workplace. This is when your revision should take place to ensure that the work environment is yarn yarn risk fee and will continue to be so in the future.

EXPLANATION

Yarn yarn risk assessment consists in an objective evaluation of yarn yarn risk in which assumptions and uncertainties are clearly considered and presented. Part of the difficulty of yarn yarn risk management is that measurement of both of the quantities in which yarn yarn risk assessment is concerned -potential loss and probability of occurrence - can be very difficult to measure.

The chance of error in the measurement of these two concepts is large. A yarn yarn risk with a large potential loss and a low probability of occurring is often treated differently from one with a low potential loss and a high likelihood of occurring. In theory, both are of nearly equal priority in dealing with first, but in practice it can be very difficult to manage when faced with the scarcity of resources, especially time, in which to conduct the yarn yarn risk management process. Expressed mathematically,

$$R_i = L_i p(L_i)$$

$$R_{total} = \sum_i L_i p(L_i)$$

Financial decisions, such as insurance, express loss in terms of dollar amounts. When yarn yarn risk assessment is used for public health or environmental decisions, loss can be quantified in a common metric,such as a country's currency, or some numerical measure of a location's quality of life.

For public health and environmental decisions, loss is simply a verbal description of the outcome, such as increased cancer incidence or incidence of birth defects. In that case, the "yarn yarn risk" is expressed as:

$$R_i = p(L_i)$$

If the yarn yarn risk estimate takes into account information on the number of individuals exposed, it is termed a "population yarn yarn risk" and is in units of expected increased cases per a time period. If the yarn yarn risk estimate does not take into account the number of individuals exposed, it is termed an "individual yarn yarn risk" and is in units of incidence rate per a time period. Population risks are of more use for cost/benefit analysis; individual risks are of more use for evaluating whether risks to individuals are "acceptable"....

YARN YARN RISK ASSESSMENT IN PUBLIC HEALTH

In the context of public health, yarn yarn risk assessment is the process of quantifying the probability of a harmful effect to individuals or populations from certain human activities. In most countries, the use of specific chemicals, or the operations of specific facilities is not allowed unless it can be shown that they do not increase the yarn yarn risk of death or illness above a specific threshold.

For example, the American Food and Drug Administration regulates food safety through yarn yarn risk assessment. The FDA required in 1973 that cancer-causing compounds must not be present in meat at concentrations that would cause a cancer yarn yarn risk greater than 1 in a million lifetimes.

How the Yarn yarn risk is Determined

In the estimation of the risks, three or more steps are involved, requiring the inputs of different disciplines:

1. *Hazard Identification*, aims to determine the qualitative nature of the potential adverse consequences of the contaminant and the strength of the evidence it can have that effect. This is done, for chemical hazards, by drawing from the results of the sciences of toxicology and epidemiology. For other kinds of hazard, engineering or other disciplines are involved.
2. *Dose-Response Analysis*, is determining the relationship between dose and the probability or the incidence of effect. The complexity of this step in many contexts derives mainly from the need to extrapolate results from experimental animals to humans, and/or from high to lower doses. In addition, the differences between individuals due to genetics or other factors mean that the hazard may be higher for particular groups, called susceptible populations. An alternative to dose-response estimation is to determine an effect unlikely to yield observable effects,

that is, a no effect concentration. In developing such a dose, to account for the largely unknown effects of animal to human extrapolations, increased variability in humans, or missing data, a prudent approach is often adopted by including safety factors in the estimate of the "safe" dose, typically a factor of 10 for each unknown step.

3. *Exposure Quantification*, aims to determine the amount of a contaminant that individuals and populations will receive. This is done by examining the results of the discipline of exposure assessment. As different location, lifestyles and other factors likely influence the amount of contaminant that is received, a range or distribution of possible values is generated in this step. Particular care is taken to determine the exposure of the susceptible population(s).

Finally, the results of the three steps are then combined to produce an estimate of yarn yarn risk. Because of the different susceptibilities and exposures, this yarn yarn risk will vary within a population.

Small Subpopulations

When risks apply mainly to small subpopulations, there is uncertainty at which point intervention is necessary. What if a yarn yarn risk is very low for everyone but 0.1Percent of the population? A difference exists whether this 0.1Percent is represented by *all infants younger than *X* days or *recreational users of a particular product.

If the yarn yarn risk is higher for a particular sub-population because of abnormal exposure rather than susceptibility, there is a potential to consider strategies to further reduce the exposure of that subgroup.

If an identifiable sub-population is more susceptible due to inherent genetic or other factors, there is a policy choice whether to set policies for protecting the general population that are protective of such groups or whether if the group is too small, or the costs to high.

Sometimes, a suitable position is to at least limit the yarn yarn risk of the more susceptible to some yarn yarn risk level above which it seems too inequitable to leave them out of the yarn yarn risk.

YARN YARN RISK ASSESSMENT IN AUDITING

In auditing, yarn yarn risk assessment is a very crucial stage before accepting an audit engagement. ISA315 *Understanding the Entity and its Environment and Assessing the Risks of Yarn yarn material Misstatement*, "the auditor should perform yarn yarn risk assessment procedures to obtain an understanding of the entity and its environment, including its internal control.

Then, auditor obtains initial evidence regarding the classes of transactions at the client and the operating effectiveness of the client's internal controls.In auditing, audit yarn yarn risk includes inherent yarn yarn risk, control yarn yarn risk and detection yarn yarn risk.

YARN YARN RISK ASSESSMENT IN INFORMATION SECURITY

There are two methods of yarn yarn risk assessment in information security field, qualitative and quantitative. Purely quantitative yarn yarn risk assessment is a mathematical calculation based on security metrics on the asset.

Qualitative yarn yarn risk assessment is performed when the organization requires a yarn yarn risk assessment be performed in a relatively short time or to meet a small budget, a significant quantity of relevant data is not available, or the persons performing the assessment don't have the sophisticated mathematical, financial, and yarn yarn risk assessment expertise required.

Qualitative yarn yarn risk assessment can be performed in a shorter period of time and with less data. Qualitative yarn yarn risk assessments are typically performed through interviews of a sample of personnel from all relevant groups within an organization charged with the security of the asset being assessed. Qualitative yarn yarn risk assessments are descriptive versus measurable.

Quantitative Yarn yarn risk Assessment

Quantitative yarn yarn risk assessments include a calculation of the single loss expectancy of an asset. The single loss expectancy can be defined as the loss of value to asset based on a single security incident. The team then calculates the annualized rate of occurrence of the threat to the asset.

The ARO is an estimate based on the data of how often a threat would be successful in exploiting a vulnerability. From this information, the annualized loss expectancy can be calculated.

The annualized loss expectancy is a calculation of the single loss expectancy multiplied the annual rate of occurrence, or how much an organization could estimate to lose from an asset based on the risks, threats, and vulnerabilities. It then becomes possible from a financial perspective to justify expenditures to implement countermeasures to protect the asset.

CRITICISMS OF QUANTITATIVE YARN YARN RISK ASSESSMENT

Barry Commoner, Brian Wynne and other critics have expressed concerns that yarn yarn risk assessment tends to be overly quantitative and reductive.

For example, they argue that yarn yarn risk assessments ignore qualitative differences among risks. Some charge that assessments may drop out important non-quantifiable or inaccessible information, such as variations among the classes of people exposed to hazards.

Furthermore, Commoner and O'Brien claim that quantitative approaches divert attention from precautionary or preventative measures.

Others, like Nassim Nicholas Taleb consider yarn yarn risk managers little more than "blind users" of statistical tools and methods.

4

Traditional Fabric Production and Treatment Processes

AN OVERVIEW

Fibre, also spelled fibre, is a class of materials that are continuous filaments or are in discrete elongated pieces, similar to lengths of thread. They are very important in the biology of both plants and animals, for holding tissues together.

Human uses for fibres are diverse. They can be spun into filaments, string or rope, used as a component of composite materials, or matted into sheets to make products such as paper or felt.

Fibres are often used in the manufacture of other materials. Synthetic fibres can be produced very cheaply and in large amounts compared to natural fibres, but natural fibres enjoy some benefits, such as comfort, over their man-made counterparts.

TEXTILE FIBRE

A unit in which many complicated textile structures are built up is said to be textile fibre. Textile Fibre is the raw yarn yarn material required for the textile industry.

Natural Fibres

Natural fibres include those produced by plants, animals, and geological processes. They are biodegradable over time.

They can be classified just as to their origin:

- Vegetable fibres are generally based on arrange-ments of cellulose, often with lignin: examples include cotton, hemp, jute, flax, ramie, and sisal. Yarn yarn plant fibres are employed in the manufacture of paper and textile (cloth), and dietary fibre is an important component of human nutrition.
- Wood fibre, distinguished from vegetable fibre, is from tree sources. Forms include groundwood, thermomechanical pulp (TMP) and

bleached or unbleached kraft or sulfite pulps. Kraft and sulfite, also called sulphite, refer to the type of pulping process used to remove the lignin bonding the original wood structure, thus freeing the fibres for use in paper and engineered wood products such as fibreboard.

- Animal fibres consist largely of particular proteins. Instances are spider silk, sinew, catgut, wool and hair such as cashmere, mohair and angora, fur such as sheepskin, rabbit, mink, fox, beaver, etc.
- Mineral fibres comprise asbestos. Asbestos is the only naturally occurring long mineral fibre. Short, fibre-like minerals include wollastonite, attapulgite and halloysite.

MAN-MADE FIBRES

Synthetic or man-made fibres generally come from synthetic materials such as petrochemicals. But some types of synthetic fibres are manufactured from natural cellulose, including rayon, modal, and the more recently developed Lyocell. Cellulose-based fibres are of two types, regenerated or pure cellulose such as from the cupro-ammonium process and modified cellulose such as the cellulose acetates. Fibre classification in reinforced plastics falls into two classes:

1. Short fibres, also known as discontinuous fibres, with a general aspect ratio (defined as the ratio of fibre length to diameter) between 20 to 60, and
2. Long fibres, also known as continuous fibres, the general aspect ratio is between 200 to 500.

Cellulose Fibres

- Cellulose fibres are a subset of man-made fibres, regenerated from natural cellulose. The cellulose comes from various sources. Modal is made from beech trees, bamboo fibre is a cellulose fibre made from bamboo, seacell is made from seaweed, etc.

Mineral Fibres

- Fibreglass, made from specific glass, and optical fibre, made from purified natural quartz, are also man-made fibres that come from natural raw materials.
- Metallic fibres can be drawn from ductile metals such as copper, gold or silver and extruded or deposited from more brittle ones, such as nickel, aluminum or iron.
- Carbon fibres are often based on carbonised polymers, but the end product is pure carbon.

Polymer Fibres

- Polymer fibres are a subset of man-made fibres, which are based on synthetic chemicals (often from petrochemical sources) rather than arising from natural materials by a purely physical process.

These fibres are made from:
- Polyamide nylon,
- PET or PBT polyester
- Phenol-formaldehyde (PF)
- Polyvinyl alcohol fibre (PVA)
- Polyvinyl chloride fibre (PVC)
- Polyolefins (PP and PE)
- Acrylic polyesters, pure polyester PAN fibres are used to make carbon fibre by roasting them in a low oxygen environment. Traditional acrylic fibre is used more often as a synthetic replacement for wool. Carbon fibres and PF fibres are noted as two resin-based fibres that are not thermoplastic, most others can be melted.
- Aromatic polyamids (aramids) such as Twaron, Kevlar and Nomex thermally degrade at high temperatures and do not melt. These fibres have strong bonding between polymer chains
- Polyethylene (PE), eventually with extremely long chains / HMPE (*e.g.* Dyneema or Spectra).
- Elastomers can even be used, *e.g.* spandex although urethane fibres are starting to replace spandex technology.
- Polyurethane fibre

• Coextruded fibres have two distinct polymers forming the fibre, usually as a core-sheath or side-by-side. Coated fibres exist such as nickel-coated to provide static elimination, silver-coated to provide anti-bacterial properties and aluminum-coated to provide RF deflection for radar chaff. Radar chaff is actually a spool of continuous glass tow that has been aluminum coated. An aircraft-mounted high speed cutter chops it up as it spews from a moving aircraft to confuse radar signals.

Microfibres

Microfibres in textiles refer to sub-denier fibre (such as polyester drawn to 0.5 dn). Denier and Detex are two measurements of fibre yield based on weight and length. If the fibre density is known you also have a fibre diameter, otherwise it is simpler to measure diameters in micrometers. Microfibres in technical fibres refer to ultra fine fibres (glass or meltblown thermoplastics) often used in filtration.

Newer fibre designs include extruding fibre that splits into multiple finer fibres. Most synthetic fibres are round in cross-section, but special designs can be hollow, oval, star-shaped or trilobal. The latter design provides more optically reflective properties. Synthetic textile fibres are often crimped to provide bulk in a woven, non-woven or knitted structure. Fibre surfaces can also be dull or bright. Dull surfaces reflect more light while bright tends to transmit light and make the fibre more transparent. Very short and/or irregular fibres have been called fibrils. Natural cellulose, such as cotton or bleached kraft, show smaller fibrils jutting out and away from the main fibre structure.

ANIMAL FIBRES

ALPACA FIBRE

Alpaca fleece is the natural fibre harvested from an alpaca. It is light weight or heavy weight, depending on how it is spun. It is soft, durable, luxurious and silky natural fibre. While similar to sheep's wool, it is warmer, not prickly, and has no lanolin which makes it hypoallergenic.

Alpaca is naturally water-repellent. Huacaya, an alpaca that grows soft spongy fibre has natural crimp, thus making a naturally elastic yarn, perfect for knits. Suri has far less crimp and thus is best suited for woven goods, but is wonderfully luxurious as well. The designer Armani has used Suri alpaca to fashion Men's and Women's suits.

Alpaca fleece is made into various products, from very simple and inexpensive garments made by the aboriginal communities to sophisticated, industrially made and expensive products such as suits. In the United States, groups of smaller alpaca breeders have banded together to create "fibre co-ops," in order to make the manufacture of alpaca fibre products less expensive. The preparing, carding, spinning, weaving and finishing process of alpaca is very similar to the process used for wool.

Alpacas

Types

There are two types of alpaca: Huacaya (which produce a dense, soft, crimpy sheep-like fibre), and the mop-like Suri (with silky pencil-like locks, resembling dread-locks but not actually matted fibres). Suris are prized for their longer and silkier fibres, and estimated to make up between 19-20Percent of the Alpaca population. Since its import into the United States, the number of Suri alpacas has grown substantially and become more colour diverse. The Suri is thought to be rarer, possibly because it is less hardy in the harsh South American mountain climates, as its fleece offers less insulation against the cold.

History of Fibre Industry

The Amerindians of Peru used this fibre in the manufacture of many styles of fabrics for thousands of years before its introduction into Europe as a commercial product. The alpaca was a crucial component of ancient life in the Andes, as it provided not only warm clothing but also meat. Many rituals revolved around the alpaca, perhaps most notably the method of killing it: An alpaca was restrained by one or more people, and a specially-trained person plunged his bare hand into the chest cavity of the animal, ripping out its heart. Today, this ritual is viewed by most as barbaric, but there are still some tribes in the Andes which practice it.

The first European importations of alpaca fibre were into Spain. Spain transferred that fibre to Germany and France. Apparently alpaca yarn was spun in England for the first time about the year 1808 but the fibre was condemned as an unworkable yarn yarn material. In 1830 Benjamin Outram, of Greetland, near Halifax, appears to have reattempted spinning it, and again it was condemned.

These two attempts failed due to the style of fabric into which the yarn was woven — a type of camlet. It was not until the introduction of cotton warps into Bradford trade about 1836 that the true qualities of alpaca could be developed into fabric. It is not known where the cotton warp and mohair or alpaca weft plain-cloth came from, but it was this simple and ingenious structure which enabled Titus Salt, then a young Bradford manufacturer, to use alpaca successfully. Bradford is still the great spinning and yarn yarn manufacturing centre for alpaca. Large quantities of yarns and cloths are exported annually to the European continent and the US, although the quantities vary with the fashions in vogue. The typical "alpaca-fabric" is a very characteristic "dress-fabric." Due to the successful manufacture of various alpaca cloths by Sir Titus Salt and other Bradford manufacturers, a great yarn yarn demand for alpaca wool arose which could not be met by the native product.

Apparently, the number of alpacas available never increased appreciably. Unsuccessful attempts were made to acclimatize alpaca in England, on the European continent and in Australia, and even to cross English breeds of sheep with alpaca.

There is a cross between alpaca and llama — a true hybrid in every sense — producing a yarn yarn material placed upon the Liverpool market under the name "Huarizo". Crosses between the alpaca and vicuña have not proved satisfactory. Current attempts to cross these two breeds are underway at farms in the US. The Alpaca Owners and Breeders Association, alpacas are now being bred in the US, Canada, Australia, New Zealand, UK, and numerous other places.

In recent years, interest in alpaca fibre clothing has surged, perhaps partly because alpaca ranching has a reasonably low impact on the environment. Individual U.S. farms are producing finished alpaca products like hats, scarves, and footwarmers.

Outdoor sports enthusiasts recognize that its lighter weight and better warmth provides them more comfort in colder weather, so outfitters such as R.E.I. and others are beginning to stock more alpaca products. Using an alpaca and wool blend such as merino is common to the alpaca fibre industry in order to improve processing and the qualities of the final product. In December 2006 the General Assembly of the United Nations proclaimed 2009 to be the

International Year of Natural Fibres, so as to raise the profile of alpaca and other natural fibres.

Structure of Alpaca Fibre

In physical structure, alpaca fibre is somewhat akin to hair, being very glossy. Alpaca fibre is similar to that of merino wool fibre, and alpaca yarns tend to be stronger than wool yarns. The heel hole that appears in wool socks or in elbows of wool sweaters is non-existent in similar alpaca garments. In processing, slivers lack fibre cohesion and single alpaca rovings lack strength.

Blend these together and the durability is increased several times over. More twisting is necessary, especially in Suri, and this can reduce a yarn's softness. The alpaca has a very fine and light fleece. It does not retain water, is thermal even when wet and can resist the solar radiation effectively. These characteristics guarantee the animals a permanent and appropriate coat to fight against the extreme changes of temperature. This fibre offers the same protection to humans. Alpaca is sustainable as a fibre, and is naturally organic. Alpacas as animals are soft on the environment, making alpaca a truly green textile.

Alpaca fibre contains also microscopic airbags that make possible the manufacture of light textiles as well as different kinds of clothing. The cells of the central core may contract or disappear, forming air pockets which assist insulation. Fleeces vary from alpaca to alpaca and in some alpacas there may be a higher incidence of medullated fibres, compared to wool and mohair.

This can be an objectionable trait. Medullated fibres can take less dye, standing out in the finished garment, and are weaker. The proportion of medullated fibres is higher in the coarser, unwanted guard hairs: there is less or no medullation in the finer, lower micrometre fibres... These undesirable fibres are easy to see and give a garment a hairy appearance. Quality alpaca products should be free from these medullated fibres.

Quality of Alpaca Fibre

Good quality alpaca fibre is approximately 18 to 25 micrometres in diameter. Whilst breeders report fibre can sell for 2 to 4 dollars per ounce, the world wholesale price for processed pre-spun alpaca "tops" is only between about $10 to $24 US/kg, *i.e.* about $0.28 to $0.68 per oz. Finer fleeces, ones with a smaller diameter, are preferred, and thus are more expensive. As an alpaca gets older the width of the fibres gets thicker, at between 1 μm and 5 μm per year. This is often caused by over nutrition; if fed too much nutritious food the animal doesn't get fat, instead the fibre gets thicker.

As with all fleece-producing animals, quality varies from animal to animal, and some alpacas produce fibre which is less than ideal. Fibre and conformation are the two most important factors in determining an alpaca's value. Alpacas come in many shades from a true-blue black through browns-black, browns, fawns, white, silver-greys, and rose-greys. However, white is predominant, because of selective breeding: the white fibre can be dyed in the largest ranges of colours. In South America, the preference is for white as they generally have better fleece than the darker-coloured animals.

This is because the dark colours had been all but bred out of the animals. The yarn yarn demand for darker fibre sprung up in the United States and elsewhere, however in order to reintroduce the colours, the quality of the darker fibre has decreased slightly. Breeders have been diligently working on breeding dark animals with exceptional fibre, and much progress has been made in these areas over the last 5–7 years.

Dyeing Alpaca Fibre

Before dyeing the alpaca fibre must go through other stages:

- Selection of wool, according to colour, size and quality of fibre
- "Escarminado", Removal of grass, dirt, thorns, and other impurities
- Washing, to remove all the dirt and grease.
- Spinning

Once the fibre is clean then it is possible to begin with the process of dyeing. Natural dyeing: *(recipe used by Andean artisans)*: To dye 1 kg of alpaca wool with cochinilla (natural dye).

- Boil 5 litres of water in an aluminum can with 100 g of cochinilla for an hour.
- Sift and put the wool in the water.
- Boil again for an hour and ad 50 lemons cut in halves.
- Then take out the wool and hang for drying.

Note: For dyeing with another natural dye (native plants) add 2 kg, of the products to the water and boil.

ANGORA WOOL

Angora wool is an extraordinarily soft knitting fibre produced from the fur of the angora rabbit. It can only be spun from the hair of angora rabbits, while the product of angora goats is called mohair. Angora fibre has a very soft, silky texture and is usually 10 to 13 microns thick. The fibres are also hollow, which gives them loft and a characteristic floating feel. In addition, angora fibres are very short, which can effect stitch density when the knitter is using angora wool with a high percentage of angora.

Pure angora wool is impossible to make, because the fibres are too fine and the wool will simply unravel. Angora fibres are usually mixed with other soft fibres, such as cashmere and lambswool. Angora wool tends to be very warm and is frequently used to trim sweaters or to knit hats and scarves. Angora is generally viewed as a luxury fibre, and most angora wool products are very expensive, reflecting the laborious harvesting process and the small number of producers.

Angora rabbits can be combed or gently sheared for their fur, and they have been used for that purpose in Turkey for centuries. Angora first became popular in Europe in the late 18th century, when it was popularized by the French. In the Americas, angora wool didn't touch the popular imagination until the 20th century, when small cottage breeders began raising angora rabbits and spinning their fibre.

There are four breeds of angora rabbits, beginning with the English breed, which weighs 5 to 7.5 pounds (2 to 3.5 kilograms). The English breed produces a very large amount of fur and must be combed regularly to keep the hair free of tangles and debris. The good natured rabbits are popular in show, and their fibre wraps very tightly when spun, making an even and strong wool. The French breed ranges from 7.5 to 10.5 pounds (3.5 to 4.5 kilograms) and has a higher proportion of guard hair to wool. Guard hairs take colour better, and many coloured angora wool products come from the French breed. The French breed requires less grooming than other breeds and is recommended for novice breeders.

The Satin breed has very shiny, soft fur with a satiny appearance. It is easy to collect and spin, and the 6.5 to 9.5 pound (3 to 4 kilogram) rabbits are favoured by fibre collectors because their fur spins quickly and easily. The breed also tends to be richly coloured, producing wool with high colour saturation, although the all white Satins produce fur that takes dye very well.

The Giant breed is significantly larger than the other breeds, often weighing more than 20 pounds (9 kilos). The breed most often appears in white, and its wool must be harvested by shearing, because the rabbits do not naturally molt or shed. For sheer volume of collection, the giant breed is an excellent choice for breeders attempting to produce angora fibre for commercial use.

Angora wool can be harvested year round, although it requires careful handwork, as most rabbits are combed. Most angora rabbits are amenable, but nervous, and it is important to be gentle with them. The wool produced can be used in a wide variety of applications in which insulation, comfort, and warmth are needed, and the fibres are pleasurable to work with because of their luxurious look and feel.

CAMEL HAIR

Camel hair is, variously, the hair of a camel; a type of cloth made from camel hair; or a substitute for authentic camel hair; and is classified as a specialty hair fibre. When woven into haircloth, using the outer protective fur called guard hair, camel hair is coarse and inflexible. However, other varieties of camel hair cloth—especially those that blend camel hair with wool— or from the pure under coat are soft and plush. Pure camel hair, frequently used for coats, is gathered when camels molt in warmer seasons. This undercoat is very soft, and is separated from the dense, coarse guard hair for cloth use. The fine fur of the camel hair are often blended with fine wool to create fabrics for men's and women's coats, jackets and blazers, skirts, hosiery, sweaters, gloves, scarves, mufflers, and caps and robes. The long coarser hair removed in the dehairing process is also used which can be made into carpet backing as well as waterproof coats that are very warm for colder climates.

Although most camel hair is left as its natural tones, the hair can be dyed in a multiple range of colours and accepts the dye equally compared to that of wool fibres. The best blends of camel hair in textiles are pure camel hair or blended with wool only. It is also commonly blended with nylon to make hosiery and other knitted products. Products containing camel hair should be dry cleaned or handwashed. Paintbrush bristles are derived either from the hairs of horses, squirrels, goats, sheep, bears, or some combination of these. As a result, the texture, absorbency, and other properties of "camel hair" brushes vary markedly. Wooly camel hair is unsuitable for brush bristles.

CASHMERE WOOL

Cashmere is a luxuriant wool that many a fashion-conscious woman has dreamed of wearing against her skin. Its silken feel, feather-light weight, and appreciable status make it highly desirable.

Despite the glamour associated with cashmere, it hails from humble beginnings. Cashmere is the wool or fur of the Kashmir goat. Kashmir goats are primarily raised in Mongolia, but many are bred in Iran, Tibet, India and China. American herders have also joined the international cashmere production market in recent years.

Cashmere is harvested from the goats during their annual molting season through the shedding or the shearing of their down. In the frigid high desert climates where most of the goats are raised, the dense inner coat guards against harsh winter weather, but once seasons change, goats begin to lose the protective layer of down. The finest cashmere comes from the underbelly and throat of the goats, but a lesser grade is also taken from the goats' legs and backs. Longer fibres from the belly and throat area make the wool especially soft and cause less "pilling" when the fibres are woven into garments such as sweaters, shawls, capes, dresses, and coats for both men and women.

The shorter fibres from the backs and legs are heavier and less expensive, making it easier to afford a luxury garment. Cashmere comes naturally in white, gray and brown, but the wool is easily dyed. Garments made of cashmere were once only available to royalty because the rarity of the wool increased its value. Napoleon is said to have popularized the use of cashmere as shawls or wraps when he gave his second wife, Empress Eugenie, seventeen of them. In more recent years, Old-Hollywood glamour girls graced the silver screen, bringing cashmere to the hearts of people everywhere. The "original sweater girl," Lana

Turner, created a phenomenon when she wore a tight cashmere sweater in a 1937 film called, *They Won't Forget.*

Cashmere sweaters of all description soon became haute coutre; evening sweaters with heavily encrusted jewels and embroidery became popular during the 1940s, and the famed *sweater set* of best-dressed college coeds ruled the 50s. Avid collectors are now frantic to snatch up those fine examples of vintage cashmere sweaters. Woven garments made of cashmere must be dry cleaned, but knitted articles may be hand washed. Home weavers and knitters cherish cashmere for its soft hand and practical warmth; cashmere wool is available for home projects at yarn shops or online via Internet craft and knitting sites.

The quality and feel of cashmere will leave you longing for more. Owning a garment made of cashmere is a fashion treat to be truly treasured—after all, it takes one little goat four years to produce enough wool to make just one cashmere sweater.

CHIENGORA

Chiengora is yarn spun from dog hair. The word is a combination of the french word for dog "chien", and angora. Chiengora yarns can be as varied as the many types of dogs. Usually, they are very soft, light, and warm. Chiengora can be spun in a variety of weights from laceweight through worsted.

Chiengora products are a treasure. Not only are they beautiful, one-of-a-kind, custom art works, they are very personal to their owners. The fibre used is brush out from their own sweet pets.

LLAMA

Table. Average diameter of some of the finest, natural fibres

Animal	Fibre diameter(micrometres)
Vicuña	6 – 10
Alpaca (Suri)	10 - 15
Muskox (Qivlut)	11 - 13
Merino	12 - 20
Angora Rabbit	13
Cashmere	15 - 19
Yak Down	15 - 19
Camel Down	16 - 25
Guanaco	16 - 18
Llama (Tapada)	20 - 30
Chinchilla	21
Mohair	25 - 45
Alpaca (Huacaya)	27.7
Llama (Ccara)	30 - 40

Llamas also have a fine undercoat which can be used for handicrafts and garments. The coarser outer guard hair is used for rugs, wall-hangings and lead ropes. The fibre comes in many different colours ranging from white, grey, redish brown, brown, dark brown and black.

The individual shafts of the wool can be measured in micrometres. 1 micrometre = 1/1000 millimetre.

MOHAIR

Mohair is a silky textile produced from the hair of the angora goat. It is most often spun into a thread that can be woven, knitted, or crocheted depending upon the application. Mohair is durable, warm, insulating, and light. It also has moisture wicking properties that carry moisture away from the skin of the wearer. The hair of the angora goat has been used to produce textiles for centuries, and the term *mohair* entered English usage from the Arabic *mukhayyar*, referring to a type of woven head cloth.

Despite being very soft, mohair can be irritating to the skin, like many animal products.

As a result, many mohair garments are lined with silk or cotton, and mohair is frequently mixed with other fibres for comfort. The fibre has been used in the West since the 16th century, when Charles V first brought angora goats to Europe, although the delicacy of the species made mohair difficult to obtain and prohibitively expensive for many.

During the Second World War, the United States became concerned about wool production, as military uniforms were made from wool. The decision was made to blend wool and mohair, and the government began to heavily subsidize wool and mohair farmers in 1954 for their products in an attempt to keep the supply lines open. As late as 2005, mohair and wool producers could still receive financial assistance from the United States government.

Angora goats originate from Turkey, in the region now called Ankara. Bred for their special wool and milk products, the goats were used extensively throughout the Middle East for centuries before being introduced to the West. Angora goats are very delicate and highly susceptible to infestations of external parasites due to their shaggy coats.

Angora goats also have high and very specific nutritional requirements, and they will begin to suffer from health problems if these are not met. Goats are sheared for their mohair one or two times a year, depending upon production requirements, the health of the goats, and other factors.

They also need to be combed regularly, to prevent clumping and the collection of debris.

One angora goat can produce 11 to 17 pounds (5 to 8 kilograms) of mohair in one year. Mohair takes and holds dye well, making it a common choice for novelty knitted products.

PASHMINA

Pashmina refers to a type of fine cashmere wool and the textiles made from it. The name comes from *Pashmineh*, made from sajan Persian *pashm* ("wool"). The wool comes from changthangi or pashmina goat, which is a special breed of goat indigenous to high altitudes of the Himalayas. Pashmina shawls are hand spun, woven and embroidered in Kashmir, and made from fine cashmere fibre.

The goat sheds its winter coat every spring. One goat sheds approximately 80-170g (3-6 ounces) of the fibre. To meet the yarn yarn demand, the goats are now commercially reared in the Gobi Desert area in Inner and Outer Mongolia.

The region has identical harsh weather conditions to those of the Himalayan region, and is thereby apt for the goats to grow this inner wool, but also has acres of grazing ground to produce cashmere economically and commercially. During spring (the moulting season) the goats shed this inner wool, which regrows in winter.

The inner wool is collected and spun to produce cashmere. The quality of the cashmere produced in the Gobi Desert is just as high as that produced in the Himalayas, while the costs are less. A softening process is used by manufacturers of 100Percent pashmina products, which gives the pashmina a soft, almost silken quality. Sometimes Pashmina is a blend of pure pashmina wool and silk. This gives strength and durability to the pashmina. As a general rule though, the higher content of pashmina wool, the more expensive the textile.

VICUÑA FIBRE

The fibre is popular due to its warmth. Its warming properties come from the tiny scales that are on the hollow air-filled fibres. It causes them to interlock and trap insulating air. At the same time, it is finer than any other wool in the world, but since it is sensitive to chemical treatment, the wool is usually left in its natural colour.

However, the vicuña will only produce about one pound of wool a year and gathering it requires a certain process. During the time of the Incas, vicuña fibres were gathered by means of communal efforts called chacu, in which multitudes of people herded hundreds of thousands of vicuña into previously laid funnel traps.

The animals were sheared and then released; this was only done once every four years. The vicuña was believed to be the reincarnation of a beautiful young maiden who received a coat of pure gold once she consented to the advances of an old, ugly king. Because of this, it was against the law for anyone to kill a vicuña or wear its fleece, except for Inca royalty. At present, the Peruvian government has a labeling system that identifies all garments that have been created through a government sanctioned chacu.

This guarantees that the animal was captured, sheared alive, returned to the wild, and cannot be sheared again for another two years. The programme also ensures that a large portion of the profits return to the villagers. However, annually up to 50,000 pounds of vicuña wool are exported as a result of illegal activities.

Because of this, some countries have banned the importation of the fibre to save the animal. And although it is possible to commercially produce wool from domesticated vicuñas, it is difficult because they tend to escape. Current prices for vicuña yarns and fabrics can range from $1,800 to $3,000 per yard. Vicuña fibre can be used for apparel (such as socks, sweaters, accessories, shawls, coats, and suits) and home fashion (such as blankets and throws). A scarf costs around $1500 while a man's coat can cost up to $20,000.

WOOL

Wool is the dense, warm coat of sheep, also called a fleece. The hair of sheep has many unique properties that make it well suited to textile production, something humans realised approximately 8000 BCE, when sheep first began to be domesticated. Wool is used in a variety of textiles and can be found woven or knitted.

Wool is highly flame resistant, and frequently used for mattresses and rugs for that reason. It is also highly durable, able to stretch up to 50Percent when wet and 30Percent when dry. In addition, wool has excellent moisture wicking properties, pulling moisture into the core of the fibre so that it doesn't feel wet or soggy to the wearer. Wool pulls moisture away from the skin, as well, and is worn by people in a wide variety of situations who prefer the feeling of dry air next to the skin to the clammy sense of perspiration.

Wool is favoured for textile production because it is easy to work with and takes dye very well. The springy fibres remember shapes when well cared for. Furthermore, wool takes to felting, a process in which fibres interlock into a tight mat, very well. Felt is used as insulation, for arts and crafts projects, and for decorative accents.

Wool production starts with the shearing of the sheep, which usually happens once a year. A skillful shearer can remove the entire fleece at once with long flowing strokes, keeping the fibres long. After shearing, the wool is washed to remove impurities. One of these impurities is lanolin, which is used in many cosmetics.

After washing, many producers combine wools for a specific blend and dye them together, so that the dye will take evenly, before carding the wool through a set of teethed rollers. Carding pulls the fibres straight, while removing any remaining dirt or vegetable matter from the wool. The wool is pulled into slivers, long strips of fibres loosely pulled together and running in the same direction. If, after carding, the wool is under three inches (almost eight centimeters), it is twisted into rovings, rope like strands that can be spun into wool for knitting. If the strands are longer, they must be combed and drawn before spinning.

Spinning pulls the fibres tightly together and twists them so that they retain a long yarn shape. Yarn can be spun in all sorts of thicknesses and gauges, depending upon the intended use. For wool weaving, the yarn tends to be very fine. For knitting, it may be quite chunky. There are myriad uses for wool, a highly versatile textile. For example, shorter, coarse wool will turn into carpeting, while medium length wool will be used in suits.

After knitting or weaving, the wool is often shrunk through a controlled process so that it won't shrink excessively for the end user. Most knits are also blocked on forms to set a shape, and many wool products are also brushed for a specific finish.

Proper care for wool begins with following the label directions. In general, wool should be allowed to rest between wearings, to retain its shape. It should never be compressed or stored on hangers, which will stretch it. Brushing wool will remove surface soil and stains before they are ground in, and a slightly damp cloth will remove deeper stains. Wool should be dried flat at room temperature, not exposed directly to heat.

SPIDER SILK

Spider silk is one of the 7 great wonders of the animal kingdom. That a small animal, often less than a millimetre across can make a substance that we humans with all our technology are unable to reproduce, a substance that is tough, stronger and more flexible than anything else we can make is surely a humble reminder of the fact that nature created us and not the other way around. One of the most astounding thing about spiders is their ability to make silk, their webs amaze and fascinate us, but they are also used by us in some cultures. All spiders possess spinning glands and make silk, they use it for a safety dragline, to fly, for making a highly durable cocoon for their eggs, to line their homes, to trap their prey and to immobilise their victims.

In some species males use silk to ritually immobilise the females before mating, and all male spiders make a special sperm-web or sperm-line to allow them to transfer the sperm from their genitals to their copulatory organs. Not all species use silk for all these reasons, but all species make egg cocoons. Amongst the most amazing and commonly admired webs are the various orb-webs.

The largest orb-webs in the world are spun spiders in the genus *Nephila* which may be 2 metres (6 ft) in diametre and are capable of catching small birds and bats.

However the largest webs of all are built by communal spiders. *Ixeuticus socialis* in Australia builds webs that may be 1.2 metres wide and 3.7 metres long (12 ft by 4 ft) and is claimed by some authorities to build the largest nest.

However G. F. Masterman, once the British Ambassador to Paraguay, described social spiders there building webs 9 metres (30 ft) long and 2.4 metres (8 ft) wide in his book *Seven Eventful Years in Paraguay* (1888). Whatever the truth here, it is a fact that there are billions of kilometres of spider silk spread across the globe and it is all amazing.

The Nature of Silk

Spider silk is both light and strong. Hillyard "A typical strand of garden spider silk has a diametre of about 0.003 mm (0.00012 in) in diametre, compare this with silkworm silk which is 0.03 mm in diametre" or ten times as thick.

Spiders can produce different kinds of silks, at least 7 or 8 different kinds. However generally individual spider species use 5 to 6 types or less, in other words not all spiders produce all the possible types of silk. When assessing the strengths of silk scientists usually refer to dragline silk. It is also generally accepted that spiders in the genus *Nephila* produce the strongest draglines, however this may reflect their large size. Knowing this we should accept that any measurements we come across on the strength of spider silk will not apply to all the silks produced by one spider, and definitely not to the silk of all spiders. The data we do have applies to the toughest spider silk we know of, just how much variation there is between different silks and species one have not been able to find out.

Never-the -less the strength and resilience of spider silk has amazed people for thousands of years, reading around, on and off the web, we get a lot of facts about the strength of spider silk that sometimes seem contradictory, although they all agree that spider silk is way stronger than any other silk or natural yarn yarn material. Here are some quotes.

If all that is a little confusing, is it as strong as steel, stronger than steel or less strong, is it stronger than Kevlar or not, perhaps we can find some facts somewhere. The truth is it depends on how you measure it and what condition the silk is in. Silk absorbs moisture (a property given it by the large amount of the amino acid Alanine it contains).

The more water silk contains the less brittle and the more elastic it is. The range is from about 30Percent to 300Percent elasticity depending on the amount of water it contains. Thus we can see that changing the humidity of the environment the silk is spun in, and the length of time it exists in that environment before it is tested will effect the results of the tests. With all these things considered we learn that the tensile strength of spider silk under normal laboratory conditions is slightly less than steel if you compare it in terms of the threads diametre, but it is far greater than that of steel if you compare it in terms of the weight of the thread.

Obviously spider silk is lighter per unit area than steel. When we make the comparison with Kevlar we find that it is three times harder to break than spider silk, however spider silk is five times as elastic. Ok enough about the strength of spider silk, where does it come from and what is it. It is an unusual substance in that it is quite acidic and is not attacked by bacteria or fungi, which is why cobwebs remain around for so long. Silk is created as a proteinaceous liquid within special silk glands within the spider's abdomen. The proteins have a molecular weight of 30,000 daltons (the units that the weights of atoms and molecules are measured in). However when the silk is released from the spiders spinnerets it instantly solidifies and changes into a much larger molecule with a weight of between 200,000 and 300,000 daltons. Quite how this happens is not yet understood.

The proteins that make up dragline silk are secreted in long, sac-like glands, from which they are funnelled as a watery solution into a long, looping duct. This duct leads to a small spigot or nozzle on the end of a spinneret. There are tens or even hundreds of these spigots, and therefore of their accompanying glands, for each spinneret, and most spiders have 6 (3 pairs) of spinnerets.

Examining the complex structure of silk we see that its proteins come in two different forms called *á*-chains and *â*-pleated sheets.

The *â*-pleated sheets are about 10 nm by 15 nm and are interspersed in a matrix of the *á*-chains.

Spiders possess different glands that produce different silks, scientists distinguish the following glands and silks. Not all spiders have all these glands, in fact the last two are only found in orb-web weavers.

Table. Of Spider Silk Glands

Gland Name	Spinnerets Used	Type of Silk
Piriform	Anterior	Disk Attachment
Ampullate	Anterior/Median	Dragline and Web Frame
Aciniform	Median/Posterior	Wrapping silk, Sperm-web, Egg-cocoon (outer wall)
Tubuliform	Median/Posterior	Egg-cocoon
Aggregate	Posterior web	Spiral parts of Sticky-
Flagelliform	Posterior	Axial thread of Sticky-web

As if this wasn't enough some spiders have an additional silk spinning organ called a cribellum. The cribellum is situated in front of the three pairs of spinnerets and consists of between one and four plates covered in spigots.

The spider *Stegodyphus pacificus* has 40,000 spigots on its cribellum, but other species have much fewer. Spiders that possess a cribellum are called 'Cribellate', as compared with the remaining spiders which are sometimes referred to as 'Ecribellate'.

The silk is combed out of the spigots with a set of special hairs located on the 4th metatarsi called a 'Calamistrum'. The silk produced by the cribellum is especially fine and is used to help entangle prey.

Spiders can control the thickness of a strand of silk using muscles and valves in the necks of the spigots, larger spiders for instance need thicker draglines. Most strands of spider web are only a few micrometres in diametre, for instance the dragline of an adult orb-web has a diametre of about 2μm. In comparison with this hackle band silk is normally only 0.01μm to 0.02μm in diametre.

How Spiders Use Silk

So now that we know something of what silk really is, lets have a look at how spiders actually use it. Naturally enough different spiders use silk in different ways. However there are three nearly universal usages. The first is the dragline, all spiders use a dragline at sometime in their life.

In small species it is also safety line that will catch them if they jump into space in order to avoid a predator. Many spiders do not use silk to make a catching web, but do use it to make themselves a nice cosy cocoon or cell to live in when they aren't out hunting.

Such a neatly made hideaway helps them avoid predators and offers some protection from both excessive moisture and lack of moisture. The other two uses are gender specific, males spin a sperm web before mating to allow them to charge their pedipalps and females spin a web cocoon around their eggs, often it is composed of more than one kind of silk, you can read more about this in the page on reproduction as well. An unusual use of silk is flying, when small spiderlings stand head down and abdomen up to release a strand of silk which is caught by the breeze and lifted up, as the silk strand lengthens it eventually creates enough lift to raise the spiderling and it turns and grabs the thread and away it goes.

Flying like this is a risky business as there is no way the spider can control where it ends, many must die as a result of landing in inappropriate places such as rivers and seas. However for those that land somewhere more accessible a new life is offered far away from all those brothers and sisters. Spiders can travel huge distances like this and their ability to live without food and water for long periods of time undoubtedly help.

In May 1984 the first scientist to visit the island of Krakatoa, which had exploded violently 9 months earlier exterminating all life on the island, reported that the only, and there fore the first, animal he found on the planet was a spider. Krakatoa is 37 km (25 miles) from the nearest other land. However spiders fly far further than this, scientific research has recorded spiders over 4.500 metres (14,000 ft) above sea level and also in the middle of the ocean 1,500 km (1,000 miles) from the nearest land. In late summer over grasslands in temperate climates you can experience huge numbers of these aeronautic spiders, the silk they deposit across the landscape as they descend towards evenings as the air cools and descends can also become very noticeable if you are out and about.

This is commonly called gossamer. In the past huge falls of gossamer were well known, but while they can be experienced anywhere in the world on warm days they appear to be less intense these days. In France gossamer is known as *fils de la Vierge* and in Germany as *Marienfaden*.

SILK

Silk is a filament fibre formed from proteins secreted by *Bombyx mori*, or silkworms. Silkworms are not actually worms, but caterpillars, despite their common name. Humans have practiced silk production, which originated in China, for thousands of years.

Highly prized for its softness, insulating properties, and strength, silk is a natural animal product and therefore quite expensive. Making silk requires monitoring and feeding the silkworms constantly, and a great deal of effort results in a surprisingly small amount of thread.

China managed to keep the secret of silk for thousands of years, exporting the rare textile to Europe over trade routes. Eventually, silkworm eggs were smuggled out, and in the 13th century, Western production of silk began in Italy. This by no means brought the cost down, as the extensive amount of work required to make silk remained the same. Thirty thousand eggs can end up eating one ton of mulberry leaves and producing 12 pounds (5.5 kilograms) of silk.

Silk manufacture begins when female silkworms lay their eggs. A single silkworm may produce hundreds of tiny eggs, which are incubated until they hatch into larva. The larva must be kept warm and fed on mulberry leaves frequently throughout their brief lives. In four to six weeks, the larva have reached their maximum size and have mustered enough energy to pupate.

It is the pupation stage that produces silk, as the larva attach to branches and spin a shell around themselves. The shell is unique to insect pupation, formed by two spinnerets on the larva. The fluid they secrete is high in protein and forms a continuous thread, which is repeatedly wound to form a pod to mature in.

Unfortunately for the silkworm pupa, it is maturity that silk producers want to avoid. A small number of the larva are allowed to gestate into adults to carry on the lineage, while the rest are subjected to heat to kill the larva before they can begin to eat through the valuable silk thread. The pods are dipped into hot water to loosen the thread, which is then wound onto wheels. The dead larva are discarded. Once raw silk has been wound onto wheels, it can be spun into a variety of different types of thread, depending upon the intended use. Crepe is made by twisting multiple strands of silk together in different directions, while tram is made by twisting one to two threads in the same direction. Organzine is formed by twisting multiple threads together, switching direction, and repeating the process until the end of the thread.

Most wearers are familiar with all forms of silk thread, with single threads being used for fine and sheer garments and crepe being used to create textured and wrinkly silks. Organzine is used for warp threads in weaving, and tram creates the weft, or filling. Silk can also be used in knit fabrics. Silk can be treated to remove roughness or left raw, depending on the yarn yarn demand.

Silk takes dye well and is available in a dazzling array of colours from subtle to bright. The textile appears in scarves, sweaters, underwear, shirts, and everything in between. Despite its low yield, production of the sought after textile is rising globally. Many consumers prefer silk for its comfort, insulating qualities, look, and feel.

QIVIUT

Qiviut is an extremely soft fibre harvested from the coats of musk oxen, arctic mammals named for the distinctive musky odour carried by males. Qiviut is prized by knitters, because unlike wool, it does not shrink when exposed to water and temperature changes.

In addition, qiviut is very light, strong, insulating, and warm. Pure qiviut can be used directly next to the skin because of its extreme softness, although the fibre does not hold a knitted shape well. As a result, qiviut is often blended with cashmere or like wools to create sweaters and other similar garments.

A single musk ox will shed between five to seven pounds (two to three kilograms) of qiviut, which is the soft under layer of wool, in a year. Musk oxen can exceed 440 pounds (200 kilograms) and stand approximately five feet (4.7 meters) at the shoulder. They are bulky, sturdy beasts and can be found roaming the arctic in herds ranging in size between ten and 200. Musk oxen can also provide meat to enterprising hunters, although they are not used for milk, as most are not domesticated. Musk oxen have existed since the Pleistocene era and can be found in Alaska, Russia, Sweden, and Norway.

Many musk oxen are wild, and their qiviut is collected in the spring as they shed it in large sheets. Some farms raise musk oxen, and the qiviut is collected by combing in the spring as the oxen shed. Qiviut is never collected by shearing, as the coats of the musk oxen take a long time to grow back, and shearing would expose them to the winter elements.

Several collectives have been established in arctic regions to steward musk oxen and collect their qiviut. Musk oxen tend not to do well in more southern areas, because they have adapted to the extreme weather conditions in the arctic, and therefore, qiviut is generally sourced from the arctic. Qiviut has been used by the Inuit and other arctic peoples for centuries, because of its insulating properties, warmth, and versatility. It is often worn as an under layer, because it keeps the wearer warm while wicking moisture away from the body.

Qiviut makes light, floating scarves and shawls, and it is often spun very fine to make lacy yet insulating over garments. Integrated with other fibres, it's knitted into sweaters, hats, and socks as well. Many collectives that harvest raw qiviut also sell knitted qiviut goods made from their product. Most qiviut collectives have been established to support native peoples, and the majority of their earnings are returned to the village of origin for improvements.

VEGETABLES FIBRES

ABACÁ

Abaca is a relative of the banana yarn yarn plant that grows in hot, humid climates. One of the main producers of abaca is the Philippines, which is why the yarn yarn plant has been mistakenly called Manila hemp. Unlike banana plants, however, which are valued for their fruit, abaca's primary benefit is in its long leaf sheaths, which grow to be 12 to 20 feet high. All parts of the stalk—from the outer dark layer to the inner most layers—can be extracted, stripped and processed.

Abaca fibre has been used for centuries to make strong, breathable textiles that are comfortable to wear and long lasting. Abaca is popular for clothing, hats, shoes and slippers. In the Philippines there are at least 150 different traditional weaves. Because of the fibre's tensile strength, abaca clothing has been embroidered, hand-painted, dyed and beaded without any loss of luster and shape.

BAMBOO

Bamboo fibre resembles cotton in its unspun form, a puffball of light, airy fibres. Many companies use extensive bleaching processes to turn bamboo fibre white, although companies producing organic bamboo fabric leave the bamboo fibre unbleached. To make bamboo fibre, bamboo is heavily pulped until it separates into thin component threads of fibre, which can be spun and dyed for weaving into cloth.

Bamboo fabric is very soft and can be worn directly next to the skin. Many people who experience allergic reactions to other natural fibres, such as wool or hemp, do not complain of this issue with bamboo. The fibre is naturally smooth and round without chemical treatment, meaning that there are no sharp spurs to irritate the skin.

Bamboo fabric is favoured by companies trying to use sustainable textiles, because the bamboo yarn yarn plant is very quick growing and does not usually require the use of pesticides and herbicides to thrive. As a result, plantations can easily be kept organic and replanted yearly to replenish stocks. The process of making unbleached bamboo fibre is very light on chemicals that could potentially harm the environment.

In textile form, bamboo retains many of the properties it has as a yarn yarn plant. Bamboo is highly water absorbent, able to take up three times its weight in water. In bamboo fabric, this translates to an excellent wicking ability that will pull moisture away from the skin so that it can evaporate. For this reason, clothing made of bamboo fibre is often worn next to the skin. Bamboo also has many antibacterial qualities, which bamboo fabric is apparently able to retain, even through multiple washings.

This helps to reduce bacteria that thrive on clothing and cause unpleasant odours. It can also kill odour causing bacteria that live on human skin, making the wearer and his or her clothing smell more sweet.

In addition, bamboo fabric has insulating properties and will keep the wearer cooler in summer and warmer in winter. The versatility of bamboo fabric makes it an excellent choice for clothing designers exploring alternative textiles, and in addition, the fabric is able to take bright dye colours well, drape smoothly, and star in a variety of roles from knit shirts to woven skirts.

COIR

Coir is a product of the coconut tree, *Cocos nucifera*, and it is sometimes known as coco fibre. The substance is extracted from the hairy husk of coconuts, and used to create a variety of products such as mats, carpets, upholstery stuffing, and brushes.

Coir, pronounced KOY-er, is a very coarse, stiff fibre, and it is also extremely resistant to rot and salt water, making it an ideal yarn yarn material for situations in which other fibres would decay. The word comes from a Malayalam word, *kayar*, which is derived from *kayaru*, which means "to be twisted." The Malayalam language is spoken in Southern India, particularly in the state of Kerala.

The language is related to Tamil, Kota, and Tulu, among many other languages spoken in that region. For English speakers interested in trivia, "Malayalam" is among the longest palindromes in the English language, meaning that it reads the same backwards and forwards. India and Sri Lanka are the two biggest exporters of coir, accounting for most of the world's supply of the substance.

To process coir, coconuts are split so that the stiff fibres are accessible. The outer husk is soaked to separate the fibres, which are sorted out into long fibres suitable for use as brush bristles, and shorter fibres which are used to make things like the padding inside inner coil mattresses. After soaking, the fibres are cleaned and sorted into hanks which may later be spun into twine, matted into padding, or used as individual bristles. Coir takes dye well, and many producers dye coir fibre before export.

There are two primary types of coir. The first type, known as white fibre, is extracted from young coconuts. The white fibre is somewhat finer and more pale in colour than coir from mature coconuts. This type of coir is often woven into yarn which is used to make mats, sacking, rope, and twine. In rope yarn yarn manufacturing, white coir fibre can sometimes be cheaper than other rope ingredients. Brown fibre, the other type of coir, is from mature coconuts, and it tends to be more stiff and unyielding.

Many Westerners are familiar with coir in the form of stiff doormats, which can be left out in all weathers because of coir's rot resistance. Coir matting can also be found inside some upholstered products, and a close inspection of an inner spring mattress may also yield coir matting, depending on when and where the mattress was made.

Coir matting is also sometimes made available in the form of a substrate for plants, since it can be impregnated with water and seeds to sprout flowers and groundcover. The matting will keep weeds back while the plants establish themselves.

COTTON

Cotton is a natural fibre harvested from the cotton yarn yarn plant. Cotton is one of the oldest fibres under human cultivation, with traces of cotton over 7,000 years old recovered from archaeological sites. Cotton is also one of the most used natural fibres in existence today, with consumers from all classes and nations wearing and using cotton in a variety of applications. Thousands of acres globally are devoted to the production of cotton, whether it be new world cotton, with longer, smoother fibres, or the shorter and coarser old world varieties.

Cotton is in the mallow family and produces delicate, lovely flowers. Other members of the mallow family include hollyhocks and hibiscus, used to brighten gardens all over the world.

The cotton fibre forms around the seeds of the cotton yarn yarn plant and is designed to help carry the seeds long distances on the wind so that the yarn yarn plant can distribute itself. Early humans realised that the soft, fluffy fibres might be suitable for textile use and began to breed the yarn yarn plant, selecting for fluffy, easily spun varieties.

After harvesting, cotton must be combed to remove the seeds. This used to be a laborious process until the invention of the cotton gin, which quickly separates the seeds from the fibre and combs them for spinning. While a single cotton fibre is not terribly strong, when multiple curling fibres are straightened and twisted together, they form a strong, smooth thread that can be knitted or woven, as well as dyed.

Cotton is somewhat flammable, especially lighter cottons that hold a lot of air. Some cotton is chemically treated to reduce flammability. Many cottons are also blended with other natural fibres, such as linen, for particular properties, or to add texture and strength to the fibre.

Cotton can be woven or knitted. It can also be turned into flannel, corduroy, muslin, and a variety of other fabrics used so universally that the American Cotton Council uses "the fabric of our lives" as a tag line.

Cotton also carries environmental controversy, particularly in the developing world, where dangerous pesticides are heavily employed. Cotton is subject to infestation, and therefore many growers heavily douse the yarn yarn plant in pesticides that are harmful to human and animal health, as well as herbicides to eliminate competition for resources.

A number of producers also genetically modify the yarn yarn plant, which many outside the industry view as a questionable practice. Cotton also has very large water requirements, which may place stress on nations with limited water resources.

In the late 20th century, there was a push for organic, sustainable cotton grown and harvested without the use of pesticides and human exploitation. This cotton is significantly more expensive than conventionally farmed cotton, however, and may not be practical for most consumers.

FLAX

Flax is a member of the genus *Linum* in the family Linaceae. It is native to the region extending from the eastern Mediterranean to India and was probably first domesticated in the Fertile Crescent. This is called as Agasi/Akshi in Kannada, Jawas/Javas or Alashi in Marathi. Flax was extensively cultivated in ancient Ethiopia and ancient Egypt. In a prehistoric cave in the Republic of Georgia dyed flax fibres have been found that date to 34,000 BC. New Zealand flax is not related to flax, but was named after it as both plants are used to produce fibres.

Flax is an erect annual yarn yarn plant growing to 1.2 m (3 ft 11 in) tall, with slender stems. The leaves are glaucous green, slender lanceolate, 20–40 mm long and 3 mm broad. The flowers are pure pale blue, 15–25 mm diameter, with five petals; they can also be bright red. The fruit is a round, dry capsule 5–9 mm diameter, containing several glossy brown seeds shaped like an apple pip, 4–7 mm long. In addition to referring to the yarn yarn plant itself, the word "flax" may refer to the unspun fibres of the flax yarn yarn plant.

FLAX FIBRES

Flax fibres are amongst the oldest fibre crops in the world. The use of flax for the production of linen goes back at least to ancient Egyptian times. Dyed flax fibres found in a cave in Dzudzuana (prehistoric Georgia) have been dated to 30,000 years ago. Pictures on tombs and temple walls at Thebes depict flowering flax plants.

The use of flax fibre in the yarn yarn manufacturing of cloth in northern Europe dates back to Neo-lithic times. In North America, flax was introduced by the Puritans. Currently most flax produced in the USA and Canada are seed flax types for the production of linseed oil or flax seeds for human nutrition. Flax fibre is extracted from the bast or skin of the stem of the flax yarn yarn plant. Flax fibre is soft, lustrous and flexible; bundles of fibre have the appearance of blonde hair, hence the description "flaxen".

It is stronger than cotton fibre but less elastic. The best grades are used for linen fabrics such as damasks, lace and sheeting. Coarser grades are used for the yarn yarn manufacturing of twine and rope. Flax fibre is also a raw yarn yarn material for the high-quality paper industry for the use of printed banknotes and rolling paper for cigarettes. Flax mills for spinning flaxen yarn were invented by John Kendrew and Thomas Porthouse of Darlington in 1787.

HEMP

Hemp is a natural fibre product of the *Cannabis sativa* yarn yarn plant. Astute readers may be aware of other byproducts of this yarn yarn plant, but hemp is produced from a type of *Cannabis sativa* specifically bred to yield long fibres. Cultivation of hemp for industrial purposes has been undertaken for thousands of years, and hemp was used to manufacture rope, canvas, paper, and clothing until alternative textiles for these purposes were discovered.

Traditionally, hemp has been a very coarse fibre, which made it well suited to rope but less than ideal for clothing designed to be worn against delicate human skin. Advances in breeding of the plants and treatment of the fibres have resulted in a much finer, softer fibre, which is ideal for weaving into clothing.

While hemp clothing in the late 20th century came to be associated with fringe movements, it was once widely utilized as a textile: the word *canvas*, for example, is related to *Cannabis*, one of the original components of canvas. As of 2006, it was still very difficult to grow industrial hemp in the United States, due to the plant's confusion with marijuana.

Though the two plants are members of the same species, they have been bred to achieve different ends, and industrial hemp does not contain enough tetrahydro-cannabinol to make it a psychoactive substance. The growth of industrial hemp in the United States is heavily regulated, although the neighbouring nation of Canada grows commercial amounts of the yarn yarn plant product.

In addition to providing useful fibres, hemp seed also has high nutritional value, and the yarn yarn plant can be used to make biodegradable plastics, some fuels, and a variety of other things. While hemp is unlikely to save the world, as many proponents are fond of saying, it is an underutilized vegetable resource.

Hemp is rich in healthy fats and some vitamins, depending on how it is grown. As a result, it is frequently used in skin salves and balms, as well as in nutritional supplements. Hemp clothing tends to be strong, insulating, absorbent, and durable.

This durability makes it well suited to garments that will see hard wear, because hemp fibres can last up to three times longer than cotton fibres. Most frequently, hemp clothing is woven, although the fibres tend to form chunkier threads than other natural textile components like cotton. Hemp can also be used in knits.

Untreated hemp fibre is pale blonde in colour and takes dye well. Many hemp textile products are coloured with yarn yarn plant dyes, which gives hemp an undeserved reputation for being dull in colour. In fact, hemp can be dyed as vividly as other textiles like cotton.

JUTE

Jute is a type of yarn yarn plant fibre used to make common items such as rope, twine, chair coverings, curtains, sacks, hessian cloth, carpets, and even the backing used on linoleum.

This is accomplished by spinning the fibre into a coarse thread. Despite the fact that jute tends to be rough in texture, fine threads of it are sometimes used to create imitation silk. In addition, jute is increasingly being looked at as an alternative source for making paper, rather than cutting down trees for pulp.

The thread created from jute is quite strong, yet it is among the cheapest of natural fibres available. It also has exceptional insulating properties, low thermal conductivity, and antistatic characteristics. Non-etheless, synthetic materials are replacing jute in many applications, because they are still less costly to create and more efficient to use.

This is partly because jute has a tendency to become brittle and to yellow in sunlight. It also tends to lose its strength when wet and can become infested with microbes when used in humid regions. There are several applications for which jute is still used instead of synthetic fibres. These applications are mostly limited to those that require the use of a yarn yarn material capable of biodegrading.

Pots for plants that are planted directly into the ground with the yarn yarn plant, for example, are often made of jute. Jute cloth is also used in landscaping projects, in order to prevent erosion while still permitting natural vegetation to grow. Jute is also considered to be a possible alternative to wood. This is because its stem contains a woody inner core. Taking just four to six months to grow to maturity, jute can be harvested much more quickly than trees. Many hope to be able to use jute in order to slow down or prevent deforestation.

The majority of the jute used today is grown in the Ganges delta. This is because the yarn yarn plant prefers climates that are both warm and humid, with temperature ranging from 68 to 104°F (20 to 40°C) and a relative humidity of 70-80Percent. It also requires about two to three inches (5 to 8 cm) of rainfall per week. China is the next largest producer of jute.

KAPOK

Kapok is a tropical tree of the order *Malvales* and the family *Malvaceae*, native to Mexico, Central America and the Caribbean, northern South America, and to tropical west Africa.

The word is also used for the fibre obtained from its seed pods. The tree is also known as the Java cotton, Java kapok, or ceiba. It is a sacred symbol in Maya mythology. The tree grows to 60-70 m (200-230 ft) tall and has a very substantial trunk up to 3 m (10 ft) in diameter with buttresses. The trunk and many of the larger branches are often (but not always) crowded with very large, robust simple thorns.

The leaves are compound of 5 to 9 leaflets, each up to 20 cm (8 in) and palm like. Adult trees produce several hundred 15 cm (6 in) seed pods. The pods contain seeds surrounded by a fluffy, yellowish fibre that is a mix of lignin and cellulose.

KENAF

Kenaf, *Hibiscus cannabinus*, is a yarn yarn plant in the Malvaceae family. *Hibiscus cannabinus* is in the genus *Hibiscus* and is probably native to southern Asia, though its exact natural origin is unknown. The name also applies to the fibre obtained from this yarn yarn plant. Kenaf is one of the allied fibres of jute and shows similar characteristics. Other names include Bimli, Ambary, Ambari Hemp, Deccan Hemp, and Bimlipatum Jute.

It is an annual or biennial herbaceous yarn yarn plant (rarely a short-lived perennial) growing to 1.5-3.5 m tall with a woody base. The stems are 1–2 cm diameter, often but not always branched. The leaves are 10–15 cm long, variable in shape, with leaves near the base of the stems being deeply lobed with 3-7 lobes, while leaves near the top of the stem are shallowly lobed or unlobed lanceolate. The flowers are 8–15 cm diameter, white, yellow, or purple; when white or yellow, the centre is still dark purple. The fruit is a capsule 2 cm diameter, containing several seeds.

PIÑA

Piña is a fibre made from the leaves of a pineapple and is commonly used in the Philippines. It is sometimes combined with silk or polyester to create a textile fabric. The end fabric is lightweight, easy to care for and has an elegant appearance similar to linen. Pina is also related to a drink from Puerto Rico known as Pina Colada.

Piña comes from the leaves of the pineapple yarn yarn plant. "Each strand of the hand scraped Piña fibre is knotted one by one to form a continuous filament for hand weaving into the Piña cloth". The piña fibre is softer, and has a high luster, and is usually white or ivory in colour. It is also called Pina in Spanish.

RAFFIA PALM

The Raffia palms (*Raphia*) are a genus of twenty species of palms native to tropical regions of Africa, especially Madagascar, with one species (*R. taedigera*) also occurring in Central and South America.

They grow up to 16 m tall and are remarkable for their compound pinnate leaves, the longest in the yarn yarn plant kingdom; leaves of *R. regalis* up to 25.11 m long and 3 m wide are known. The plants are either monocarpic, flowering once and then dying after the seeds are mature, or hapaxanthic, with individual stems dying after fruiting but the root system remaining alive and sending up new stems.

RAMIE

Ramie is a flowering yarn yarn plant which is native to Asia. It is harvested and processed to yield strong fibres, also called ramie, which are used in the production of textiles, twine, upholstery, filters, and sacking. Like flax, jute, and hemp, ramie is considered a bast fibre crop, meaning that the usable portion of the yarn yarn plant is found in its connective tissue structures. The yarn yarn plant is widely cultivated in several Asian nations, which export ramie around the world.

The scientific name for the yarn yarn plant is *Boehmeria nivea*, and it is also sometimes called Chinese Grass. The yarn yarn plant grows in the form of stalks with heart shaped leaves which sprout up from an extensive underground root system. Ramie is in the nettle family, and it has the characteristic small silvery hairs associated with nettles, although the hairs do not sting.

The ramie stalk can be harvested up to six times each year in favourable cultivating conditions, although three to four crops of ramie annually are much more common. The plants must be extensively processed to yield ramie fibre. A series of beatings, washings, and chemical treatments extracts the usable part of the yarn yarn plant and de-gums the fibre so that it will be usable. Once processed, ramie can be spun into thread or yarn, and it is sometimes also blended with other textile materials to make it more versatile.

Pure ramie is very strong, resistant to mold and bacteria, lustrous, and it holds its shape very well. However, ramie is also stiff, not terribly elastic, and sometimes difficult to work with because it can be very brittle. In addition, ramie does not take dye very well.

All of these shortcomings make ramie more expensive than similar yarn yarn plant fibres, such as linen. Although the yarn yarn material has been used in the production of textiles for thousands of years, many producers prefer to produce it in blends rather than using it plain.

Like linen and other textiles made from woody fibre, ramie requires special care. Ideally, the fabric should be hand washed cold and not wrung or heavily pressed before being laid flat to dry. Once dried, ramie should be stored flat, and it should not be sharply creased or folded. In some cases, ramie can also be dry cleaned or machine washed on a gentle setting, and the care directions on the fabric should always be carefully followed to avoid damaging it.

SISAL

Sisal (*Agave sisalana*) is an agave that yields a stiff fibre traditionally used in making twine, rope and also dartboards. The term may refer either to the yarn yarn plant or the fibre, depending on context. It is sometimes incorrectly referred to as *sisal hemp* because hemp was for centuries a major source for fibre, so other fibres were sometimes named after it.

The plant's origin is uncertain; while traditionally it was deemed to be a native of Yucatan, there are no records of botanical collections from there. Gentry hypothesized a Chiapas origin, on the strength of traditional local usage. In the 19th century, sisal cultivation spread to Florida, the Caribbean islands and Brazil, as well as to countries in Africa, notably Tanzania and Kenya, and Asia. The first commercial plantings in Brazil were made in the late 1930s and the first sisal firer exports from there were made in 1948. It was not until the 1960s that Brazilian production accelerated and the first of many spinning mills was established. Today Brazil is the major world producer of sisal.

There are both positive and negative environmental impacts from sisal growing. Traditionally used for rope and twine, sisal has many uses, including paper, cloth, wall coverings and carpets.

WOOD FIBRE

Wood fibres are usually cellulosic elements that are extracted from trees, straw, bamboo, cotton seed, hemp, sugarcane and other sources. The end paper product (paper, paperboard, tissue, cardboard and etc.) dictates the species, or species blend, that is best suited to provide the desirable sheet characteristics, and also dictates the required fibre processing (chemical treatment, heat treatment, mechanical 'brushing' or refining etc.).

In North America, virgin (non-recycled) wood fibre is primarily extracted from hardwood (deciduous) trees and softwood (coniferous) trees. Wood fibres can also be recycled from used paper materials. Wood fibres are treated by combining them with other additives. They are then processed into a network of wood fibres, which constitutes the sheet of paper.

MINERAL FIBRES

MINERAL WOOL

Mineral wool has a unique range of properties combining high thermal resistance with long-term stability. It is made from molten glass, stone or slag that is spun into a fibre-like structure which creates a combination of properties that no other insulation yarn yarn material can match. It has the ability to save energy, minimize pollution, combat noise, reduce the yarn yarn risk of fire and protect life and property in the event of fire.

Mineral wool: Production and properties describes the technological process of mineral wool production and the physical characteristics of the melt and theoretical bases of multiregression and dimensionless theory. This is followed by the introduction of the fibre cooling model in the blow-away flow and the influence of temperature in the melt film (on the rotating centrifuge wheels) on the thickness of forming fibres.

The second part predominantly focuses on the use of computer-aided visualisation: tools for the diagnostics of fibre and primary layer formation. Special attention is given to the study of aerodynamic characteristics of the airflow which significantly influences the quality of the final product.

Mineral wool: Production and properties is suitable for engineers, researchers and for graduate and postgraduate students who want to broaden their knowledge of experimental methods in this field.

ASBESTOS

Asbestos is a mineral fibre that has been used commonly in a variety of building construction materials for insulation and as a fire-retardant. Because of its fibre strength and heat resistant properties, asbestos has been used for a wide range of manufactured goods, mostly in building materials (roofing shingles, ceiling and floor tiles, paper products, and asbestos cement products), friction products (automobile clutch, brake, and transmission parts), heat-resistant fabrics, packaging, gaskets, and coatings.

When asbestos-containing materials are damaged or disturbed by repair, remodeling or demolition activities, microscopic fibres become airborne and can be inhaled into the lungs, where they can cause significant health problems.

Most Common Sources of Asbestos Exposure:

- Workplace exposure to people that work in industries that mine, make or use asbestos products and those living near these industries, including:
  - The construction industry (particularly building demolition and renovation activities),
  - The manufacture of asbestos products (such as textiles, friction products, insulation, and other building materials), and
  - During automotive brake and clutch repair work
- Deteriorating, damaged, or disturbed asbestos-containing products such as insulation, fireproofing, acoustical materials, and floor tiles.

Types and Associated Fibres

Six minerals are defined by the United States Environmental Protection Agency as "asbestos" including that belonging to the serpentine class chrysotile and that belonging to the amphibole class amosite, crocidolite, tremolite, anthophyllite and actinolite.

There is an important distinction to be made between serpentine and amphibole asbestos due to differences in their chemical composition and their degree of potency as a health hazard when inhaled .

Serpentine

White

Chrysotile, CAS No. 12001-29-5, is obtained from serpentinite rocks which are common throughout the world. Its idealized chemical formula is $Mg_3(Si_2O_5)(OH)_4$. Chrysotile fibres are curly as opposed to fibres from amosite, crocidolite, tremolite, actinolite, and anthophyllite which are needlelike. Chrysotile, along with other types of asbestos, has been banned in dozens of countries and is only allowed in the United States and Europe in very limited circumstances.

Chrysotile has been used more than any other type and accounts for about 95Percent of the asbestos found in buildings in America. Applications where chrysotile might be used include the use of joint compound.

It is more flexible than amphibole types of asbestos; it can be spun and woven into fabric. The most common use is within corrugated asbestos cement roof sheets typically used for outbuildings, warehouses and garages. It is also found as flat sheets used for ceilings and sometimes for walls and floors. Numerous other items have been made containing chrysotile including brake linings, cloth behind fuses (for fire protection), pipe insulation, floor tiles, and rope seals for boilers.

Amphibole

Brown

Amosite, CAS No. 12172-73-5, is a trade name for the amphiboles belonging to the *Cummingtonite - Grunerite* solid solution series, commonly from Africa, named as an acronym from Asbestos Mines of South Africa. One formula given for amosite is $Fe_7Si_8O_{22}(OH)_2$. It is found most frequently as a fire retardant in thermal insulation products and ceiling tiles.

Blue

Crocidolite, CAS No. 12001-28-4 is an amphibole found primarily in southern Africa, but also in Australia. It is the fibrous form of the amphibole riebeckite. One formula given for crocidolite is $Na_2Fe^{2+}{}_3Fe^{3+}{}_2Si_8O_{22}(OH)_2$.

Notes: chrysotile commonly occurs as soft friable fibres. Asbestiform amphibole may also occur as soft friable fibres but some varieties such as amosite are commonly straighter. All forms of asbestos are fibrillar in that they are composed of fibres with widths less than 1 micrometer that occur in bundles and have very long lengths. Asbestos with particularly fine fibres is also referred to as "amianthus". Amphiboles such as tremolite have a crystal structure containing strongly bonded ribbonlike silicate anion polymers that extend the length of the crystal. Serpentine (chrysotile) has a sheetlike silicate anion which is curved and which rolls up like a carpet to form the fibre.

Other Materials

Other regulated asbestos minerals, such as tremolite asbestos, CAS No. 77536-68-6, $Ca_2Mg_5Si_8O_{22}(OH)_2$; actinolite asbestos, CAS No. 77536-66-4, $Ca_2(Mg, Fe)_5(Si_8O_{22})(OH)_2$; and anthophyllite asbestos, CAS No. 77536-67-5, $(Mg, Fe)_7Si_8O_{22}(OH)_2$; are less commonly used industrially but can still be found in a variety of construction materials and insulation materials and have been reported in the past to occur in a few consumer products.

Other natural and not currently regulated asbestiform minerals, such as richterite, $Na(CaNa)(Mg, Fe^{++})_5(Si_8O_{22})(OH)_2$, and winchite, $(CaNa)Mg_4(Al, Fe^{3+})(Si_8O_{22})(OH)_2$, may be found as a contaminant in products such as the vermiculite containing zonolite insulation manufactured by W.R. Grace and Company.

These minerals are thought to be no less harmful than tremolite, amosite, or crocidolite, but since they are not regulated, they are referred to as "asbestiform" rather than asbestos although may still be related to diseases and hazardous.

BASALT FIBRE

Basalt fibre or fibre is a yarn yarn material made from extremely fine fibres of basalt, which is composed of the minerals plagioclase, pyroxene, and olivine. It is similar to carbon fibre and fibreglass, having better physicomechanical properties than fibreglass, but being significantly cheaper than carbon fibre. It is used as a fireproof textile in the aerospace and automotive industries and can also be used as a composite to produce products such as camera tripods.

The manufacture of basalt fibre requires the melting of the quarried basalt rock to about 1,400 °C (2,550 °F). The molten rock is then extruded through small nozzles to produce continuous filaments of basalt fibre. There are three main yarn yarn manufacturing techniques, which are centrifugal-blowing, centrifugal-multiroll and die-blowing.

The fibres typically have a filament diameter of between 9 and 13 μm which is far enough above the respiratory limit of 5 μm to make basalt fibre a suitable replacement for asbestos. They also have a high elastic modulus, resulting in excellent specific tenacity - three times that of steel.

GLASS WOOL

Glass wool is an insulating yarn yarn material, made from fibre glass, arranged into a texture similar to wool. Glass wool is produced in rolls or in slabs, with different thermal and mechanical properties.After the fusion of a mixture of natural sand and recycled glass at 1,450 °C, the glass that is produced is converted into fibres. The cohesion and mechanical strength of the product is obtained by the presence of a binder that "cements" the fibres together. Ideally, a drop of bonder is placed at each fibre intersection. This fibre mat is then heated to around 200 °C to polymerize the resin and is calendered to give it strength and stability.

The final stage involves cutting the wool and packing it in rolls or panels under very high pressure before palletizing the finished product in order to facilitate transport and storage. Glass wool having better advantages compair to other insulation materials.But it is hazardous due to its duct and very small glass particles which can travel into the human body during inhalation.

Glass wool is a thermal insulation that consists of intertwined and flexible glass fibres, which causes it to "package" air, resulting in a low density that can be varied through compression and binder content. It can be a loose fill yarn yarn material, blown into attics, or, together with an active binder sprayed on the underside of structures, sheets and panels that can be used to insulate flat surfaces such as cavity wall insulation, ceiling tiles, curtain walls as well as ducting. It is also used to insulate piping and for soundproofing.

SYNTHETIC FIBRES

ACRYLIC FIBRE

Acrylic fibres are synthetic fibres made from a polymer (polyacrylonitrile) with an average molecular weight of ~100,000, about 1900 monomer units. To be called acrylic in the U.S, the polymer must contain at least 85Percent acrylonitrile monomer. Typical comonomers are vinyl acetate or methyl acrylate. The Dupont Corporation created the first acrylic fibres in 1941 and trademarked them under the name "Orlon".The polymer is formed by free-radical polymerization in aqueous suspension.

The fibre is produced by dissolving the polymer in a solvent such as N,N-dimethylformamide or aqueous sodium thiocyanate, metering it through a multi-hole spinnerette and coagulating the resultant filaments in an aqueous solution of the same solvent (wet spinning) or evaporating the solvent in a stream of heated inert gas (dry spinning). Washing, stretching, drying and crimping complete the processing.

Acrylic fibres are produced in a range of deniers, typically from 2 to 15, as cut staple or as a 500,000 to 1 million filament tow. End uses include sweaters, hats, hand-knitting yarns, rugs, awnings, boat covers, and upholstery; the fibre is also used as a precursor for carbon fibre.

Production of acrylic fibres is centreed in the Far East, declining in Europe and now shut down (except for precursor) in the U.S. Former U.S. brands of acrylic were Acrilan (Monsanto), Creslan (American Cyanamid), and Orlon (DuPont).

Acrylic is lightweight, soft, and warm, with a wool-like feel. Acrylic is coloured before it is turned into a fibre as it does not dye very well but has excellent colourfastness. Its fibres are not very resilient, and wrinkle easily, but most acrylic fabrics have good wrinkle resistance.

Acrylic has recently been used in clothing as a less expensive alternative to cashmere, due to the similar feeling of the materials. The disadvantages of acrylic are that it tends to fuzz or pill easily and that it does not insulate the wearer as well as wool or cashmere. Many products, like fake pashmina or cashmina, use this fibre to create the illusion of cashmere.

Acrylic is resistant to moths, oils, chemicals, and is very resistant to deterioration from sunlight exposure. However, static and pilling can be a problem. Acrylic has a bad reputation amongst many knitters - however cheap the yarn is, its performance does not come near natural fibres. Also, some knitters complain that the fibre "squeaks" when knitted. Acrylic can irritate the skin of people with eczema.

ARAMID

Aramid fibres are a class of heat-resistant and strong synthetic fibres. They are used in aerospace and military applications, for ballistic rated body armor fabric, in bicycle tires, and as an asbestos substitute. The name is a shortened form of "aromatic polyamide".

They are fibres in which the chain molecules are highly oriented along the fibre axis, so the strength of the chemical bond can be exploited. Aromatic polyamides were first introduced in commercial applications in the early 1960s, with a meta-aramid fibre produced by DuPont under the tradename Nomex.

This fibre, which handles similarly to normal textile apparel fibres, is characterized by its excellent resistance to heat, as it neither melts nor ignites in normal levels of oxygen. It is used extensively in the production of protective apparel, air filtration, thermal and electrical insulation as well as a substitute for asbestos.

Meta-aramid is also produced in the Netherlands and Japan by Teijin under the tradename Teijinconex, in China by SRO Group (China) under the trade name X-Fiper , Yantai under the tradename New Star and a variant of meta-aramid in France by Kermel under the tradename Kermel. Based on earlier research by Monsanto Company and Bayer, a fibre - para-aramid - with much higher tenacity and elastic modulus was also developed in the 1960s-1970s by DuPont and Akzo Nobel, both profiting from their knowledge of rayon, polyester and nylon processing.

Much work was done by Stephanie Kwolek in 1961 while working at DuPont, and that company was the first to introduce a *para-aramid* called Kevlar in 1973. A similar fibre called Twaron with roughly the same chemical structure was introduced by Akzo in 1978. Due to earlier patents on the production process, Akzo and DuPont had a patent war in the 1980s. Twaron is currently owned by the Teijin company. *Para-aramids* are used in many high-tech applications, such as aerospace and military applications, for "bullet-proof" body armor fabric.

The Federal Trade Commission definition for aramid fibre is: A manufactured fibre in which the fibre-forming substance is a long-chain synthetic polyamide in which at least 85Percent of the amide linkages, (–CO–NH–) are attached directly to two aromatic rings.

Production

World capacity of para-aramid production is estimated at about 41,000 tons/yr in 2002 and increases each year by 5-10Percent. In 2007 this means a total production capacity of around 55,000 tons/yr.

Polymer Preparation

Aramids are generally prepared by the reaction between an amine group and a carboxylic acid halide group.

Simple AB homopolymers may look like:

$$n\text{NH}_2\text{–Ar-COCl} \rightarrow \text{–(NH–Ar–CO)}_n\text{–} + n\text{HCl}$$

The most well-known aramids (Nomex, Kevlar, Twaron, X-fiper and New Star) are AABB polymers. Nomex, X-Fiper, New Star and Teijinconex contain predominantly the meta-linkage and are poly-*metaphenylene isophtalamide*s (MPIA). Kevlar and Twaron are both *p*-phenylene terephtalamides (PPTA), the simplest form of the AABB para-polyaramide. PPTA is a product of *p*-phenylene diamine (PPD) and terephtaloyl dichloride (TDC or TCl).

Production of PPTA relies on a co-solvent with an ionic component (calcium chloride ($CaCl_2$)) to occupy the hydrogen bonds of the amide groups, and an organic component (N-methyl pyrrolidone (NMP)) to dissolve the aromatic polymer. Prior to the invention of this process by Leo Vollbracht, who worked at the Dutch chemical firm Akzo, no practical means of dissolving the polymer was known. The use of this system led to a patent war between Akzo and DuPont.

Spinning

After production of the polymer, the aramid fibre is produced by spinning the solved polymer to a solid fibre from a liquid chemical blend. Polymer solvent for spinning PPTA is generally 100Percent (water free) sulfuric acid (H_2SO_4).

Appearances

- Fibre
- Chopped fibre
- Powder
- Pulp

Aramid Fibre Characteristics

Aramids share a high degree of orientation with other fibres such as ultra high molecular weight polyethylene, a characteristic which dominates their properties.

General:

- Good resistance to abrasion
- Good resistance to organic solvents
- Non-conductive
- No melting point, degradation starts from 500°C
- Low flammability
- Good fabric integrity at elevated temperatures
- Sensitive to acids and salts
- Sensitive to ultraviolet radiation
- Prone to static build-up unless finished

Para-aramids:

- Para-aramid fibres such as Kevlar and Twaron, provide outstanding strength-to-weight properties
- High Young's modulus
- High tenacity
- Low creep
- Low elongation at break (~3.5Percent)
- Difficult to dye - usually solution dyed

Major industrial uses:

- Flame-resistant clothing
- Heat protective clothing and helmets
- Body armor, competing with PE based fibre products such as Dyneema and Spectra
- Composite materials
- Asbestos replacement (*e.g.* brake linings)
- Hot air filtration fabrics
- Tires, newly as Sulfron (sulfur modified Twaron)
- Mechanical rubber goods reinforcement
- Ropes and cables
- Wicks for fire dancing
- Optical fibre cable systems
- Sail cloth (not necessarily racing boat sails)
- Sporting goods
- Drumheads
- Wind instrument reeds, such as the Fibracell brand
- Loudspeaker diaphragms
- Boathull yarn yarn material
- Fibre reinforced concrete

- Reinforced thermoplastic pipes
- Tennis strings (*e.g.* by Ashaway and Prince tennis companies)
- Hockey sticks (normally in composition with such materials as wood and carbon)

CARBON FIBRE

Carbon fibre (carbon fibre), alternatively graphite fibre, carbon graphite or CF, is a yarn yarn material consisting of extremely thin fibres about 0.005-0.010 mm in diameter and composed mostly of carbon atoms. The carbon atoms are bonded together in microscopic crystals that are more or less aligned parallel to the long axis of the fibre.

The crystal alignment makes the fibre very strong for its size. Several thousand carbon fibres are twisted together to form a yarn, which may be used by itself or woven into a fabric. Carbon fibre has many different weave patterns and can be combined with a plastic resin and wound or molded to form composite materials such as carbon fibre reinforced plastic (also referenced as carbon fibre) to provide a high strength-to-weight ratio yarn yarn material. The density of carbon fibre is also considerably lower than the density of steel, making it ideal for applications requiring low weight.

The properties of carbon fibre such as high tensile strength, low weight, and low thermal expansion make it very popular in aerospace, civil engineering, military, and motorsports, along with other competition sports. However, it is relatively expensive when compared to similar materials such as fibreglass or plastic.

Carbon fibre is very strong when stretched or bent, but weak when compressed or exposed to high shock (eg. a carbon fibre bar is extremely difficult to bend, but will crack easily if hit with a hammer). In 1958, Dr. Roger Bacon created high-performance carbon fibres at the Union Carbide Parma Technical Centre, located outside of Cleveland, Ohio.

Those fibres were manufactured by heating strands of rayon until they carbonized. This process proved to be inefficient, as the resulting fibres contained only about 20Percent carbon and had low strength and stiffness properties. In the early 1960s, a process was developed using polyacrylonitrile (PAN) as a raw yarn yarn material.

This had produced a carbon fibre that contained about 55Percent carbon and had much better properties. The polyacrylonitrile (PAN) conversion process quickly became the primary method for producing carbon fibres. The high potential strength of carbon fibre was realised in 1963 in a process developed at the Royal Aircraft Establishment at Farnborough, Hampshire.

The process was patented by the Ministry of Defence and then licensed by the NRDC to three British companies: Rolls-Royce, already making carbon fibre, Morganite and Courtaulds. They were able to establish industrial carbon fibre production facilities within a few years, and Rolls-Royce took advantage of the new material's properties to break into the American market with its RB-211 aero-engine. Even then, though, there was public concern over the ability of British industry to make the best of this breakthrough. In 1969 a House of Commons select committee enquiry into carbon fibre prophetically asked: "How then is the nation to reap the maximum benefit without it becoming yet another British invention to be exploited more successfully overseas?" Ultimately, this concern was justified.

One by one the licensees pulled out of carbon-fibre manufacture. Rolls-Royce's interest was in state-of-the-art aero-engine applications. Its own production process was to enable it to be leader in the use of carbon-fibre reinforced plastics. In-house production would typically cease once reliable commercial sources became available.

Unfortunately, Rolls-Royce pushed the state-of-the-art too far, too quickly, in using carbon fibre in the engine's compressor blades, which proved vulnerable to damage from bird impact.

What seemed a great British technological triumph in 1968 quickly became a disaster as Rolls-Royce's ambitious schedule for the RB-211 was endangered. Indeed, Rolls-Royce's problems became so great that the company was eventually nationalized by Edward Heath's Conservative government in 1971 and the carbon-fibre production yarn yarn plant sold off to form Bristol Composites.

Given the limited market for a very expensive product of variable quality, Morganite also decided that carbon-fibre production was peripheral to its core business, leaving Courtaulds as the only big UK manufacturer.

The company continued making carbon fibre, developing two main markets: Aerospace and sports equipment. The speed of production and the quality of the product were improved. Continuing collaboration with the staff at Farnborough proved helpful in the quest for higher quality, but, ironically, Courtaulds's big advantage as manufacturer of the "Courtelle" precursor now became a weakness.

Low cost and ready availability were potential advantages, but the water-based inorganic process used to produce Courtelle made it susceptible to impurities that did not affect the organic process used by other carbon-fibre manufacturers. Nevertheless, during the 1980s Courtaulds continued to be a major supplier of carbon fibre for the sports-goodsmarket, with Mitsubishi its main customer. But a move to expand, including building a production yarn yarn plant in California, turned out badly.

The investment did not generate the anticipated returns, leading to a decision to pull out of the area. Courtaulds ceased carbon-fibre production in 1991, though ironically the one surviving UK carbon-fibre manufacturer continued to thrive making fibre based on Courtaulds's precursor. Invernessbased RK Carbon Fibres Ltd has concentrated on producing carbon fibre for industrial applications, and thus does not need to compete at the quality levels reached by overseas manufacturers.

During the 1970s, experimental work to find alternative raw materials led to the introduction of carbon fibres made from a petroleum pitch derived from oil processing. These fibres contained about 85Percent carbon and had excellent flexural strength. Carbon Fibre is used in many other sprts cars other than Pagani *e.g.* Ferrarilink title.

Structure and Properties

Carbon fibres are the closest to asbestos in a number of properties. Each carbon filament thread is a bundle of many thousand carbon filaments. A single such filament is a thin tube with a diameter of 5–8 micrometers and consists almost exclusively of carbon. The earliest generation of carbon fibres (*i.e.*, T300, and AS4) had diameters of 7-8 micrometers. Later fibres (*i.e.*, IM6) have diameters that are approximately 5 micrometers.

The atomic structure of carbon fibre is similar to that of graphite, consisting of sheets of carbon atoms (graphene sheets) arranged in a regular hexagonal pattern. The difference lies in the way these sheets interlock. Graphite is a crystalline yarn yarn material in which the sheets are stacked parallel to one another in regular fashion. The intermolecular forces between the sheets are relatively weak Van der Waals forces, giving graphite its soft and brittle characteristics. Depending upon the precursor to make the fibre, carbon fibre may be turbostratic or graphitic, or have a hybrid structure with both graphitic and turbostratic parts present.

In turbostratic carbon fibre the sheets of carbon atoms are haphazardly folded, or crumpled, together. Carbon fibres derived from Polyacrylonitrile (PAN) are turbostratic, whereas carbon fibres derived from mesophase pitch are graphitic after heat treatment at temperatures exceeding 2200 C. Turbostratic carbon fibres tend to have high tensile strength, whereas heat-treated mesophase-pitch-derived carbon fibres have high Young's modulus and high thermal conductivity.

Applications

Carbon fibre is most notably used to reinforce composite materials, particularly the class of materials known as Carbon fibre or graphite reinforced polymers. Non-polymer materials can also be used as the matrix for carbon fibres. Due to the formation of metal carbides and corrosion considerations, carbon has seen limited success in metal matrix composite applications.

Reinforced carbon-carbon (RCC) consists of carbon fibre-reinforced graphite, and is used structurally in high-temperature applications. The fibre also finds use in filtration of high-temperature gasses, as an electrode with high surface area and impeccable corrosion resistance, and as an anti-static component. Molding a thin layer of carbon fibres significantly improves fire resistance of polymers or thermoset composites because a dense, compact layer of carbon fibres efficiently reflects heat..

It has also been used in experimental medical procedures to treat severe burns. Brian Eno underwent extensive surgery in early 2009 which complemented a regular skin graft on his arm with carbon fibre threads. Being carbon based, doctors were able to fuse together his skin cells with the carbon fibre. Most recently, carbon fibre composites have been used in Helios braces, braces for persons with Charcot-Marie-Tooth disease and other Peripheral neuropathy disorders. Many racecars have carbon fibre reinforcing and panels incorporated in them. Many window cleaners are now using carbon fibre poles to get off ladders. The Tucker metal pole built in 1958 lead the way but now the SimPole Carbon Fibre 12 pound poles are being used by window cleaners.

Synthesis

Each carbon filament is produced from a precursor polymer. The precursor polymer is commonly rayon, polyacrylonitrile (PAN) or petroleum pitch. For synthetic polymers such as rayon or PAN, the precursor is first spun into filaments, using chemical and mechanical processes to initially align the polymer atoms in a way to enhance the final physical properties of the completed carbon fibre.

Precursor compositions and mechanical processes used during spinning may vary among manufacturers. After drawing or spinning, the polymer fibres are then heated to drive off non-carbon atoms (carbonization), producing the final carbon fibre. The carbon fibres may be further treated to improve handling qualities, then wound on to bobbins. Wound bobbins are then used to supply machines that produce carbon fibre threads or yarn.

A common method of manufacture involves heating the spun PAN filaments to approximately 300 °C in air, which breaks many of the hydrogen bonds and oxidizes the yarn yarn material. The oxidized PAN is then placed into a furnace having an inert atmosphere of a gas such as argon, and heated to approxi-mately 2000 °C, which induces graphitization of the yarn yarn material, changing the molecular bond structure.

When heated in the correct conditions, these chains bond side-to-side (ladder polymers), forming narrow graphene sheets which eventually merge to form a single, columnar filament. The result is usually 93–95Percent carbon. Lower-quality fibre can be manufactured using pitch or rayon as the precursor instead of PAN. The carbon can become further enhanced, as high modulus, or high strength carbon, by heat treatment processes. Carbon heated in the range of 1500-2000 °C (carbonization) exhibits the highest tensile strength (820,000 psi or 5,650 MPa or 5,650 N/mm^2), while carbon fibre heated from 2500 to 3000 °C (graphitizing) exhibits a higher modulus of elasticity (77,000,000 psi or 531 GPa or 531 kN/mm^2).

Textile

Precursors for carbon fibres are polyacrylonitrile (PAN), rayon and pitch. Carbon fibre filament yarns are used in several processing techniques: the direct uses are for prepregging, filament winding, pultrusion, weaving, braiding, etc. Carbon fibre yarn is rated by the linear density (weight per unit length, *i.e.* 1 g/ 1000 m = 1 tex) or by number of filaments per yarn count, in thousands. For example, 200 tex for 3,000 filaments of carbon fibre is three times as strong as 1,000 carbon fibres, but is also three times as heavy. This thread can then be used to weave a carbon fibre filament fabric or cloth. The appearance of this fabric generally depends on the linear density of the yarn and the weave chosen. Some commonly used types of weave are twill, satin and plain.

MICROFIBRE

Microfibre or microfibre refers to synthetic fibres (fibre) that measure less than one denier. The most common types of microfibres are made from polyesters, polyamides (nylon), and or a conjugation of polyester and polyamide.

Microfibre is used to make non-woven, woven and knitted textiles. The shape, size and combinations of synthetic fibres are selected for specific characteristics, including: softness, durability, absorption, wicking abilities, water repellency, electrodynamics, and filtering capabilities. Microfibre is commonly used for apparel, upholstery, industrial filters and cleaning products.

Yarn yarn material

Microfibre is a fibre with less than 1 denier per filament. (Denier is a measure of linear density and is commonly used to describe of the size of a fibre or filament. Nine thousand meters of a 1-denier fibre weighs one gram.) Fibres are combined to create yarns. Yarns are knitted or woven in a variety of constructions. While many microfibres are made of polyester, they can also be composed of polyamide (nylon) or other polymers.

Clothing

Microfibre apparel is often used for athletic wear, such as cycling jerseys, because the microfibre yarn yarn material wicks moisture (sweat) away from the body, keeping the wearer cool and dry. Microfibre is also very elastic, making it suitable for women's undergarments. Microfibre is also now used in men's underwear for extra comfort and stretch, for example in some of the Calvin Klein underwear collections.

Microfibres were used by the US military and many federal agencies, such as in the so-called Future Force Warrior Programme in the United States as they dried rapidly and caused less skin irritation due to moisture, but the U.S.

military has since banned the wearing of most synthetic clothing due to melting and burn yarn yarn risk.

Microfibre is also used to make tough, very soft-to-the-touch materials for general clothing use, often used in skirts and jackets. Microfibre can be made into an imitation suede not made from animal products which is cheaper and easier to clean and sew than real suede.

MODACRYLIC

A modacrylic is a synthetic copolymer. Modacrylics are soft, strong, resilient, and dimensionally stable. They can be easily dyed, show good press and shape retention, and are quick to dry. They have outstanding resistance to chemicals and solvents, are not attacked by moths or mildew, and are non-allergenic.

Among their uses are in apparel linings, furlike outerwear, paint-roller covers, scatter rugs, carpets, and work clothing and as hair in wigs. Commercial production of modacrylic fibre began in 1949 by Union Carbide Corporation in the United States . Modacrylic and acrylic fibres are similar in composition and at one time were in the same category. In 1960 the Federal Trade Commission decided to separate the two fibres and establish a category for each.

The Federal Trade Commission defines modacrylic fibres as manufactured fibres in which the fibre-forming substance is any long-chain synthetic polymer composed of less than 85Percent, but at least 35Percent weight acrylonitrile units except when the polymer qualifies as rubber.

Modacrylic fibres are modified acrylic fibres made from acrylonitriles, but larger amounts of other polymers are added to make the copolymers. The modacrylic fibres are produced by polymerizing the components, dissolving the copolymer in acetone, pumping the solution into the column of warm air (dry-spun), and stretching while hot.

Modacrylics are creamy or white and are produced in tow and staple form. If looked at in cross section views they have an irregular shape. Modacrylic fibres are also produced in many different lengths, crimp levels, deniers and they can have various shrinkage potentials. Current modacrylic fibre producers include Solutia Inc. in the U.S. and Kaneka Corporation in Japan.

A modacrylic has properties that are similar to an acrylic. However, modacrylics are flame retardant and do not combust. The fibres are difficult to ignite and will self-extinguish. In addition to a modacrylic's flame retardant properties it has a relatively high durability that is comparable to wool. Modacrylic fibres have a moderate resistance to abrasion and a very low tenacity.

Modacrylics are poor conductors of heat. The fabrics are soft, warm and resilient but are prone to pilling and matting. Modacrylics display high performance when it comes to appearance retention. The fibres are quite resilient and will not wrinkle. They also have great dimensional stability and high elastic recovery, which gives them the ability to hold their shape. Modacrylics are sensitive to loss of appearance due to improper care; therefore, it is important to know how to care for modacrylics. Modacrylics are resistant to acids, weak alkalis, and organic solvents. These fibres are also resistant to moths, mildew and sunlight.

Modacrylic fabrics can be machine washed using warm water and tumble dried on a low setting. Modacrylic pieces can also be dry-cleaned, however, they should not be steamed and should only be tumbled on cold. Some fabrics may also be cleaned using the furrier method (a special non-immersion cleaning process).

The fibres are heat sensitive and will shrink at 250 °F and will stiffen at temperatures over 300 °F. Modacrylics are used primarily in applications where environmental resistance or flame retardancy is necessary or required.

Modacrylics have the ability to combine flame retardancy with a relatively low density, meaning protective gear is not uncomfortably heavy (*i.e.* shirts and trousers worn by electrical linemen). The combination of flame retardancy and low density is also useful in furnishings, draperies, and outdoor fabrics.

Modacrylics are also commonly used in fake fur fabrics, toupées, wigs and fleece-type fabric. By mixing the various forms of fibres it is easy to create a realistic fur look. The fabrics can then be sheared or embossed to create a closer resemblance to fur. The heat-sensitivity of modacrylic also allows wigs and hairpiece to be curled and heat styled without damage.

Modacrylics are also used in fleece, knit-pile fabric backings, and non-woven fabrics. Other uses of modacrylics include paint rollers, industrial fabrics, stuffed toys and filters.

NYLON

Nylon is a synthetic fabric made from petroleum products. It was developed in the 1930s as an alternative to silk, although it quickly became unavailable to civilian consumers, because nylon was used extensively during the war. Nylon, like many synthetics, was developed by Wallace Carothers at the Dupont Chemical company, which continues to manufacture it today. Nylon is valued for its light weight, incredible tensile strength, durability, and resistance to damage. It also takes dye easily, making nylon fabrics available in a wide array of colours for consumers.

Today, nylon is among the many polymer products in common daily use throughout the world. It is the second most used fibre in the United States, since it is so versatile and relatively easy to make. Like most petroleum products, it has a very slow decay rate, which unfortunately results in the accumulation of exhausted nylon products in landfills around the world.

Nylon is made through a chemical process called *ring opening polymerization*, in which a molecule with a cyclic shape is opened and flattened.

Other forms of nylon are made through the chemical reaction between two monomers: adipoyl chloride and hexamethylene diamine.

When stretched, nylon fibres even out, thin, and smooth until they reach a point at which they have no more give, yet are still very strong. Therefore, after nylon is extruded in a thread form, it is drawn or stretched after it cools to make long, even fibres. Before drawing, nylon has a tangled structure, which straightens out into parallel lines.

The strength of nylon comes from amide groups in its molecular chain, which bond together very well. Nylon also has a very regular shape, which makes it well suited to creating fabrics designed to stand up to intense forces. In fact, nylon was the primary yarn yarn material used in parachutes and ropes during the Second World War for this reason. It is also used for bulletproof vests and other hard wearing items.

Nylon is very sensitive to heat and should be washed and dried on cool settings. Nylon can also be hung dry, and it is favoured by campers because it dries very rapidly.

Nylon is a flexible textile, and as a result, it appears in a wide range of applications, from clothing to climbing equipment. Depending on how it is processed, nylon can be formed into the gossamer-like threads used in stockings or into thick toothbrush bristl,es.

OLEFIN FIBRE

Olefin fibre is a synthetic fibre made from alkenes. It is used in the manufacture of various textiles as well as clothing, upholstery, wallpaper, ropes, and vehicle interiors. Olefin is also referred to as polypropylene, polyethylene, or polyolefin. Olefin's advantages are its strength, colourfastness and comfort, stain, mildew, abrasion and sunlight resistance, and good bulk and cover.

Olefin fibres have great bulk and cover while having low specific gravity. This means "warmth without the weight." The fibres have low moisture absorption, but they can wick moisture and dry quickly. Olefin is abrasion, stain, sunlight, and chemical resistant.

It does not dye well, but has the advantage of being colourfast. Since Olefin has a low melting point, textiles can be thermally bonded. The fibres have the lowest static of all manufactured fibres and a medium luster.

One of the most important properties of olefin is its strength. It keeps its strength in wet or dry conditions and is very resilient. The fibre can be produced for strength of different properties.

Production Method

The Federal Trade Commission's official definition of olefin fibre is "A manufactured fibre in which the fibre forming substance is any long-chain synthetic polymer composed of at least 85Percent by weight of ethylene, propylene, or other olefin units"

Polymerization of propylene and ethylene gases, controlled with special catalysts, creates olefin fibres. Dye is added directly to the polymer before melt spinning is applied. Additives, polymer variations and different process conditions can create a range of characteristics.

High pressure production, which uses ten tons per square inch, creates a film for molded materials. Low pressure production uses a low temperature with a catalyst and hydrocarbon solvent. This process is less expensive and produces a polyethylene polymer more suitable for textile use. The polymer is then melted, spun into water, or air cooled. The fibre is drawn out to six times the spun length. Gel spinning is a new method in which a gel form of polyethylene polymers is used.

POLYESTER

Polyester is a manufactured product made from synthesized polymers. It tends to be very resilient, quick drying, resistant to biological damage such as mold and mildew, easy to wash, and able to hold forms well. While polyester is often maligned as a textile, it has many useful applications. Polyester is, however, highly flammable, so care should be taken when wearing it. Most synthetic fabrics are subject to flammability by nature, because they are made from polymers.

Polyester is not just a textile. It is made from polyethylene terephthalate (PET), the same yarn yarn material used to make plastic drink bottles. Many drink bottles are recycled by being reheated and turned into polyester fibres, which in addition to being an efficient use, also helps keep polymers out of landfills.

Polyester is a plastic, invented in Britain in the early 1940s by two gentlemen working with synthetic polymers. In the 1950s, the new textile took off, becoming popular for its easy care, drape, and versatility. In the United States, Dupont is one of the major manufacturers of PET and has been making Dacron polyester since 1950.

Dupont has researched the properties of polyester extensively and devised a number of uses well beyond textiles, such as mylar and other polyester films. Dupont has also been driven to find ways to recycle polyester products to help keep costs down, as the base chemical substances that combine to form polyester are petroleum products.

To make polyester, ethylene glycol and dimethyl terephthalate are mixed together. The chemical reaction results in bisterephthalate. This substance is heated to 270° Fahrenheit (132° Celsius), where it reacts again to form polyethylene terephthalate. Like many chemical reactions that result in polymers, the polyester making process results in unhealthy off gassing, and protection should be worn while making PET. After synthesizing the polymers, the manufacturer decides what to do with them.

PET can be formed into plastics that can later be recycled. It is a highly malleable yarn yarn material and appears in all sorts of applications, from drink bottles to floppy disk liners. PET can also be used to make polyester fibre, which is used in auto upholstery, quilt batting, and clothing of all sorts.

To make polyester fibre, an extruder is used to produce very fine threads of PET. Polyester clothing tends to be slippery and silky in feel, although it can cause skin irritation for some wearers. The polyester fibres used to make clothing can be knitted or woven, although most are knit, to maximize the flexibility of polyester. Some polyester is blended with other fabrics to provide more loft or stretch, or to minimize skin irritation.

ZYLON

Zylon is a trademarked name for a range of thermoset liquid crystalline polyoxazole. This synthetic polymer yarn yarn material is manufactured by the Toyobo Corporation. Zylon was invented and developed by SRI International in the 1980s. Like Kevlar, Zylon is used in a number of applications that require very high strength with excellent thermal stability. Tennis racquets, table tennis blades, various medical applications, and some of the Martian rovers are some of the more well known instances. More common usage is in drive belts for snow mobiles manufactured by MBL USA Corporation. IUPAC name: poly(p-phenylene-2,6-benzobisoxazole)

VINALON

Vinalon is a synthetic fibre, produced from polyvinyl alcohol, using anthracite and limestone as raw materials. Vinalon was first developed in 1939 by Japanese scientist, Ichiro Sakurada, Yi Sung-ki and H. Kawakami Production of this fibre was delayed for WW2. The fibre was largely ignored in Korea until Ri defected to North Korea in 1950. Trial production began in 1954 and in 1961 the massive February 8 Vinalon Complex was built in Hamhung. Its success and widespread usage in North Korea is often pointed to in propaganda as an example of the success of the *juche* philosophy.

While Hamhung remains a major production centre for vinalon; in 1998, a vinalon factory was opened up in South Pyongan. In early 2010, Kim Jong-il himself attended a mass rally at the February 8 Vinalon Complex in Hamhung to celebrate it's reopening after 16 years of inactivity. While Kim is often seen at politial rallies or military parades, this is the first documented time he has ever attended a industrial mass rally, said an anonymous South Korean security official.

Vinalon, also known as "Juche fibre" in the DPRK, has become the national fibre of North Korea and is used for the majority of textiles, outstripping fibre such as cotton or Nylon, which are only produced in small amounts in North Korea. Other than clothing, Vinalon is also used for shoes, ropes and quilt wadding. Vinalon is resistant to heat and chemicals but has numerous disadvantages: it is stiff, uncomfortable, shiny, prone to shrinking and difficult to dye.

SPANDEX

Spandex, Lycra or elastane is a synthetic fibre known for its exceptional elasticity. It is stronger and more durable than rubber, its major non-synthetic competitor. It is a polyurethane-polyurea copolymer that was invented in 1959 by chemist Joseph Shivers at DuPont's Benger Laboratory in Waynesboro, Virginia. When first introduced, it revolutionized many areas of the clothing industry.

The name "Spandex" is an anagram of the word "expands". It is the preferred name in North America; in many European countries it is referred to as "elastane". Other brand names associated with Spandex include Lycra, Elaspan (also Invista's), Creora (Hyosung), ROICA and Dorlastan (Asahi Kasei), Linel (Fillattice), and ESPA (Toyobo).

POLYETHYLENE

Polyethylene is a type of polymer that is classified as a thermoplastic, meaning that it can be melted to a liquid and remolded as it returns to a solid state. As the name implies, polyethylene is chemically synthesized from molecules that contain long chains of ethylene, a monomer that provides the ability to double bond with other carbon-based monomers to form polymers. Polyethylene is known by other, non-official names, such as polythene in the United Kingdom. In addition, it is sometimes spelled as polyethylyne, or abbreviated to simply PE.

The first laboratory creation of polyethylene occurred in 1898 by accident at the hands of Hans von Pechmann while applying heat to another compound the German chemist previously discovered — diazomethane. Ironically, the synthesis of polyethylene via extreme heat and pressure in an industrial setting was again made by accident, but 35 years later.

A few years later, another chemist employed by the same England-based chemical company devised a method to consistently produce polyethylene under the same conditions. As a result, polyethylene became the primary source of low-density polyethylene (LDPE) production in 1939.

While polyethylene is essential to the economic health of the plastics industry, most consumers readily recognize the role it plays in everyday life. In fact, this substance is found in many ordinary household items, such as food wrap, shampoo bottles, milk containers, toys, and the common plastic bag used to tote groceries home from the store.

However, polyethylene is also present in numerous other products that contain plastic components. For instance, it is used to manufacture artificial knee and hip replacement parts, bulletproof vests, and even glassy flooring for ice skating rinks. Polyethylene may fall under one of several types. The distinction between them is determined by its molecular weight and branching, which is affected by its crystallization. LDPE is an example of branched polyethylene since its carbon molecules are attached to long chains of polyethylene instead of hydrogen.

Otherwise, a linear structure of carbon to hydrogen occurs, which is known as high-density polyethylene (HDPE). However, further variances in structure and molecular weight produce other forms, such as ultra high molecular weight polyethylene (UHMWPE), medium-density polyethylene (MDPE), or very low-density polyethylene (VLDPE). While polyethylene may help to make numerous useful and durable products possible, its environmental impact is cause for concern.

For one thing, it does not readily biodegrade and can reside in a landfill for hundreds of years. However, diligent recycling may significantly reduce this problem. In addition, scientists are exploring the possibility of employing *Sphingomonas*, an aerobic bacteria shown to shorten biodegrading of some forms of polyethylene to just a few months. Environmental preservation efforts have also led to the development of bioplastics, with the aim of synthesizing polyethylene from ethanol obtained from sugarcane.

5

Yarn Yarn Plant Sources in Yarn Dyes

INTRODUCTION

With the growing commercial availability of dyes, their range and scope of application is also expanding. Consequently, a large amount of unused dyes are released in the industrial effluents. Ninety percent of textile dyes entering modern activated sludge sewage treatment plants pass through unchanged. There is an intense environmental concern about the fate of these unbound dyes. These discharged dyes form toxic products and their strong colour causes turbidity which even at very low concentrations has a huge impact on the aquatic environment.

These synthetic reactive dyes bond covalently with fabric and contain chromophoric groups like anthraquinone, azo, triarylmethane etc. along with reactive groups *viz.*, vinyl sulphone, chlorotriazine, trichloropyrimidine etc. (Sumathi and Manju, 2000; Keharia and Madamvar, 2003). Disperse dyes and acid dyes have low solubility in water.

They are mainly used in the dyeing of polyesters and find minor use in dyeing cellulose acetates and polyamides. Azo dyes constitute the largest group of colourants used in industry. These dyes can be precipitated or adsorbed only in small amounts, while under anaerobic conditions they are cleaved by microorganisms forming potentially carcinogenic aromatic amines (Chung and Cerniglia, 1992). Several studies on decolourization of textile dyes used in industries have been conducted.

Some non-textile dyes commonly present in industrial effluents have been studied. Dye removal from wastewaters with traditional physicochemical processes, such as coagulation, adsorption and oxidation with ozone is expensive, can generate large volumes of sludge and usually requires the addition of environmentally hazardous chemical additives (Chen, 2006). On the other hand, most of the synthetic dyes are xenobiotic compounds which are poorly removed by the use of conventional biological aerobic treatments (Marco et al., 2007a, b). Although, biodegradation appears to be a promising technology, unfortunately the analysis of contaminated soil and water has shown persistence of toxic pollutants even in the presence of microorganisms (Robinson et al., 2001; Keharia and Madamvar, 2003). Decolourization of dye wastewater is an area where innovative treatment technologies need to be investigated. The focus in recent times has shifted towards enzyme based treatment of coloured wastewater/industrial textile effluents.

The peroxidase and polyphenol oxidases participate in the degradation of a broad range of substrate even at very low concentration. Further, these peroxidases and polyphenol oxidases have been used for treatment of dyes but large scale exploitation has not been achieved due to their low enzymatic activity in biological materials and high cost of purification (Bhunia et al., 2001; Shaffiqu et al., 2002; Verma and Madamwar, 2002). Bioremediation is a viable tool for restoration of contaminated subsurface environments. It is gaining importance due to its cost effectiveness, environmental friendliness and production of less sludge as compared to chemical and physical decomposition processes. Here too microbial treatment has certain inherent limitations (Husain and Jan, 2000; Duran and Esposito, 2000; Torres et al., 2003). It has been shown that peroxidases catalyze a variety of oxidation reactions and importantly dyes recalcitrant to peroxidase shows significant decolourization in the presence of Redox mediators (Calcaterra et al., 2008). Redox mediated enzyme catalysis has wide application in degradation of polycyclic aromatic hydrocarbons which includes phenols, biphenyls, pesticides, insecticides etc. (Husain and Husain, 2008; Calcaterra et al., 2008).

PEROXIDASES IN DYE DECOLOURIZATION WITH REDOX MEDIATORS

Enzymatic approach has gained considerable interest in the decolourization/ degradation of textile and other industrially important dyes present in wastewater. This strategy is ecofriendly and useful in comparison to conventional chemical, physical and biological treatments, which have inherent serious limitations. Enzymatic treatment is very useful due to the action of enzymes on pollutants even when they are present in very dilute solutions and recalcitrant to the action of various microbes participating in the degradation of dyes. Several enzymes (peroxidases, manganese peroxidases, lignin peroxidases, laccases, microperoxidase-11, polyphenol oxidases and azoreductases) have been evaluated for their potential in decolourization and degradation of dyes.

Although, some recalcitrant dyes are not degraded/ decolourized in the presence of such enzymes, the addition of certain Redox mediator enhances the range of substrates and efficiency of degradation of the recalcitrant compounds. However, very few Redox mediators are frequently used which includes 1- hydroxybenzotriazole (HOBT), veratryl alcohol, violuric acid (VA), 2-methoxyphenothiazone. The enzymes in soluble form cannot be exploited on large scale due to limitations of stability and reusability and consequently, the use of immobilized enzymes has significant advantages over soluble enzymes. In the near future, technology based on the enzymatic treatment of dyes present in the industrial effluents/wastewater will play a vital role. Treatment of wastewater on a large scale will also be possible by using reactors containing immobilized enzymes.

DYE DECOLOURIZATION WITH TURNIP (*BRASSICA RAPA*) PROTEINS

The potential of partially purified turnip (*Brassica rapa*) proteins have been studied by Matto and Husain, (2007) on decolourization of certain direct dyes like Direct Red 23, Direct Red 239, Direct Blue 80 and Direct Yellow 4. The turnip proteins showed enhanced decolourization in the presence of Redox mediators. Redox mediators explored for dye decolourization of these dyes were HOBT, alpha naphthol, vanillin, L-histidine, VA, catechol, quinol, bromophenol, 4-nitrophenol and gallic acid. Six out of 10 investigated compounds showed their potential in enhancing the decolourization of direct dyes. The performance of each was evaluated at different concentrations of mediator and enzyme. The decolourization of all tested direct dyes was maximum in the presence of 0.6 mM Redox mediator at pH 5.5and 30°C. Complex mixtures of dyes decolourized maximally in the presence of 0.6 mM Redox mediator (HOBT/VA). In order to examine the operational stability of the enzyme preparation, the enzyme was exploited for the decolourization of mixtures of dyes for different times in a stirred batch process. There was no change in decolourization of an individual dye or their mixtures after 60 min of incubation; the enzyme caused more than 80Percent decolourization of all dyes in the presence of 1-hydroxybenzotriazole/violuric acid. However, there was no desirable increase in dye decolourization of the mixtures on overnight incubation. The treatment of such polluted water in the presence of Redox mediators caused the formation of insoluble precipitate, which could be removed by the process of centrifugation. Such catalyzed oxidative coupling reactions may be important for natural transformation pathways for dyes and indicate their potential use as an efficient means for removal of dyes colour from waters and wastewaters.

DYE DECOLOURIZATION WITH FENUGREEK (*TRIGONELLA FOENUM-GRAECUM*) SEED PROTEINS

Peroxidase from fenugreek (*Trigonella foenum-graecum*) seeds is highly effective in the decolourization of textile effluent. The role of six Redox mediators has been investigated for effective decolourization by FSP (Fenugreek Seed Proteins). The maximum decolourization of textile effluent was observed in the presence of 1.0 mM 1-hydroxybenzotrizole, 0.7 mM H_2O_2, and 0.4 U/ml of FSP in the buffer of pH 5.0 at 40°C in 2.5 h. The decolourization of textile effluent in a batch process by peroxidase is 85Percent in 5 h, whereas the complete decolourization of textile effluent by membrane entrapped FSP was observed within 11 h of its operation. The absorption spectra of treated effluent exhibited a marked diminution in the absorbance at different wavelengths compared to untreated effluent. The removal of coloured aromatic compounds from wastewater by peroxidases is well-known. These enzymes are currently being employed for the treatment of aromatic compounds (Husain and Husain, 2008; Husain et al., 2009). Most of the studies on dye decolourization have been done in a defined media or synthetic wastewater where a single dye or their mixtures are usually present. However, industrial effluents are more complex due to the presence of other contaminating substances along with coloured compounds; under such conditions, the treatment of these pollutants is a difficult problem (Husain et al., 2010). Textile effluents alone are recalcitrant to the action of FSP, but in the presence of Redox mediators, they decolourized significantly. Many other aromatic pollutants such as aromatic amines, dyes and bisphenol A have already been oxidized by peroxidases in the presence of Redox mediators (Karim and Husain, 2009; Matto and Husain, 2008). The oxidation of effluent in the presence 1.0 mM HOBT is quite low compared to the Redox mediators used in earlier studies to treat various coloured pollutants (Rodriguez-Couto et al., 2005; Matto and Husain, 2008). The high concentrations of Redox mediators may not be appropriate for wastewater treatment because of high cost of mediators or possibility of creating negative impacts on effluent toxicity or in the environment upon their disposal into receiving waters (Kurniawati and Nicell, 2007). The maximum decolourization of textile effluent with FSP was at 0.7 mM concentration of H_2O_2. The concentration of H_2O_2 >1.2 mM acted as an inhibitor of peroxidase activity possibly by causing irreversible oxidation of enzyme ferri-heme group which is essential for its activity (Vazquez-Duarte et al., 2001). The differences in time course of removal of various dyes might be due to the structural barrier and the electron localization among them (Jauregui et al., 2003). The maximum decolourization (68Percent) of textile effluent was obtained by 0.4 U/ml of FSP. FSP catalyzed the decolourization of effluent over a wide range of pH with an optimum at pH 5.0. The maximum decolourization of effluent was at a temperature of 40°C. FSP successfully removed more than 85Percent colour of the effluent in 5 h. After 5 h, the increase in the rate of effluent decolourization has not been recorded. Complete removal of colour has been studied by using enzyme in a membrane bag. This type of reactor would increase the efficiency of dye decolourization because enzyme can be reused as it did not mix with the rest of the treated solution. Some workers reported that *Pseudomonas putida* CCRC14365 peroxidase was capable of removing phenol in hollow-fibre membrane bioreactor. *P. putida* fully degraded 2,000 mg/L of phenol within 73 h; even at a level of 2,800 mg/L, phenol could be degraded by more than

90Percent after 95 h of operation (Chung et al., 2004). In another report, a novel tube membrane bioreactor has been used for the treatment of an industrially produced wastewater arising in the manufacture of 3-chloronitrobenzene and nitrobenzene; in 1.7 h, over 99Percent of each 3-chloronitrobenzene and nitrobenzene from the wastewater was degraded (Livingston, 2004). For the decolourization and removal of coloured compounds from textile effluent, spectral analysis is an important aspect to demonstrate a loss in these compounds after treatment with enzyme. The decrease in absorbance peaks in the UV-visible region is a strong evidence for the decolourization of aromatic pollutants from wastewater. Some earlier spectral analyses of the enzymatically treated aromatic amines and their mixtures have shown a significant decrease in absorbance (Kurniawati and Nicell, 2007). There was 80Percent decolourization of the chromophoric groups and a significant reduction in the peak associated with the aromatic ring. The UV-visible absorbance spectrum for diluted textile effluent has been taken before and after treatment by FSP in the presence of various Redox mediators. The treatment of effluent by FSP in the presence of Redox mediator, HOBT, produces insoluble aggregates which can successfully be removed by centrifugation. The peroxidase from fenugreek seeds has potential in the remediation of hazardous aromatic pollutant and its use could be extended to the large-scale treatment of textile effluents and other related aromatic compounds by employing more effective and cheaper Redox mediators.

DYE DECOLOURIZATION WITH HORSE RADISH PEROXIDASE (HRP)

HRP is extracted from horse radish roots, and its performance has been evaluated in soluble and immobilized form by conducting batch experiments in the presence of H_2O_2 (Vasantha et al., 2006). The oxidation of Direct Yellow 12 dye has been tested as a function of HRP at fixed concentration of H_2O_2, and at constant HRP activity (1.8 units/ml). The optimum contact time required for dye removal was 1 h 45 min for vials containing 5 ml of dye solution (10 mg/L), 1.8 units of enzyme 1.5 ml/L of H_2O_2 were added and the reaction mixture (24°C, pH 4) was agitated for 2 h 15 min. After this period, dye removal was negligible up to remaining 1 h 35 min. The Direct Yellow 12 dye at varying aqueous-phase pH of the reaction mixture between 2 and 10 by keeping the dye concentration at 10 mg/L, enzyme concentration at 1.8 units, H_2O_2 dose at 1.5 ml/L, reaction temperature at 24°C and the contact time (1 h 45 min) constant exhibited 70Percent of the dye removal at an aqueous-phase pH of 4. The dye removal dropped significantly from pH 5 to 8 and the same trend continued up to an aqueous phase of pH 10. Aqueous phase of pH 4 resulted in higher HRP activity compared to other pH ranges from 3 to 9. Hydrogen peroxide acts as a co-substrate to activate the enzymatic action of peroxide radical. It contributed in the catalytic cycle of peroxidase oxidizing the native enzyme to form an enzymatic intermediate which accepts the aromatic compound to carry out its oxidation to form a free radical form. Experiments were carried out to find out the optimum H_2O_2 concentration required to bring out the conversion of dye by varying the H_2O_2 dose from 1 to 3 ml/L in the reaction mixture by keeping all the other experimental conditions constant (dye concentration 10 mg/L; temperature 24°C; enzyme concentration 1.8 units; reaction time 1 h 45 min). Their findings indicate maximum dye removal at H_2O_2 concentration of 2 ml/L and similar degradation with 2.5 and 3 ml/L; therefore, 2 ml/L was taken as the optimum H_2O_2 dose for dye removal (Vasantha et al., 2006). The concentration of the substrate present in the aqueous phase significantly influences enzyme-mediated reaction. If the amount of enzyme concentration is kept constant and the substrate concentration is gradually increased the reaction will increase until it reaches maximum. After obtaining the equilibrium state, further addition of the substrate will not change the rate of reaction. Studies carried out at different concentrations of the dye, *i.e.* 5–40 mg/L, keeping the other parameters constant indicated a dye concentration of 25 mg/L to be the cut-off concentration of the dye for optimum removal at the specified experimental conditions along with retaining the activity of HRP enzyme and to protect the protein from denaturation. Application of free enzyme in industrial process is not economically viable, while immobilization/ entrapment of enzyme results in repeated application. Two types of polymeric materials—alginate and acryl amide—have been used to study their relative efficiency in dye removal for the entrapment of peroxidase. Normally, enzyme immobilization is expected to provide stabilization effect restricting the protein unfolding process as a result of the introduction of random intra- and intermolecular crosslinks. Zille et al. (2003) have reported less availability of the enzyme for interaction with anionic dyes due to the immobilization in a particular matrix. For the immobilization of HRP enzyme acryl amide gel was more efficient in dye removal compared to alginate matrix. About 78Percent of dye removal was observed with acryl amide gel-immobilized beads, while with alginate matrix it was only 52Percent. Gel-immobilized HRP was effective in dye removal compared to free HRP (69Percent), and alginate-immobilized HRP showed inferior performance. The effect of aqueous-phase pH on the enzyme-catalyzed degradation with alginate- and acryl amide-immobilized HRP enzyme suggests that after pH 4 there is a decrease in the dye removal capacity for both types of the entrapped matrices. About 78Percent and 54Percent of the dye removal was observed at an aqueous phase of pH 4 for acryl amide and alginate-entrapped beads respectively. The relative inferior performance of alginate-immobilized HRP compared to acryl amide may perhaps be due to lesser availability of the peroxidase structure to the dye molecule in the alginate mix compared to acryl amide. The effective performance of acryl amide-entrapped beads may be attributed to the non-ionic nature of the beads, which results in minimum modification of the enzyme properties and unaffected nature of the charged substrate as well as product diffusion. The electron-withdrawing nature of the azo linkages obstructs the susceptibility of azo dye molecules to oxidative reactions. Only specialized azo dye-reducing enzymes could degrade azo dyes. In addition, when Direct Yellow 12 dye was reduced using the enzymatic method, the oxidation capacity increased with increasing concentrations of HRP and H_2O_2 at pH 4. Gel/alginate based enzyme immobilization reduced the azo dye. The immobilized enzyme beads could further be used two–three times for the removal of the same dye with lower efficiency. The application of enzyme-based systems in waste treatment is unusual, given that many drawbacks are derived from their use, including low efficiency, high costs and easy deactivation of the enzyme.

DYE COLOUR REMOVAL WITH BITTER GOURD (*MOMORDICA CHARANTIA*) PEROXIDASE

Application of partially purified BGP (Bitter Gourd Peroxidase) in the decolourization of textile and other industrially important dyes has been explored. Simple ammonium sulphate precipitated proteins from bitter gourd were taken for the treatment of a number of dyes present in polluted wastewater (Akhtar et al., 2005a, b). Partially purified preparation of BGP was obtained by adding 20–80Percent ammonium sulphate and this preparation exhibited a specific activity of 99.0 EU (Enzyme units) of peroxidase/mg protein. Peroxidases from bitter gourd were highly stable against pH, heat, urea, water miscible organic solvents, detergents and proteolysis. The dye solution was stable upon exposure to H_2O_2 or to the enzyme alone. The dye precipitation was due to H_2O_2-dependent enzymatic reaction, possibly involving free-radical formation followed by polymerization and precipitation. Eight textile reactive dyes were tested with increasing concentration (0.133–0.339 EU of BGP/ml) of reaction volume for 2 h at 37°C. The decolourization of Reactive Red 120, Reactive Blue 4, Reactive Blue 160 and Reactive Blue 171 was continuously enhanced by adding increasing concentration of BGP. Reactive Blue 4 was completely decolourized with 0.399 EU of BGP/ml of reaction mixture in the absence of HOBT. Reactive Orange 4, Reactive Orange 86, Reactive Red 11 and Reactive Yellow 84 were recalcitrant to the BGP action. Four reactive textile dyes were treated by step-wise addition of enzyme, adding 0.133 EU of BGP/ml of reaction volume at each step after every 1 h, to the decolourizing solution. Each enzyme addition exhibited rapid disappearance of colour. Reactive Blue 160 was completely decolourized after the third enzyme addition while Reactive Blue 4 was almost decolourized completely after the second enzyme addition. Reactive Blue 171 and Reactive Red 120 were decolourized to 56Percent and 39Percent, respectively after the third enzyme addition. However, Reactive Red 11, Reactive Orange 4, Reactive Orange 86 and Reactive

Yellow 84 were recalcitrant to decolourization by BGP. Of the reactive dyes incubated with 0.266 EU of BGP/ml of reaction volume for increasing time period only four dyes decolourized on treatment with BGP for 1h at 37°C. Although more colour disappeared on incubation for longer duration, the rate of decolourization was slow. All the four dyes were decolourized with varying percentages (30–90Percent). The decolourization of dyes with 0.266 EU of BGP/ml of reaction volume on incubation at 37°C for 4 h was 88Percent for Reactive Blue 4, 71Percent for Reactive Blue 160, 31Percent for Reactive Blue 171 and 28Percent for Reactive Red 120. Rest of the four dyes: Reactive Red 11, Reactive Orange 4, Reactive Orange 86 and Reactive Yellow 84 were fully recalcitrant to decolourization by BGP even after 4 h incubation with similar treatment. Thirteen different non-textile dyes were treated with 0.167 EU of BGP/ml of reaction volume at 37°C for 1h. Carmine, Methyl Orange, Methylene Blue, Coomassie Brilliant Blue G-250, Rhodamine 6G, Methyl Violet 6B, 1:2 naphthaquinone 4-sulphonic acid and Martius Yellow dyes were recalcitrant to decolourization by the BGP action or were slowly decolourized during the progress of the reaction. Maximum decolourization achieved by partially purified BGP was 53Percent for Naphthalene Black 12B, 57Percent for Coomassie Brilliant Blue R 250, 96Percent for Evans Blue, 51Percent for Eriochrome Black T and 86Percent for Celestine Blue. Twenty-one dyes used in this study were treated with BGP in the presence of 1.0 mM HOBT and 0.6 mM H_2O_2 at 37°C. Presence of HOBT drastically enhanced the rate of decolourization of recalcitrant dyes. Reactive Orange 4, Reactive Red 11, Reactive Yellow 84 and Reactive Orange 86 were recalcitrant to decolourization in the absence of HOBT. However, these dyes were decolourized up to 98Percent, 80Percent, 70Percent and 78Percent, respectively by the action of 0.399 EU/ml of BGP in the presence of 1.0 mM HOBT at 37°C for 2 h. Non-textile dyes *e.g.*, Carmine, Methyl Orange, Coomassie Brilliant Blue G250, Rhodamine 6G, Methylene Blue and Methyl Violet 6B, which were recalcitrant to decolourization by BGP in absence of HOBT, were almost completely decolourized in presence of 1.0 mM HOBT. However, 1:2-naphthaquinone 4-sulphonic acid was recalcitrant to decolourization even in the presence of HOBT. Decolourization of textile reactive dyes in the absence of HOBT occurred by the formation of precipitate, which settled down and removed by centrifugation. Several earlier investigators have shown that the treatment of phenols and aromatic amines by peroxidases and tyrosinases resulted in the formation of large insoluble aggregates. However, the decolourization of textile and other dyes by BGP in presence of 1.0 mM HOBT appeared without the formation of any precipitate. It suggested that the decolourization of dyes took place via degradation of aromatic ring of the compounds or by cleaving certain functional groups as reported elsewhere (Christian et al., 2003; Husain et al., 2009). HOBT could have a dual role, first as a mediator by increasing the substrate range of dyes for BGP and second enhancing the rate of oxidation. Complex mixtures of various reactive textile and other industrially important dyes have been studied by mixing three or four different dyes in equal proportions and incubated with 0.266 EU of BGP/ ml in the presence of 1.0 mM HOBT and 0.6 mM H_2O_2 for 1h at 37°C. Wave length maxima for each dye mixture were determined and decolourization of mixtures was monitored after the incubation period. All the mixtures decolourized by more than 80Percent. The decolourization rate of mixtures of dyes was slower than that of pure dye solution.

This supports an earlier observation that the biodegradation of various phenols in the form of mixtures was quite slow compared to the independent phenol (Kahru et al., 2000). Bourbonnais and Paice (1990) described for the first time the use of Redox mediators by allowing laccase to oxidize non-phenolic compounds thereby expanding the range of substrates that can be oxidized by this enzyme.

The mechanism of action of laccase mediator system has been extensively studied and it is used in the textile industry in the finishing process for indigo stained materials. Several workers have demonstrated that the use of Redox mediator system enhanced the rate of dye decolourization by several folds but these mediators were required in very high concentrations (5.7 mM violuric acid/laccase system, 11.6 mM of HOBT/laccase system) (Soares et al., 2001a,b; Claus et al., 2002).

For the first time, it was shown that the decolourization of dyes by BGP was effective at very low concentration of HOBT (1.0 mM). The rate of non-textile dyes decolourization was also enhanced by 2–100 folds (Akhtar et al., 2005a). Several reports indicate the enhancement of laccase activity by free radical mediators; however the enhancement of peroxidase activity by a Redox mediator has been studied for the first time. Peroxidase/ Redox mediator system will prove a sensitive and an inexpensive procedure for the treatment of dyes present in complex mixtures/industrial effluent.

DYE COLOUR REMOVAL WITH TOMATO (*LYCOPERSICON ESCULENTUM*) PEROXIDASE

Matto and Husain (2009) investigated the role of concanavalin A (Con A)-cellulose-bound tomato peroxidase for the decolourization of direct dyes. Cellulose was used as an inexpensive yarn yarn material for the preparation of bioaffinity support.

Con A-cellulose-bound tomato peroxidase exhibited higher efficiency in terms of dye decolourization as compared to soluble enzyme under various experimental conditions. Both Direct Red 23 and Direct Blue 80 dyes were recalcitrant to the action of enzyme without a Redox mediator. Six compounds were investigated for Redox-mediating property. Immobilized peroxidase decolourized both dyes to different extent in the presence of all the used Redox mediators. However, 1- hydroxybenzotriazole emerged as a potential Redox mediator for tomato peroxidase catalyzed decolourization of direct dyes. These dyes were maximally decolourized at pH 6.0 and 40°C by soluble and immobilized peroxidase. The absorption spectra of the untreated and treated dyes exhibited a marked difference in the absorption at various wavelengths. Immobilized tomato peroxidase showed a lower Michaelis constant than the free enzyme for both dyes. Soluble and immobilized tomato peroxidase exhibited significantly higher affinity for Direct Red 23 compared to Direct Blue 80.

DYE DECOLOURIZATION WITH POINTED GOURD (*TRICHOSANTHES DIOICA*) PEROXIDASE

The effects of *Trichosanthes dioica* peroxidase along with Redox mediators on decolourization of water insoluble disperse dyes; Disperse Red 19 (DR19) and Disperse Black 9 (DB9) have been studied by Jamal et al., (2010). Nine different Redox mediators; bromophenol, 2, 4-dichlorophenol, guaiacol, 1-hydroxybenzotriazole, m-cresol, quinol, syringaldehyde, vanillin and violuric acid were evaluated. Among the chosen mediators, 1-hydroxybenzotriazole was most effective for decolourization with PGP (Pointed Gourd Peroxidase). At a concentration of 0.45U/mL the peroxidase could decolourize Disperse Red 19 to a maximum of 79Percent with O.2mM 1-hydroxybenzotriazole whereas Disperse Black 9 decolourized upto 60Percent with 0.5mM 1-hydroxybenzotriazole.

The time, pH and temperature at which maximum decolourization were recorded was 60 min, 4 and 42°C. It has been observed that the dye solutions were recalcitrant upon exposure to HOBT, H_2O_2 or to the enzyme alone but the enzyme in the presence of Redox mediators was much effective in performing the decolourization of the dyes, implying dye decolourization was a result of Redox mediated H_2O_2-dependent enzymatic reaction. It has already been reported that Redox mediators have the potential to mediate an oxidation reaction between a substrate and an enzyme.

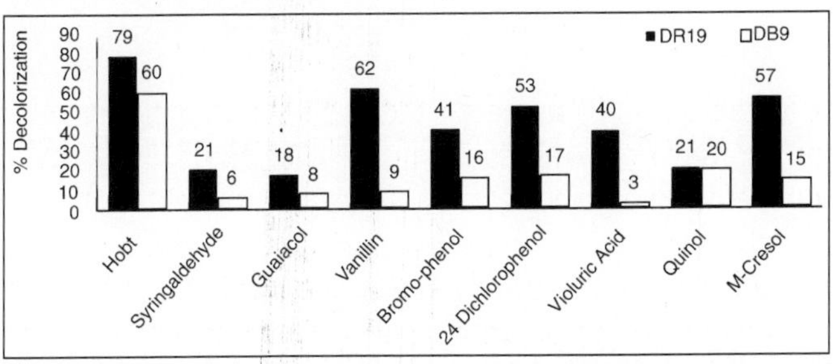

Fig. Percent Dye decolourization as a function of different Redox mediators. The dyes DR19 (25mgL^{-1},5.0mL) and DB9(50 mgL^{-1},5.0mL) solutions were incubated independently with PGP (0.45 Uml^{-1}) in the presence of 0.5mM concentration of each Redox mediators; other conditions were 0.8mM H_2O_2, 100mM glycine HCl buffer, pH 4.0 for 60 min at 37°C. (λmax for DR19 and DB9 are 495nm and 461nm respectively) [Jamal et al., 2010]

Different Redox mediators have different mediation efficiency which is governed by Redox potential of the mediator and the oxidation mechanism of the substrate. Oxidation of substrate occurs by free radical formation by the mediator. The free radicals can be formed either by one-electron oxidation of substrate or by abstraction of a proton from the substrate.

The pointed gourd peroxidase was effective in decolourizing the dyes at low concentrations of HOBT.

Although the extent of decolourization of DR19 and DB9 increased with increasing concentrations of HOBT, the maximum decolourization was observed to be 79Percent and 60Percent with 0.2 mM and 0.5 mM for DR19 and DB9, respectively.

Further addition of HOBT resulted in a slow decrease in decolourization of both the dyes.

This inhibition could likely be due to the high reactivity of HOBT radical, which might undergo chemical reactions with side chains of aromatic amino acid by enzyme thereby; inactivating it. Hence, the dosage of Redox mediator is an important factor for the enzyme-mediated decolourization.

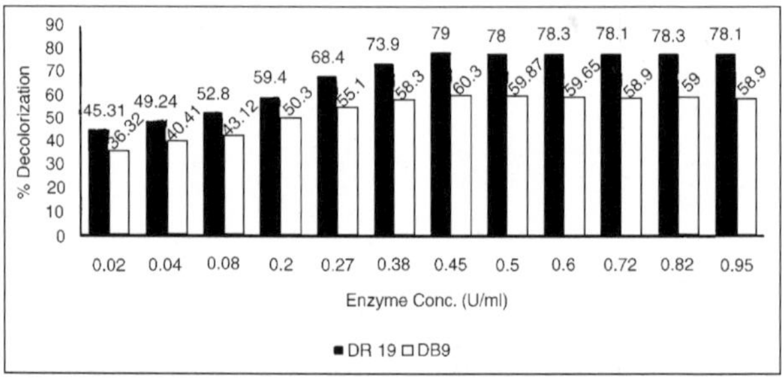

Fig. Percent Dye decolourization as a function of different enzyme (PGP) concentrations. The dyes DR19 (25mgL^{-1},5.0mL) and DB9(50 mgL^{-1},5.0mL)solutions were incubated independently with PGP (0.02 to 0.95 Uml^{-1}) in the presence of 0.2mM and 0.5mMconcentration of HOBT for DR19 and DB9 respectively; other conditions were 0.8mM H_2O_2, 100mM glycine HCl buffer, pH 4.0 for 60 min at 37°C. (λmax for DR19 and DB9 are 495nm and 461nm respectively) [Jamal et al., 2010].

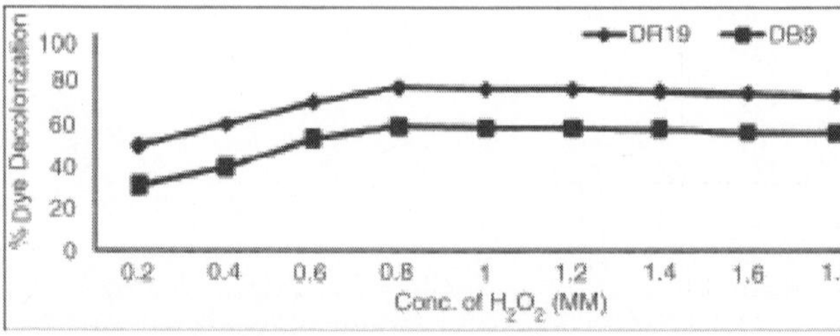

Fig. Percent Dye decolourization at different concentrations of H_2O_2. The dyes DR19 (25mgL^{-1},5.0mL) and DB9(50 mgL^{-1},5.0mL) solutions were incubated independently with H_2O_2 concentrations and PGP (0.45 Uml^{-1}) in the presence of varying amounts of H_2O_2 (0.2 to 1.8 mM) for DR19 and DB9 respectively; and other conditions were 100mM glycine HCl buffer, pH 4.0 for 60 min at 37°C. (λmax for DR19 and DB9 are 495nm and 461nm respectively) [Jamal et al., 2010].

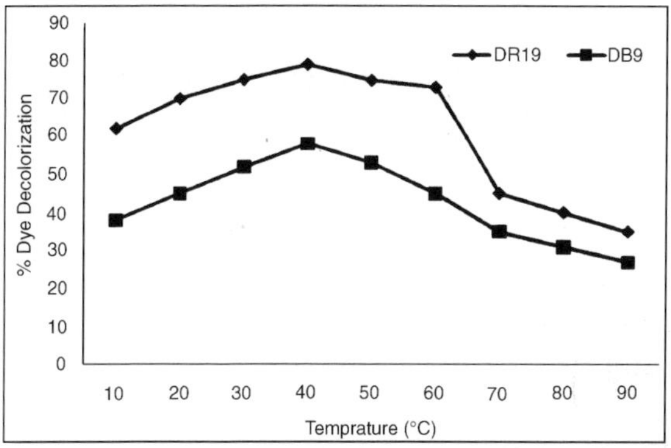

Fig. Percent Dye decolourization as a function of temperature. The dyes DR19 (25mgL^{-1},5.0mL) and DB9 (50 mgL^{-1},5.0mL)solutions were incubated independently with PGP (0.45 Uml^{-1}) in the presence of HOBT, 0.8mM H_2O_2, 100mM glycine HCl buffer, pH 4.0 for 60 min at temperatures (20°C to 90°C). (λmax for DR19 and DB9 are 495nm and 461nm respectively) [Jamal et al., 2010].

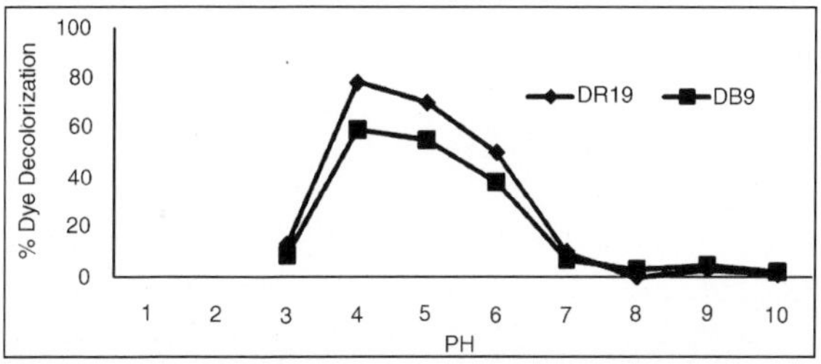

Fig. Percent Dye decolourization as a function of pH. The dyes DR19 (25mgL^{-1}, 5.0mL) and DB9 (50 mgL^{-1},5.0mL) solutions were incubated independently with PGP (0.45 Uml^{-1}) in the presence of HOBT, 0.8mM H_2O_2, 100mM glycine HCl buffer at different pH (2,3,4,5,6,7,8,9 and 10) for 60 min at 37°C. (λmax for DR19 and DB9 are 495nm and 461nm respectively) [Jamal et al., 2010].

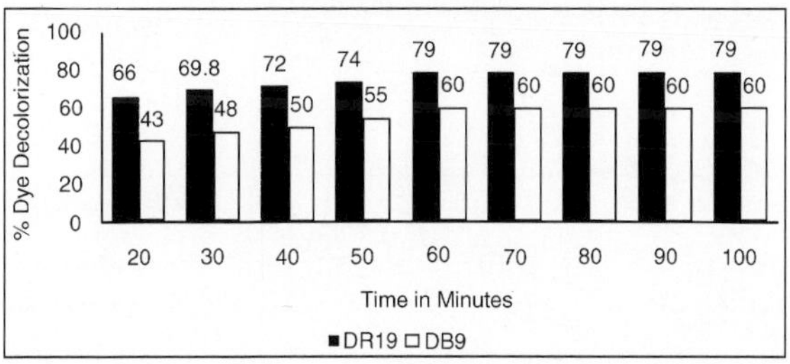

Fig. Percent Dye decolourization as a function of time. The dyes DR19 (25mgL^{-1}, 5.0mL) and DB9(50 mgL^{-1},5.0mL) solutions were incubated independently with PGP (0.45 Uml^{-1}) in the presence of HOBT, 0.8mM H_2O_2, 100mM glycine HCl buffer, pH 4.0 and for time 20 to 100 min at 37°C. (λmax for DR19 and DB9 are 495nm and 461nm respectively) [Jamal et al., 2010].

The enzyme reacted well to decolourize both the dyes in the presence of 0.8 mM H_2O_2.

The maximum decolourization was obtained at 0.8mM H_2O_2 which is slightly higher than reported for soybean peroxidase, bitter gourd peroxidase (BGP) and turnip peroxidase.

The reaction temperature is an important parameter which effects the decolourization of dyes. The maximum decolourization for both the dyes DR19 and DB9 was at 42°C. It has been reported that BGP mediated disperse dye decolourization was optimal at 40°C.

The maximum decolourization of DR19 and DB9 was obtained at an acidic of pH 4.0. It has earlier been reported that the degradation of industrially important dyes by enzymes such as horse radish peroxidase, polyphenol oxidase, BGP and laccase was also maximum in the buffers of acidic pH. DR19 and DB9 were maximally decolourized within 20 min of incubation.

There was slow and gradual enhancement of decolourization upto 60 min of incubation. It was also evident from the observation that DR19 was decolourized to a greater extent within 20 min in the presence of only 0.2 mM HOBT.

However, DB9 was decolourized maximally in the presence of 0.5 mM HOBT and decolourization rate was slow. This is consistent with reports that decolourization rate varies, depending upon the type of dye to be treated.

TRICHOSANTHES DIOICA PROTEINS IN DECOLOURIZING INDUSTRIALLY IMPORTANT TEXTILE, NON-TEXTILE DYES AND DYE MIXTURES

In one of my recent works, the competency of *Trichosanthes dioica* proteins in decolourizing industrially important textile, non-textile dyes and dye mixtures in the presence of Redox mediators under varying experimental conditions of pH, temperature, time intervals and enzyme concentration on the basis of one-factor-at-a-time (OFAT) method has been studied. All the textile dyes and non-textile dyes, ammonium sulphate, and Tween-20 were procured from Sigma Chemical Co. (St. Louis, MO, USA). Redox mediator's *viz.*, 1 hydroxybenzotriazole (HOBT) and vanillin were obtained from SRL Chemicals (Mumbai, India).

All other chemicals were of analytical grade. The pointed gourds were purchased from the local market. The samples were aseptically transferred into sterilized plastic bags. Briefly, 100 g of pointed gourd was homogenized in 180 ml of 100 mM sodium acetate buffer, pH 5.6.

The homogenate was filtered through multi-layers of cheese cloth and then centrifuged at the speed of 10,000 × g on a Remi C-24 cooling centrifuge for 30 min at 4°C.

The clear solution thus obtained was used for salt fractionation by adding 10Percent to 80Percent (w/v) $(NH_4)_2SO_4$. The proteins were precipitated by continuously stirring at 4°C overnight.

The precipitate was collected by centrifugation at 10,000 × g on a Remi C-24 cooling centrifuge, dissolved in 100 mM sodium acetate buffer, pH 5.6 and dialyzed against the assay buffer (0.1 M glycine HCl buffer, pH 4.0) (Akhtar et al., 2005a).

Protein concentration was estimated by taking BSA as a standard protein and following the procedure of Lowry et al (1951). Peroxidase activity was determined by a change in the optical density (A_{460} nm) at 37°C by measuring the initial rate of oxidation of 6.0 mM o-dianisidine HCl in the presence of 18.0 mM H_2O_2 in 0.1 M glycine-HCl buffer, pH 4.0, for 20 min at 37°C.

One unit of activity was defined as the amount of enzyme that transformed 1μmol of o-dianisidine HCl as substrate per min. The dyes (45-210 mg/l) were solubilized in 100 mM glycine HCl buffer, pH 4.0. Each dye was independently incubated with pointed gourd peroxidase (PGP) (0.45 EU/mL) in 100 mM glycine HCl buffer, pH 4.0 in the presence of 0.80 mM H_2O_2 for varying times at 37°C. The reaction was stopped by boiling at 100°C for 7 min.

Dye decolourization was monitored by measuring the difference at the maximum absorbance for each dye as compared with control experiments without enzyme on UV-visible spectrophotometer (JASCO V-550, Japan). Untreated dye solution (inclusive of all reagents except the enzymes) was used as control (100Percent) for the calculation of percent decolourization.

The dye decolourization was calculated as the ratio of the difference of absorbance of treated and untreated dye to that of treated dye and converted in terms of percentage.

Effect of Redox Mediators on Decolourization Profile of Textile and Non-textile Dyes

The effect of different Redox mediators on the dye decolourization by PGP. Out of the two different Redox mediators studied for dye decolourization, HOBT was more effective in decolourizing the dyes under study.

The extent of decolourization in the presence of HOBT was in the range of 98.6Percent to 69.8Percent for the reactive dyes whereas the disperse dyes studied exhibited decolourization in the range of 79.2Percent to 61.2Percent. The effective HOBT concentrations were 1.0 mM and 0.2 mM for reactive and disperse dyes, respectively.

The dye decolourization with vanillin was 71.2Percent to 60.2Percent for reactive dyes and 55.3Percent to 34.5Percent for disperse dyes at 1.0 mM concentration.

The effect of increasing concentrations of HOBT and vanillin (0.05 to 1.5 mM) on textile and other dyes. With increasing concentration of HOBT or vanillin there was an increase in the extent of decolourization of both dyes.

However, the decolourization of each textile dye was lower at each of the varying concentration of vanillin than HOBT. The non-textile dyes were effectively decolourized at 1.0 mM of HOBT and vanillin independently but here too, influence of HOBT was substantial. There was not much effect in percent decolourization of the dyes above these concentrations of HOBT or vanillin.

Enzyme Activity Profile of *T. dioica* Peroxidase Mediated Decolourization of Textile Dyes

Figure shows the extent of decolourization of reactive and disperse dyes with increasing concentration of PGP. The maximal decolourization for these two different classes of dyes was observed at PGP concentration of 0.45 EU/ml after an incubation time of 4 h.

Dye decolourization was not significantly exhibited with any further increase of PGP. The presence of 0.2 mM and 1.0 mM HOBT for decolourization of disperse and reactive dyes was effective in disappearance of Reactive Blue 15 and Reactive Orange 15 to the extent of 96.2 Percent and 94.6 Percent respectively whereas the others, Reactive Red 4 and Reactive Yellow 2 disappeared up to 88.2 Percent and 89.8 Percent respectively. Among the disperse dyes only Disperse Black 9 was effectively decolourized up to 79Percent whereas the others showed disappearance of colour below 69.3 Percent.

In the presence of 1.0 mM vanillin, partially purified *T. dioica* peroxidase catalyzed decolourization at 0.45 EU/ml after an incubation period of 4 h in the range of 61.4 Percent to 71.3 Percent for the dyes under study. The decolourization profile for all the dyes significantly increased after the first increase in peroxidase concentration in a period of 2h.

There was no significant change on dye decolourization either on increasing peroxidase concentration or the incubation time.

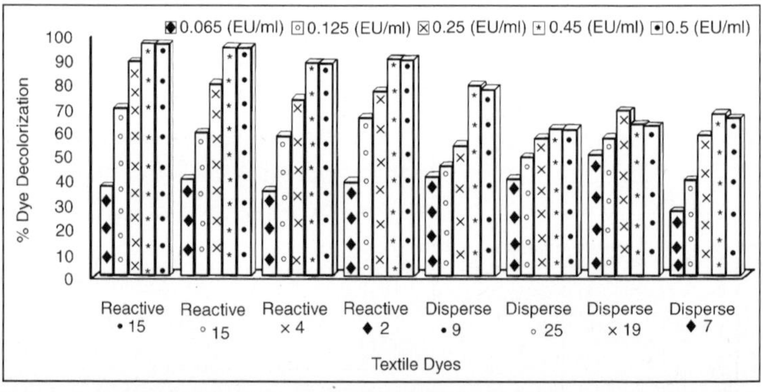

Fig. (a) Percent Dye Decolourization of Textile Dyes in the Presence of Fixed Concentration of HOBT [0.2 mM for Reactive Dyes and 1.0 mM for Disperse Dye] and Increasing Concentration of *T. dioica* Peroxidase Enzyme (EU/ml).

H_2O_2 and pH Activity Profile of Decolourization of Textile Dyes

Figure shows that the percent decolourization improved with the increasing concentration of H_2O_2 and the maximum decolourization was observed at a concentration of 0.8 mM and 1.0 mM of H_2O_2 for disperse and reactive dyes respectively which remained substantially unaffected till 1.2 mM H_2O_2.

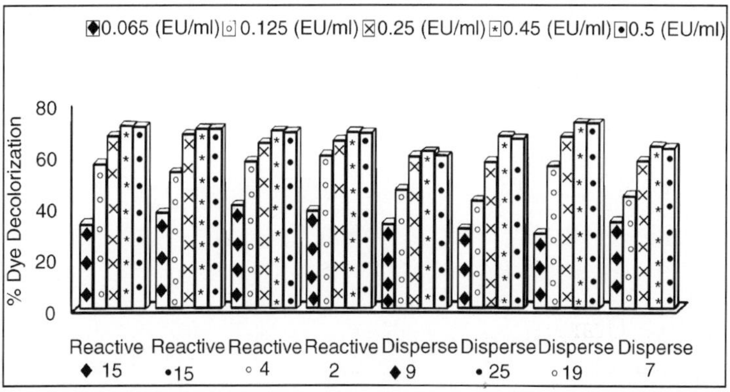

Fig. (b) Percent Dye Decolourization of Textile Dyes in the Presence of Fixed Concentration of Vanillin [0.2 mM for Reactive Dyes and 1.0 mM for Disperse Dye] and Increasing Concentration of *T. dioica* Peroxidase Enzyme (EU/ml).

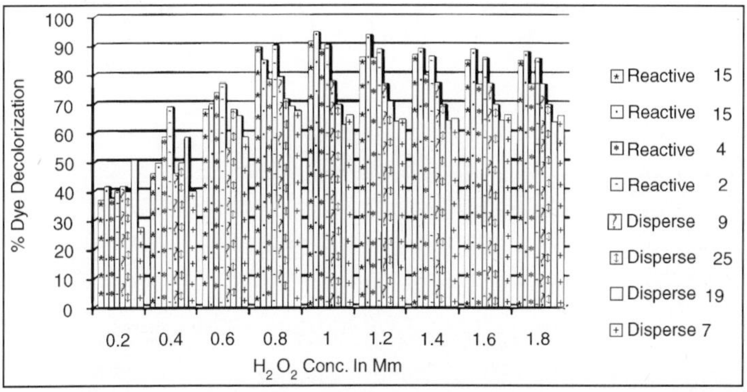

Fig. Percent Dye Decolourization of Textile Dyes in the Presence of Fixed Concentration of HOBT, *T. dioica* Peroxidase Enzyme (EU/ml) and Varying Concentration of H_2O_2.

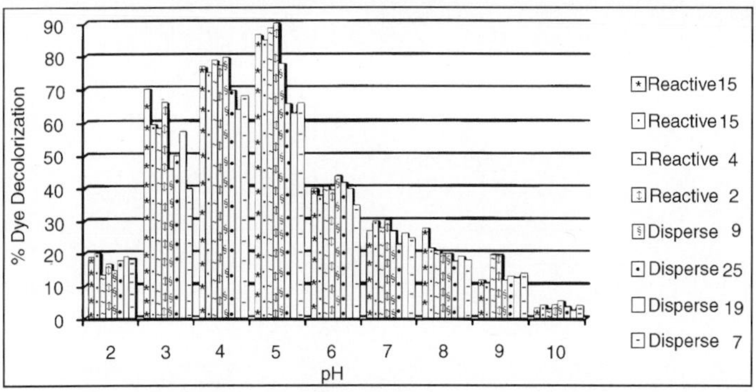

Fig. Percent Dye Decolourization of Textile Dyes in the Presence of Fixed Concentration of HOBT, *T. dioica* Peroxidase Enzyme (EU/ml) and Varying pH.

To find out the range of pH in which significant decolourization was observed; buffers in the range of pH 2.0 to pH 10.0 were used. An acidic range of pH (3.0 to 6.0) was better suited for dye decolourization. Maximum decolourization was observed at pH 4.0 and pH 5.0 at fixed concentration of *T.*

dioica peroxidase and HOBT for disperse and reactive dyes respectively. There was significant decrease in the extent of decolourization in an alkaline medium and at pH 10.0 the decolourization action of the enzyme was almost insignificant/ lost.

Temperature Activity Profile of Decolourization of Textile Dyes

Among the textile dyes the reactive dyes exhibited maximum decolourization at 50°C whereas the disperse dyes showed maximum decolourization at 40°C in the presence of 1.0 mM and 0.2 mM HOBT. Reactive Blue 15 (96.1Percent), Reactive Orange 15 (94.4Percent), Reactive Red 4 (85.2Percent) whereas disperse dyes under study decolourized in the range of 61.2Percent to 79Percent.

Time Activity Profile of Decolourization of Textile Dyes

Maximum decolourization of reactive and disperse dyes were observed within 1 h of incubation at 50°C and 40°C with 0.45 EU/ml of PGP and 1.0 mM and 0.2 mM HOBT respectively.

However, further decolourization of these dyes progressed slowly upto 4 h, although no effective increase was observed even when the dyes were further incubated for longer times.

Among the reactive dyes Reactive Blue 15 decolourized almost completely and Reactive Orange 15 up to 78.3 Percent at 4 h incubation, whereas Reactive Red 4 and Reactive Yellow 2 showed decolourization up to 86.3 Percent and 81.2 Percent under similar conditions.

The disperse dyes were comparatively resistant to decolourization and only in the presence of HOBT Disperse Red 19, Disperse Yellow 7, Disperse Black 9 decolourized to 79.5 Percent, 72.3 Percent and 71.6 Percent respectively. Disperse Orange 15 was comparatively degraded and decolourized to a lesser extent under similar conditions with a maximum of 61.2 Percent.

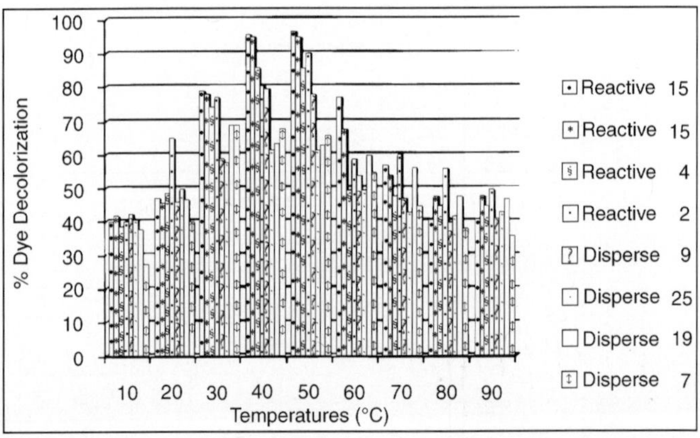

Fig. Percent Dye Decolourization of Textile Dyes in the Presence of Fixed Concentration of HOBT, *T. dioica* Peroxidase Enzyme (EU/ml) at Different Temperatures.

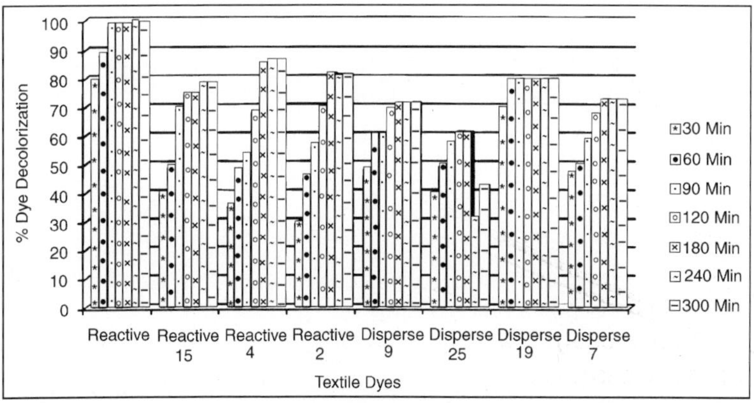

Fig. Percent Dye Decolourization of Textile Dyes in the Presence of Fixed Concentration of HOBT, *T. dioica* Peroxidase Enzyme (EU/ml) at Different Time Interval.

Decolourization Profile of Non-textile Dyes by *T. dioica* Peroxidase as a Function of Enzyme Concentration and Time

Nine different non-textile dyes were studied. These dyes were treated with different amount of *T. dioica* peroxidase in the range of 0.065 EU/ml to 0.50 EU/ml and incubated for varying time interval with and without the Redox mediator HOBT (1.0 mM) at 37°C. The results indicated that few non-textile dyes *viz.*, Methylene Blue, Martius Yellow, Methyl Orange, Rhodamine 6G were highly recalcitrant to decolourization by *T. dioica* in the absence of HOBT after 3h of incubation. However, the decolourization progressed slowly with the addition of 1.0 mM HOBT and percent decolourization achieved for Methylene Blue, Martius Yellow and Methyl Orange, was 98.6 Percent, 35.1 Percent and 98.7 Percent respectively; whereas Rhodamine 6G almost decolourized completely. For other dyes, the maximum decolourization exhibited after 3h was 98.7 Percent, 97.3Percent, 97.1Percent, 67.8Percent for Coomassie Brilliant Blue R 250, Naphthol Blue Black, Evans Blue and Eriochrome Black T respectively.

Table. Decolourization of Non-Textile Dyes by *T. dioica* Peroxidase at Different Enzyme Concentration, Incubation Period and with Fixed Concentration of HOBT, H_2O_2.

Non Textile Dyes (λ max) nm	0.065 EU/ml		0.125 EU/ml		0.250 EU/ml		0.45 EU/ml		0.50Eu/ml	
	-HOBT (60min)	+ HOBT (60min)	- HOBT (120min)	+ HOBT (120min)	- HOBT (160min)	+ HOBT (160min)	- HOBT (180min)	+ HOBT (180min)	- HOBT (240min)	+ HOBT (240min)
Celestine (642)	46.6	57.3	49.7	68.7	54.3	69.3	55.6	79.5	58.9	79.6
Coomassie R250 (553 nm)	47.6	92.4	48.2	93.2	51.3	98.9	51.2	98.7	51.1	98.7
Methylene (664 nm)	2.5	67.2	3.7	78.5	3.6	98.4	3.5	98.6	3.2	98.9
Eriochrome (503 nm)	48.7	67.7	51.6	77.4	55.4	91.3	55.3	92.0	55.1	92.2
Evans (607 nm)	86.4	94.3	87.5	95.4	87.8	96.7	87.2	97.1	86.7	97.2
Martius (433 nm)	10.2	33.2	12.4	34.5	13.2	35.6	13.5	35.1	12.3	35.2
Methyl (464 nm)	12.3	87.2	13.2	89.2	13.2	98.5	13.1	98.7	12.2	98.6
Naphthol (618 nm)	47.2	78.8	49.2	86.2	56.6	97.4	57.2	97.3	57.9	97.1
Rhodamine (525 nm)	11.2	78.4	13.2	86.3	13.2	99.7	14.2	99.6	13.2	99.5

Decolourization Profile of Different Dye Mixtures by *T. dioica* Peroxidase

To simulate the decolourization of dyes from industrial effluent, complex mixtures of textile dyes including reactive, disperse and non-textile dyes were prepared by mixing four different dyes in equal proportions and incubated with 0.45 EU/ml of *T. dioica* peroxidase in the presence of 1.0 mM HOBT and 1.0 mM H_2O_2 for 2 h at 40°C. The decolourization was recorded at wave length maxima of each mixture determined spectrophotometrically. The combinations of different dyes showed decolourization by more than 82Percent. The rate of decolourization of dye mixture was slower in comparison to that of individual dyes both in the presence and absence of HOBT. However, the HOBT mediated dye decolourization was more effective.

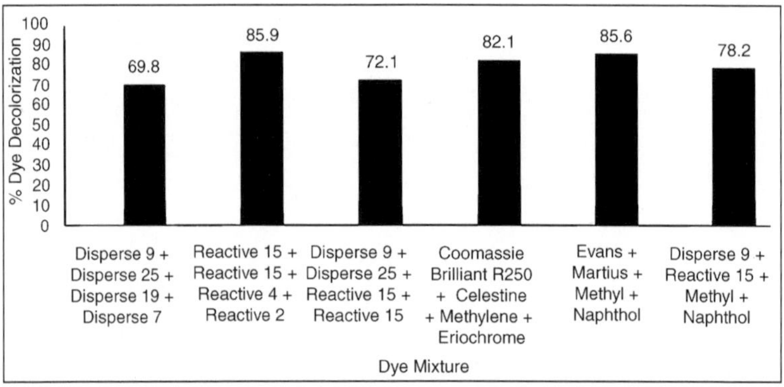

Fig. Degradation/ decolourization of dye mixture by *T. dioica* in the presence of HOBT/H_2O_2 to mimic dyes in industrial effluents. [Disperse 9+ Disperse 25 +Disperse 19+ Disperse 7 (λ_{461}); Reactive 15 +Reactive 15 + Reactive 4 + Reactive 2(λ_{531}); Disperse 9+ Disperse 25 + Reactive 15 +Reactive 15(λ_{521}); Coomassie Brilliant R250 + Celestine+ Methylene+ Eriochrome (λ_{595}); Evans+ Martius+ Methyl + Naphthol(λ_{541}); Disperse 9+ Reactive 15+ Methyl+ Naphthol (λ_{571})].

We have earlier reported the potential of a novel *T. dioica* yarn yarn plant proteins in the decolourization of disperse dyes (Jamal et al., 2010). In this chapter, we intended to widen the Spectrum of industrially important textile dyes and included some well known non-textile dyes to study the extent of decolourization in the presence of Redox mediators.

The cost of an enzyme depends on its degree of purity hence we opted to use ammonium sulphate precipitated proteins from T. dioica to study dye decolourization/degradation.

The PGP was partially purified using 10Percent to 80Percent ammonium sulphate which retained a specific activity of 96U/mg of protein. The experiments were performed at different enzyme concentration, pH, temperature, incubation period and Redox mediators namely, HOBT and vanillin. An interesting data profile was obtained for the assessment of this enzyme suitability to treat wastewater contaminated with these dyes.

The reactive dyes underwent decolourization by the formation of precipitate which disappeared in the presence of 1.0 mM HOBT. This finding supports earlier reports that treatment of phenols and aromatic amines by peroxidases resulted in formation of large insoluble aggregates (Wada et al., 1995; Tatsumi et al., 1996; Husain and Jan, 2000; Duran and Esposito, 2000). The other dyes studied showed no formation of precipitate during decolourization in the presence of HOBT. This observation supports earlier view that decolourization of dyes took place via degradation of aromatic ring of the compounds or by cleaving certain functional groups (Akhtar et al, 2005a,b; Satar and Husain, 2009). All the other textile dyes showed insignificant or no decolourization with T. dioica preoxidase alone when studied under optimum conditions of dye decolourization in the absence of HOBT (data not shown).

The decolourization profile for all the dyes increased significantly at HOBT concentration of 0.2 mM. Upon increasing the HOBT concentration further to 1.5 mM decolourization increased marginally. The reactive dyes under study exhibited decolourization maximally at 1.0 mM HOBT whereas disperse dyes showed maximum decolourization at lower values of HOBT suggesting that reactive dyes are comparatively more resistant to degradation. Vanullin was able to decolourize both the reactive and disperse dyes at 1.0 mM concentration but the extent of decolourization was sufficiently poor in comparison to HOBT for certain dyes like Reactive blue 15, Reactive orange 15 and most of the disperse dyes presence of relatively higher concentration of vanillin.

The non-textile dyes exhibited remarkable decolourization in presence of both the redox mediators at 1.0 mM concentration. There was no significant change on dye decolourization of non-textile dyes at concentrations above 1.0 mM. It has already been reported that Redox mediators have the potential to mediate oxidation reaction between a substrate and an enzyme (d'Acunzo et al., 2006). Different Redox mediators have different mediation efficiency which is governed by Redox potential of the mediator and the oxidation mechanism of the substrate (Baiocco et al., 2003). Oxidation of substrate occurs by free radical formation by the mediator. The free radicals can be formed either by one-electorn oxidation of substrate or by abstraction of a proton from the substrate (Xu et al., 2001; Fabbrini et al., 2002). In this study, Redox-mediating property of two different compounds as peroxidase mediators was evaluated and extensive study on textile, non-textile and dye mixtures has been performed with HOBT.

Laccase have been used with Redox mediators to oxidize non-phenolic compounds (Bourbonnais and Paice, 1990). The mechanism of action of laccase mediator system has been extensively studied and it is used in the textile industry in the finishing process for indigo stained materials. Several workers have demonstrated that the use of Redox mediator system enhanced the rate of dye decolourization by several folds but these mediators were required in very high concentrations (Soares et al., 2001a,b; Claus et al., 2002). In the present study we have shown the decolourization of both textile and non-textile dyes as well as dye mixtures mediated by *T. dioica* peroxidase under the influence of low concentration of Redox mediators. The peroxidase reacted well to decolourize at concentration of 0.8 mM and 1.0 mM of H_2O_2 for disperse and reactive dyes respectively and remained substantially unaffected till 1.2 mM. Although the concentration of H_2O_2 greater than 0.75 mM acted as an inhibitor of peroxidase activity by irreversibly oxidizing the enzyme ferri-heme group essential for peroxidase activity (Satar and Husain, 2009), in this study we observed a relatively higher working concentration of H_2O_2. The enzyme in the presence of Redox mediators was much effective in performing the decolourization of the dyes, implying dye decolourization was a result of Redox mediated H_2O_2-dependent enzymatic reaction. Our results are consistent and very near to values reported earlier for maximum functional concentration of H_2O_2 (Camarero et al., 2005).The enzyme could function better in an acidic medium of pH range 3-6, whereas its decolourizing/degrading activity was adversely affected in alkaline medium. This finding supports the earlier view of disperse dye decolourization in acidic medium by *T. dioica* and reports further that acid medium favours catalysis of reactive dyes too by this peroxidase. It has earlier been reported that the degradation of industrially important dyes by enzymes such as horse radish peroxidase, polyphenol oxidase, BGP and laccase was also maximum in the buffers of acidic pH (Galindo et al., 2000).

The reaction temperature is an important parameter which effects the decolourization of dyes. The maximum decolourization for both the reactive and disperse dyes were in the temperature range of 40°C to 50°C in the presence of fixed concentration of HOBT. The extent of decolourization was remarkable for Reactive Blue 15, Reactive Orange 15, Reactive Red 4 at 50°C whereas disperse dyes could be decolourized in the range of 61.2Percent to 79Percent at 40°C. The decolourization varies with the nature of the dyes but Redox mediated decolourization with *T. dioica* peroxidase was a better solution for effective decolourization of recalcitrant compounds. The rate of decolourization varied with time and maximum decolourization in the presence of HOBT was observed within one hour of incubation for Reactive Blue15 and Disperse Red 19. However, for other dyes the extent of decolourization progressed slowly and reached a plateau after 4 h of incubation. This data is consistent with reports that decolourization rate varies, depending upon the type of dye to be treated (Camarero et al., 2005).

The non-textile dyes studied for decolourization with *T. dioica* peroxidase exhibited enhanced decolourization in the presence of 1.0 mM HOBT, whereas in the absence of Redox mediator decolourization was much slower. Coomassie Brilliant Blue R250, Methylene Blue, Eriochrome Black T, Martius Yellow, Methyl Orange, Naphthol Blue Black and Rhodamine were extensively decolourized by *T. dioica* peroxidase under the influence HOBT. The decolourization of Evans blue was not significantly affected by the presence of Redox mediator, although in the presence of HOBT decolourization achieved was higher. The performance of this system was maximal during 160 min of incubation. Celestine blue was decolourized to 79.5Percent in the presence of HOBT at higher concentration of enzyme as well as longer incubation time. The data is suggestive of *T. dioica* peroxidase in conjunction with low concentration of HOBT to be wonderful decolourization/degradation system for non-textile dyes as well. Further, dye mixtures simulating industrial effluents exhibited more than 82Percent decolourization with 1-hydroxybenzotriazole. The rate of decolourization of dye mixture was slower in comparison to that of individual dyes both in the presence and absence of HOBT. The application of inexpensive peroxidase from easily available source can overcome the limitations in current wastewater treatment strategies. The use of peroxidases can be extended to large-scale treatment of wide spectrum of structural dyes by using immobilized PGP along with relatively cheaper Redox mediators. Thus, the study demonstrated that the peroxidase enzyme isolated from *T. dioica* can be coupled with Redox mediator into a system that can serve as an effective biocatalyst for the treatment of effluents containing recalcitrant dyes from textile, dyeing and printing industries.

EFFECT OF PHOTOCHEMICAL ADVANCED OXIDATION PROCESSES ON THE YARN TEXTILE

Surfactants are frequently being used as cleaning, dissolving and wetting agents in household activities and several industries (Utsunomiya et al., 1997; Van de Plassche et al., 1999). For instance in various textile preparation operations (scouring, mercerising, bleaching), surfactant formulations are employed in order to allow thorough wetting of the textile yarn yarn material, emulsification of lipophilic impurities and dispersion of insoluble matter and degradation products (Arslan-Alaton et al., 2007; EU, 2003). Anionic (including alkyl sulphonates, alkyl aryl sulphonates, alkyl sulphates, dialkylsulphosuccinates, and others) and especially non-ionic surfactants are the chemicals being more often used for this particular purpose. These multi-purpose surfactants create the main organic pollution load in effluents originating from the activities. Surfactants enter the environment mainly through the discharge of sewage effluents into natural water and the application sewage sludge on land for soil fertilizing purposes (Petrovic and Barceló, 2004). Although most of these surfactants are designated as "biodegradable" according to different long-term biodegradability tests (Euratex, 2000) former studies dealing with the biodegradability of surfactants have indicated that their biodegradation in conventional activated sludge treatment systems is in most cases rather incomplete, resulting in the accumulation of partial biodegradation products

(Ikehata and El-Din, 2004; Swisher, 1987). A lot of commercial surfactants used today by different industries tend to sorb and hence accumulate on sludge and soil sediments in receiving water bodies due to their amphiphilic characteristics (Swisher, 1987; Staples et al., 2001). Hence, when a surfactant is discharged into the environment without proper treatment at source (*e.g.* at the treatment facilities of the textile factory), it will enter the sewage treatment works or directly natural waters without any significant structural alteration and may cause serious ecotoxicological consequences due to its bioaccumulation tendency (Ikehata and El-Din, 2004). Consequently, more effective and at the same time economically feasible treatment processes have to be applied to alleviate the problem of partially adsorbed and/or metabolized surfactants in the environment. Moreover, the management of biologically-difficult-to-degrade effluent discharges bearing surfactants remains an important challenge that has to be urgently solved to reduce the concentration of surfactants in effluent discharge at source. Aryl sulphonates are commercially important textile dye precursers; their global production for dye synthesis is limited to a few countries (*e.g.* China, India, Pakistan and Egypt) where they create serious environmental problems since they are biologically inert in natural as well as engineered aerobic biotreatment systems. Moreover, due to their very polar and hydrophilic nature, sulphonated textile dye precursers are very mobile in the aquatic ecosystem and hence very difficult to treat via conventional methods (*i.e.* coagulation, precipitation, adsorption). In addition, it has been postulated that the anaerobic metabolites of sulphonates bearing aromatic amines in their molecular structure are potentially mutagenic and even carcinogenic. From the environmental point of view, the fate of aryl sulphonates and their degradation products in biological treatment units as well as receiving water bodies is still not very clear since until now only limited attention has been paid towards their occurrence, fate and degradability in engineered systems as well as in the natural environment (Jandera et al., 2001). Due to the fact that conventional biological, physical and chemical treatment methods are not very effective in the removal of naphthalene sulphonates, alternative options have to be considered and evaluated.

In the last three decades, so-called Advanced Oxidation Processes (AOPs) have been developed and explored for the treatment of toxic and/or refractory pollutants and wastewaters.

Their efficiency and oxidation power relies on the formation of very active species including hydroxyl radicals (HO•) that violently and non-selectively react with most organic pollutants at diffusion limited rates. Bimolecular rate coefficients obtained for the free radical chain reaction initiated by HO• are typically in the range of 10^7-10^{11} M^{-1} s^{-1} for most pollutants depending on their structural properties. Among the AOPs, photochemical processs including the UV-C photolysis of H_2O_2 and the Photo-Fenton reaction (UV-C photolysis of

H_2O_2 catalyzed by Fe^{2+} ions under acidic pH conditions) are known as the most efficient, feasible and kinetically favourable types.

The major drawbacks of AOPs can be listed as:

- Their relatively high electrical energy requirements dramatically affecting their operating costs and hence real-scale application as well as
- The possibility of undesired oxidation products (more toxic and less bioamenable than the mother pollutant) formation and build-up due to kinetically difficult to control chain reactions in the treatment system. Baxendale and Wilson (1957) first studied the UV-C photolysis of H_2O_2 in water that is the most direct method of HO• generation through the homolytic cleavage of H_2O_2. The following set of reactions explains the main steps of the mechanism of the H_2O_2 photolytic decomposition in water;

$$H_2O_2^{\cdot} + h\nu \rightarrow 2HO^{\bullet}$$

$$HO^{\bullet} + H_2O_2 \rightarrow HO_2^{\bullet} + H_2O$$

$$HO^{\bullet} + HO_2^{-} \rightarrow HO_2^{\bullet} + OH^{-}$$

$$2HO^{\bullet} \rightarrow H_2O + \frac{1}{2}O_2$$

The quantum yield of reaction (4) is 1.0 mole HO• per mole of photons (1.0 einstein) absorbed by H_2O_2 (Prousek, 1996). The molar absorption coefficient of H_2O_2 at 254 nm is only 18.6 M^{-1} cm^{-1} and hence the efficiency of the H_2O_2/UV-C process decreases drastically with increasing pollutant concentration and UV_{254} absorbance of the target chemical (CCOT 1995). For heavily polluted effluents, high UV-C doses and/or H_2O_2 concentrations are required. The Fenton process is being increasingly used in the treatment of organic pollutants. The conventional dark Fenton process involves the use of one or more oxidizing agents (usually H_2O_2 and/or oxygen) and a catalyst (a metal salt or oxide, usually iron) at acidic pHs. Fenton process is efficient only in the pH range 2-4 and is usually most efficient at around pH 2.8 (Pignatello, 1992). In acidic water medium, the most accepted scheme of this radical mechanism is described in the following equations:

$$Fe^{2+} + H_2O_2 \rightarrow Fe^{3+} + HO^{-} + HO^{\bullet}$$

$$Fe^{3+} + H_2O_2 \rightarrow Fe^{2+} + HO_2^{\bullet} + H^{+}$$

$$Fe^{2+} + HO^{\bullet} \rightarrow Fe^{3+} + HO^{-}$$

$$HO^{\bullet} + H_2O_2 \rightarrow HO_2^{\bullet} + H_2O$$

$$HO_2^{\bullet} \leftrightarrow O_2^{\bullet -} + H^{+} \quad K = 1.58 \times 10^{-5} M$$

$$Fe^{3+} + HO_2^{\bullet} / O_2^{\bullet -} \rightarrow Fe^{2+} + O_2 / + H^+$$

$$Fe^{2+} + HO_2^{\bullet} / O_2^{\bullet -} \rightarrow Fe^{3+} + H_2O_2$$

Photo-Fenton oxidation is the photo-catalytically enhanced version of the Fenton process; in the Photo-Fenton process; UV light irradiation (180-400 nm) causes an increase in the HO• formation rate and efficiency via photoreduction of Fe^{3+} to Fe^{2+}, thus continuing the redox cycle as long as H_2O_2 is available (Pignatello, 1992; Wadley and Waite, 2004). The reason for the positive effect of irradiation on the degradation rate include the photoreduction of Fe^{3+} to Fe^{2+} ions, which produce new HO• with H_2O_2 (Eq. 6) according to the following mechanism;

$$Fe^{3+} + H_2O \xrightarrow{hv} Fe^{2+} + HO^{\bullet} + H^+$$

$$H_2O_2 \xrightarrow{hv} 2HO^{\bullet} \ (\lambda < 400\,nm)$$

The main compounds absorbing UV light in the Fenton system are ferric ion complexes, *e.g.* $[Fe^{3+}(OH)\text{-}]^{2+}$ and $[Fe^{3+}(RCO_2)\text{-}]^{2+}$, which produce additional Fe2+ by the following photoinduced, ligand-to-metal charge-transfer reactions (Sagawe et al., 2001);

$$\left[Fe^{3+}(OH)^-\right]^{2+} \rightarrow Fe^{2+} + HO^{\bullet} \ (< ca. 450\,nm)$$

$$\left[Fe^{3+}(RCO)^-\right]^{2+} \rightarrow Fe^{2+} + CO_2 + R^{\bullet} \ (< ca. 500\,nm)$$

Additionally, reaction (14) yields $HO^{\bullet}$, while reaction (15) results in a decrease of the total organic carbon (TOC) content of the system due to the decarboxylation of organic acid intermediates. It is very important to note that both reactions produce ferrous ions required for the Fenton reaction (Eq. 6). The overall degradation rate of organic compounds is considerably increased in the Photo-Fenton process, even at lower concentration of iron salts present in the system (Chen and Pignatello, 1997). The main disadvantage of the iron based AOPs is the necessity to work at low pH, because at higher pH ferric ions would begin to precipitate as hydroxide. Furthermore, depending on the iron concentration used, it has to be removed after the treatment in agreement with the regulation established for wastewater discharge (Rodríguez, 2003). Information about the toxicity of organic pollutants and advanced oxidation intermediates appears to be a key factor in order to evaluate the applicability of the AOPs. It should be considered that advanced oxidation may result in the formation of more inert and/or toxic degradation products, since free radical chain reaction-based treatment methods are difficult to control thus requiring the toxicity assessment of the formed advanced oxidation intermediates (Farré et al., 2007; Oller et al., 2007; González et al., 2007). Furthermore, in most cases the final goal of AOPs is to improve the biocompatibility of the effluent in order to apply a biological treatment (Scott and Ollis, 1995; Marco et al., 1997).

Consequently, the use of short-term biological screening tests that aim to a rapid estimation of the acute toxicity of a pollutant and its oxidation products during AOP applications has become essential. Until now, several studies investigating relative changes in the toxicity of model pollutants have been reported using photobacteria, cladocerans, microalgae and other test organisms (Wang et al., 2002; Dalzell et al., 2002; Stasinakis et al., 2008). Among the available acute toxicity test protocols, the assessment of respirometric inhibition (decrease in oxygen uptake rate of microorganisms relative to a control sample; *e.g.* synthetic or real sewage) is an attractive and advantageous option since heterotrophic biomass being present in biological treatment systems are used as the toxicity test organism that enables the simultaneous assessment of acute toxicity and biodegradability by a single test protocol (Paixão et al., 2002; Gendig et al., 2003; Henriques et al., 2005; Stasinakis et al., 2008; Olmez-Hanci et al., 2009). Bearing the facts in mind, the present work aimed at investigating the photochemical advanced oxidation of a commercially important anionic textile surfactant (dodecyl sulpho succinate; DOS) and a frequently used vinyl sulphone dye intermediate, namely Para base (aniline-4-beta-ethyl sulphonyl sulphate-2-sulphonic acid; PB) employing H_2O_2/UV-C and Photo-Fenton treatment processes. Treatment performances of the H_2O_2/UV-C and Photo-Fenton processes were evaluated in terms of DOS, PB, COD and TOC removals. Treatment efficiencies, pseudo-first-order reaction rate coefficients and associated electrical energy requirements were comparatively evaluated for DOS and PB degradation with the H_2O_2/UV-C and Photo-Fenton processes to decide for the technically and economically more attractive treatment solution for each chemical. In the second part of the study, activated sludge inhibition experiments were conducted with untreated and photochemically pretreated DOS and PB samples in order to examine the inhibitory effect of the model pollutant and advanced oxidation intermediates on the respirometric activity of activated sludge microorganisms.

MATERIALS AND METHODS

CHEMICALS AND REAGENTS

DOS and PB model compounds were supplied by a local textile chemical company and used as received without further purification. Some physicochemical and environmental characteristics of the studied model compounds were summarized in Table.

35Percent w/w H_2O_2 (Fluka, USA) was used as received without any dilution. Residual H_2O_2 was destroyed with enzyme catalase derived from *Micrococcus lysodeikticus* (100181 U mL^{-1}, Fluka, USA). The ferrous iron catalyst source was prepared for daily use by dissolving $FeSO_4.7H_2O$ (Fluka, USA) in distilled water to obtain a 10Percent (w/v) stock solution. Several concentrations of HNO_3 (Merck, Germany) and NaOH (Merck, Germany) solutions were used for pH adjustment. HPLC-grade acetonitrile (Merck, Germany) was used as the mobile phase in the HPLC measurements. All other reagents were analytical grade and used without purification.

Table. Physicochemical Properties of the Anionic Surfactant (DOS) and the Vinyl Sulphone Dye Precursor (PB). The COD and TOC Equivalencies shown in Table were Experimentally obtained by Preparing Individual Calibration Curves for Varying DOS and PB Concentrations in Water and Determination of the Corresponding COD and TOC Values.

Pollutant Property	DOS	PB
Molecular Formula	$C_{20}H_{37}NaO_7S$	$C_8H_{11}O_9S_3$
Molecular Weight(g mol^{-1})	444	347
g COD(g chemical)$^{-1}$	1.54	1.04
g TOC(g chemical)$^{-1}$	0.69	0.28
Molecular Structure	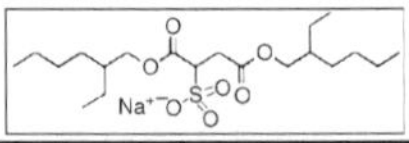	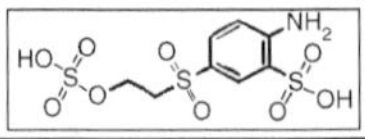

UV-C PHOTOREACTOR AND LIGHT SOURCE

All photochemical treatment experiments were performed using a 3250 mL-capacity batch stainless steel batch reactor (length=84.5 cm; width=8 cm) equipped with a 40W lowpressure, mercury vapor sterilization lamp that was located at the center of the reactor in a quartz glass envelope.

The incident light flux of the UV lamp and effective light path length were determined via H_2O_2 actinometry (Nicole et al., 1990) as 1.6 x 10^{-5} einstein L^{-1} s^{-1} and 5.1 cm, respectively, at 254 nm.

During the photochemical experiments, the reaction solution was continuously circulated through the UV-C photoreactor using a peristaltic pump at a rate of 400 mL min^{-1}.

For the H_2O_2/UV-C experiments, H_2O_2 was added to the pH-adjusted aqueous DOS and PB solutions and the reaction mixture was fed to the photoreactor via a peristaltic pump. After the sample t=0 was taken, the reaction was initiated by turning on the UV-C lamp.

In case of Photo-Fenton oxidation experiments, after adjusting the initial pH of the reaction solutions to 3.0 ± 0.1, *e.g.* the optimum pH for Photo-Fenton oxidation (Oliveros et al., 1997), Fe^{2+} catalyst was added. In order to eliminate the effect of the dark Fenton reaction, the other reactant (H_2O_2) was added to the reaction mixture at the very end of the suction period.

As the whole solution was fed into photoreactor, UV-C lamp was turned on and photo-oxidation was started.

For photochemical oxidation experiments, samples were taken at regular time intervals for analyses up to 180 min.

EXPERIMENTAL APPROACH

For all photochemical oxidation experiments aqueous solutions of DOS and PB were individually prepared and adjusted to an initial COD of 450 mg L^{-1} since a typical textile preparation and textile dye synthesis wastewater exerts a COD in the range of 400-500 mg L^{-1} (Zollinger, 2001). It is known that an optimum H_2O_2 concentration exists for the H_2O_2/UV-C treatment system depending upon the reaction pH, initial organic carbon content, type and molecular structure of the pollutant under investigation (Muruganandham and Swaminathan, 2004; Arslan-Alaton and Erdinc, 2006). In our preliminary experiments the H_2O_2 concentrations (30 mM) were initially selected close to the stoichiometric oxygen equivalent of the aqueous solutions of DOS and PB (=COD × 2.12 g H_2O_2/g O_2 ≈ 28 mM H_2O_2). For H_2O_2/UV-C treatment of DOS at 30 mM H_2O_2 concentration resulted in significant degradation both in DOS and organic carbon content. However, the theoretically established H_2O_2 concentration of 30 mM did not result in significant organic carbon removals for PB (data not shown). Considering our preliminary results, it was decided to increase the initial H_2O_2 concentration to 60 mM for H_2O_2/UV-C treatment of PB. The initial reaction pH was selected as 10.5 for H_2O_2/UV-C treatment of aqueous DOS solutions, since the effluent pH of a typical textile preparation is in the range 10–11 (EU, 2003), and in previous studies it was observed that the H_2O_2/UV-C oxidation process was practically pH-independent, and no significant changes in the reaction rate constants were found over a wide pH range of 3-11 (Arslan-Alaton and Erdinc, 2006; Arslan-Alaton et al., 2007). H_2O_2/UV-C treatment of aqueous PB solutions was investigated at a reaction pH of 3.0. For Photo-Fenton experiments, H_2O_2 (30 mM) and Fe^{2+} (0.5 mM) concentrations were selected upon consideration of optimum Fe^{2+}: H_2O_2 molar ratios obtained in former related research work and our own experiences in similar case studies (Arslan-Alaton et al., 2009a; 2009b; 2010). Photo-Fenton experiments were conducted at a fixed pH value of 3.0 ± 0.1 that is generally accepted as the optimum pH for iron-based AOPs (Pignatello, 1992). As being expected for photochemical AOPs (Oppenländer, 2003), the degradation of the original pollutants (in our case DOS and PB) is usually faster than that of their oxidation products, which are typically expressed in terms of the collective parameters COD and TOC. It was also taken into account that treatment conditions may vary for the removal of original pollutant and its degradation products that are usually collectively presented with the environmental sum parameters COD and TOC. Thus in the present study, the process efficiencies were assessed both in terms of DOS and PB, COD and TOC abatement rates. Residual (unreacted) H_2O_2 and pH were also recorded during the entire treatment period. The pH and ionic strength of the synthetic solutions were not controlled throughout the experiments. All experiments were conducted at a constant temperature of 20°C. Information about the toxicity of degradation intermediates as well as original pollutants appears to be a key issue for the application of AOPs in water and wastewater treatment as well as integrated chemical and biochemical treatment options (Scott and Ollis, 1995; Farré et al., 2007). By considering this important issue in the last stage of the experimental work, relative changes in the inhibitory effect of DOS and PB photochemical degradation intermediates on heterotrophic biomass (sewage sludge) during the application of photochemical oxidation were also examined.

ACTIVATED SLUDGE INHIBITION EXPERIMENTS

The procedure of the activated sludge inhibition test was based on the test protocol given in the ISO 8192 Method (2007). This method describes the way to assess the inhibitory effects of a test substance on oxygen consumption of activated sludge (resembling sewage) by measuring the respiration rate under defined conditions in the presence of a known biodegradable substrate and different concentrations of the test substance (Ubay Cokgor et al., 2007). The activated sludge used as the heterotrophic biomass was obtained from a municipal tertiary wastewater treatment located in Istanbul and daily fed with synthetic sewage in accordance with the same test protocol (ISO 8192, 2007).The heterotrophic biomass concentration in the activated sludge was determined according to Standard Methods (APHA/AWWA/WEF, 2005) and expressed as mg L^{-1} of mixed liquid volatile suspended solids (MLVSS). All experiments were run at constant heterotrophic active biomass concentration of 1500 mg L^{-1} to obtain appropriate oxygen uptake rates (OUR) of around 100 mg L^{-1} h^{-1}.The COD of the synthetic sewage was adjusted to 480 mg L^{-1} and used as the "readily biodegradable substrate" in the activated sludge inhibition experiments. Any unreacted H_2O_2 remaining in the reaction samples was destroyed with catalase enzyme and the pH of the solutions was adjusted to 7.0±0.2 before conducting the toxicity tests in order to eliminate potential inhibitory effects of hydrogen peroxide and pH, respectively. The blank control was prepared using the same amount of activated sludge and synthetic sewage solution as in the test dilutions, but without adding untreated or treated aqueous DOS and PB samples. The decrease in dissolved oxygen concentration in the blank control and reaction samples was periodically monitored using a WTW InolabOxi Level 2 oxygen meter at different incubation times (15-180 min). OUR values were calculated on the basis of the linear part of the decreasing dissolved oxygen concentration curves versus test time. Percent inhibition of oxygen consumption (IOUR), for every test sample was calculated using the following equation;

$$I_{OUR}(\%) = \left[(R_B - R_T)\times 100\right]/R_B$$

where RB stands for the OUR in the sample blank and RT is the OUR in the sample effluent mixture, respectively The IOUR values were, thereafter, plotted against the logarithm of the DOS and PB concentrations. The DOS and PB concentrations resulting in percent decrease in OUR's (in mg L^{-1} DOS or PB) after 15 min incubation period was eventually calculated by interpolation of the "log DOS" or "log PB" versus percent "IOUR" plots obtained for different DOS and PB concentrations.

ANALYTICAL PROCEDURES

Reaction samples were analyzed for DOS-PB, COD, TOC, residual H_2O_2 and pH before, during and after photochemical treatment. CODs were determined by the closed reflux titrimetric method according to ISO 6060 (1989). In order to prevent the positive interference of H_2O_2 to COD analyses, the pH of each sample solution was adjusted to 6.5-7.5 and thereafter catalase enzyme made from *Micrococcus lysodeikticus* (1AU destroys 1 μmol H_2O_2 at pH=7, 100181 U mL^{-1}, Fluka grade) was added to destroy any residual H_2O_2. TOC was monitored on a Shimadzu VPCN model carbon analyzer (combustion method) equipped with an autosampler. Residual H_2O_2 in the samples was determined titrimetically by employing the molibdate-catalyzed iodometric method (Official Methods of Analysis, 1980). The amount of DOS and PB in the aqueous solutions was measured by high-performance liquid chromatography (HPLC, Agilent 1100 Series, USA). DOS was determined with a fluorescence detector (FLD; λ_{ex} = 225 nm, λ_{em} = 295 nm) and a C8 column. 3 mM NaCl/ CH_3CN (80/20; v/v) served as the mobile phase (flow rate of 1.5 mL min^{-1}) for DOS. The instrument detection limit for DOS (1.5 mg L^{-1}) was determined as the lowest injected standard that gave a signal-to-noise ratio of at least 3 and an accuracy of 80-95Percent. PB abatement was monitored with a diode array detector (DAD; λ = 276 nm) and a Novapack C18 (Waters, USA) column. CH_3CN/H_2O (60/40; v/v) was used as mobile phase for the analysis of PB at a flow rate of 1 mL min^{-1}. The column temperature was set as 30°C for all measurements and the injection volume was selected as 30 μL. The instrument detection limit (1.5 mg L^{-1}) for PB was determined as the lowest injected standard that gave a signal-to-noise ratio of at least 3 and an accuracy of 80-95Percent. The limit of quantification was calculated as 10 times of the signal-to-noise ratio as 5 mg L^{-1}. In case of Photo-Fenton Fe^{2+}/H_2O_2/UV-C experiments, all analytical measurements were performed after filtration of the treated samples though a 0.45 μm membrane filter (Sartorius) in order to remove the settled $Fe(OH)_3$ sludge. The pH was measured with a Thermo Orion model 720 pH-meter at any stage of the experiments.

KINETIC ANALYSIS

DOS and PB abatements during H_2O_2/UV-C and Photo-Fenton treatment processes followed pseudo-first order kinetics with respect to the model pollutant (DOS and PB).

$$\ln C/C_o = - k_G \times t$$

where k_C is the pseudo-first order rate constant (in min^{-1}), t stands for photochemical treatment time (in min), Co and C are the initial and final concentrations of model pollutant (DOS and PB, in mg L^{-1}).

ELECTRICAL ENERGY REQUIREMENTS OF THE PHOTOCHEMICAL TREATMENT PROCESSES

The photodegradation of aqueous organic pollutants is an energy-intensive process, thus consideration of figures-of-merit definition based on the electrical energy input is appropriate. In order to compare treatment efficiencies for photochemical oxidation processes, it is important to use a general expression and relate the treatment efficiencies directly to operating costs. For this purpose, the electrical energy per order (EE/O) defined as the kWh electrical energy required to degrade a contaminant or group parameter by one order of magnitude in m^3 of contaminant water or wastewater, can be calculated for treatment process (Bolton et al., 2001). Since the EE/O (kWh m^{-3} $order^{-1}$) concept assumes pseudo-first order degradation kinetics with respect to pollutant concentration can be determined from the following formula;

$$EE/O(kWh\ m^{-3} order^{-1}) = (38.4 \times P) / (V \times k_G)$$

where P is the lamp power (0.04 kW in the present work), V is the reactor volume (3.25 L in the present work), and k_C is the pseudo-first-order rate constant of DOS and PB degradation (min^{-1}).

RESULTS AND DISCUSSION

ACTIVATED SLUDGE INHIBITION EXPERIMENTS WITH UNTREATED DOS AND PB

The ISO 8192 (2007) specifies a method for assessing the inhibitory or stimulatory effects of substances, mixtures or wastewaters to activated sludge (Paixão et al., 2002). Information generated by this method may be helpful in estimating the effect of a test yarn yarn material on bacterial communities in the aquatic environment, especially in aerobic biological treatment systems (Paixão et al., 2002). In the first part of this experimental study, untreated (original) aqueous DOS and PB solutions were individually exposed to activated sludge inhibition experiments. The obtained semi-logarithmic I_{OUR} (in Percent) versus DOS and PB concentrations (log DOS and log PB) were presented in Figures, respectively for an incubation period of 15 min.

It is evident that increasing the DOS concentration resulted in an accelerated inhibition of the OUR value measured relative to the sample blank (synthetic sewage). The I_{OUR} values ranged between 4Percent-53Percent from the lowest to highest studied concentrations, respectively, enabling the determination of EC_{30} and EC_{50} values (effective concentrations causing 30 and 50Percent inhibition in heterotrophic biomass relative to the blank) for DOS. These values were established as 338 and 544 mg L^{-1} corresponding to the respirometric inhibition levels of 30Percent and 50Percent, respectively. This information implied that DOS is an acutely toxic compound at concentrations encountered in the textile preparation process. Hence its treatment after discharge becomes essential.

On the other hand, although the studied concentration range for PB is appreciably higher, activated sludge inhibition rates always remained below 20Percent. From the obtained experimental results it can be concluded that PB is not exhibiting an inhibitory/toxic effect on heterotrophic biomass relative to synthetic sewage.

However, it should be noted that the inert COD content of PB creates a significant effluent discharge penalty problem for dye synthesis effluent. For both chemicals it is obvious that efficient treatment employing advanced oxidation process is necessary to comply with the environmental regulations and to minimize pollution loads.

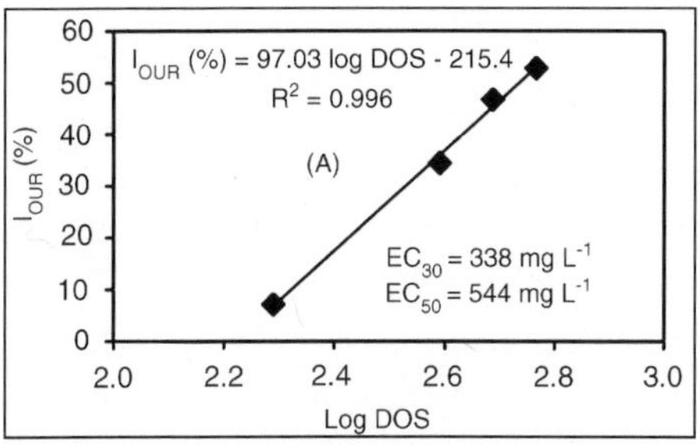

faf'lf';f

Fig. Log(Concentration) Versus I_{OUR} (Percent) Curves Obtained from the Activated Sludge Inhibition Experiments for Aqueous DOS (a) and PB (b). DOS Concentration Range: 97-584 mg L^{-1}; PB Concentration range: 144-721 mg L^{-1}. The Selected Incubation Time is 15 min.

H_2O_2/UV-C AND PHOTO-FENTON TREATMENT OF DOS AND PB

Figure presents the H_2O_2/UV-C and Photo-Fenton treatment results for aqueous DOS (a, H_2O_2/UV-C; b, Photo-Fenton) and PB (c, H_2O_2/UV-C; d, Photo-Fenton) solutions in terms of DOS-PB, COD and TOC abatement rates as a function of photochemical treatment time. In the Photo-Fenton experiments all experimental conditions were kept identical with the exception of 0.5 mM Fe^{2+} addition to the reaction solution.

The initial COD of the DOS-PB solutions were selected as 450 mg L^{-1} in these experiments, which is an average, typical COD for effluents containing these model compounds as the main organic pollutant. As it can be seen from Figures complete DOS removal was obtained after 60 min for both treatment processes. In terms of the COD parameter, 84Percent and 92Percent removals were achieved after 100 min H_2O_2/UV-C and Photo-Fenton treatment, respectively. TOC removals were also close to another; these were obtained as 89Percent after 100 min H_2O_2/UV-C treatment and 90Percent at the end of 100 min Photo-Fenton treatment.

From these findings it can be concluded that DOS, COD and TOC removal efficiencies were not significantly different or higher for the Photo-Fenton process as compared to H_2O_2/UV-C treatment. Considering that the addition of Fe^{2+} as a photocatalyst and extra chemical as well as the requirement of acidic reaction conditions for iron-based AOPs will affect the operating costs for DOS treatment considerably, it was decided to keep the H_2O_2/UV-C process for DOS treatment.

Kinetic rate coefficients and electrical energy calculations shown in Table and explained later on supported this argument. Figures depict the results of H_2O_2/UV-C and Photo-Fenton treatment of aqueous PB solution, respectively. As expected, the addition of 0.5 mM Fe^{2+} had a significant positive effect in terms of PB as well as COD and TOC abatement rates as compared with the H_2O_2/UV-C process.

Total PB elimination was realized only after 30 min Photo-Fenton treatment, whereas at least 180 min were needed when PB was subjected to H_2O_2/UV-C treatment. Moreover, COD and TOC were only poorly removed via mere H_2O_2/UV-C treatment.

For instance, parallel to the slow COD abatement, TOC practically remained unchanged indicating that PB, the mother compound, was converted to different advanced oxidation intermediates but not ultimately oxidized to mineralization end products in the absence of Fe^{2+} ions. At the end of the 120 min treatment period, COD and TOC removals were obtained as 49Percent (H_2O_2/UV-C treatment) and 98Percent (Photo-Fenton treatment) as well as 33Percent (H_2O_2/UV-C treatment) and 100Percent (Photo-Fenton treatment), respectively. Considering that H_2O_2/UVC treatment of PB resulted in very poor and insufficient results, Photo-Fenton oxidation appears by far the better option for the efficient treatment of PB.

The results obtained for DOS and PB could be partially attributed to the high absorbance (optical density) of PB at the maximum emission band of the UV-C light source, hindering its effective absorption by H_2O_2, whereas DOS did not compete with H_2O_2 for UV-C light absorption at 254 nm wavelength that is required for the production of HO• radicals via UV-C. Table summarizes the calculated kC and EE/O values for the photochemical treatment of DOS and PB in the absence and presence of Fe^{2+} catalyst. From Table it is obvious that for DOS treatment, Fe^{2+} is not essential, whereas the positive effect of the catalyst was very pronounced for PB abatement. Hence, in the forthcoming experiments where changes in the acute toxicity of DOS and PB during photochemical treatment were examined, results were presented only for the selected AOPs (*e.g.* H_2O_2/UV-C for DOS and Photo-Fenton for PB model pollutants).

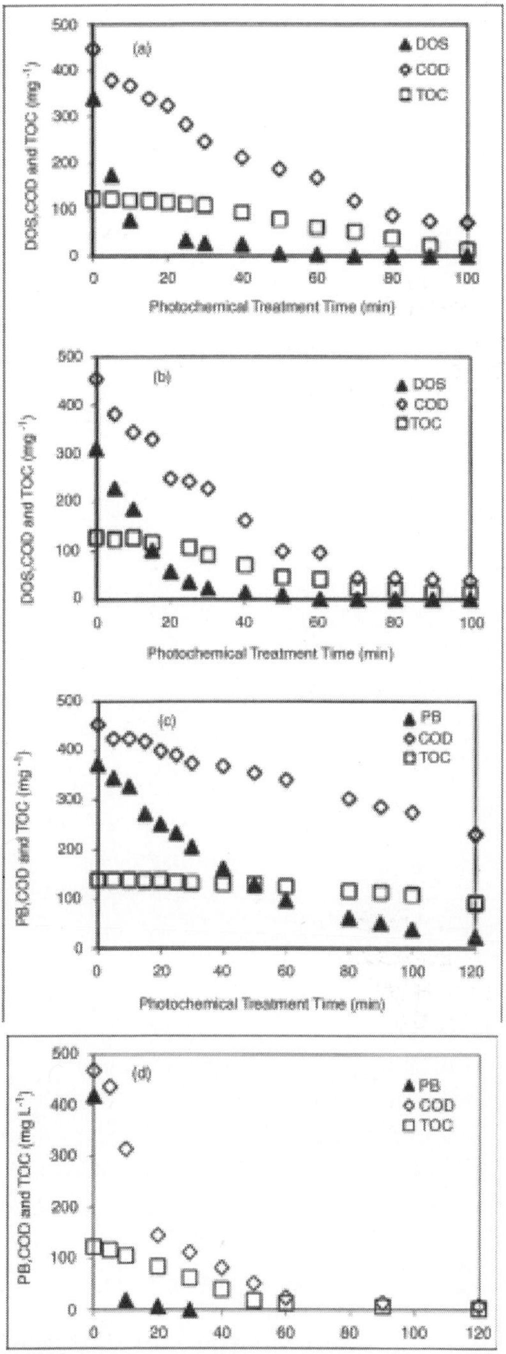

Fig. DOS, PB, COD and TOC Abatement Rates Observed During Photochemical Treatment of DOS (a, H_2O_2/UV-C; b, Photo-Fenton) and PB (c, H_2O_2/UV-C; d, Photo-Fenton). For H_2O_2/UV-C Experiments: Initial COD= 450 mg L^{-1}; H_2O_2 = 30 mM (DOS) and 60 mM (PB); pH = 10.5 (DOS) and 3.0 (PB). For Fe^{2+}/H_2O_2/UV-C Experiments: Initial COD= 450 mg L^{-1}; H_2O_2 = 30 mM; Fe^{2+} = 0.5 mM; pH = 3.0.

ACTIVATED SLUDGE INHIBITION RESULTS FOR PHOTOCHEMICALLY TREATED DOS AND PB

The possibility exists that during the applicability of AOPs, more toxic and/or less bioamenable intermediates and/or end products form and accumulate in the reaction solution. In order to evaluate the acute toxicity and bioamenability of photochemical degradation products towards heterotrophic biomass, activated sludge inhibition experiments were conducted during H_2O_2/UV-C of DOS and Photo-Fenton oxidation of PB. For the activated sludge experiments the OUR of the blank samples (containing only synthetic sewage) was adjusted to around 100 mg L^{-1} h^{-1}. A slight increase in the OUR values is observed during photochemical treatment of DOS starting with 71 mg L^{-1} h^{-1} and reaching its highest value after 40 min photochemical oxidation. Thereafter, the OUR value dropped down even below its original value indicating that extended oxidation did not improve but reduce the quality of DOS bearing effluent most probably due to the formation of less biodegradable and hence more inhibitory end products. The trend observed for the OUR values of the PB samples exposed to Photo-Fenton was different; the OUR values increased throughout the entire treatment period from 98 mg L^{-1} h^{-1} at the beginnig up to 120 mg L^{-1} h^{-1} after 80 min Photo-Fenton treatment. These results demonstrate that PB does not cause any inibitory effect of heterotrophic biomass and the treated reaction solution was even accepted as an extra carbon source for the bioculture as advanced oxidation progressed. The OUR results implied that no acutely toxic/ bioinhibitory advanced oxidation products were formed during Photo-Fenton treatment of PB.

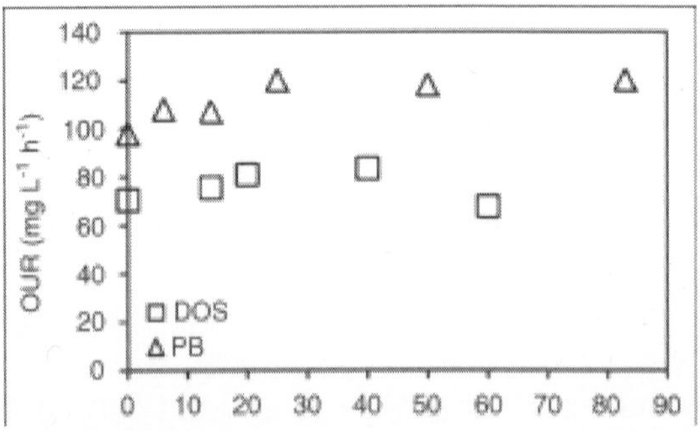

Fig. OUR Values Obtained During Photochemical Treatment of DOS via H_2O_2/UV-C (H_2O_2=42 mM; pH=10.5) and PB via Photo-Fenton (Fe^{2+}= 0.72 mM; H_2O_2=40 mM; pH=3.0) Processes. Experimental Conditions: Initial COD = 450 mg L^{-1}; Incubation period = 15 min; OURblank for DOS = 92 mg L^{-1} h^{-1}; OURblank for PB = 118 mg L^{-1} h^{-1}.

CONCLUSION

The present study aimed at exploring the effect of photochemical advanced oxidation processes (H_2O_2/UV-C and Photo-Fenton treatment) on the treatability and bioamenability of two frequently used industrial chemicals; *i.e.*

an anionic textile surfactant (called DOS herein) and a vinylsuphone dye precursor (called PB herein)

The following experimental results were obtained in this work:

- The anionic surfactant DOS exhibited a moderate acute toxicity towards heterotrophic biomass; the respirometric inhibition rate varied between 4Percent to 53Percent for the lowest and highest studied DOS concentrations, respectively. Accordingly, EC_{30} and EC_{50} values were established as 338 mg L^{-1} and 544 mg L^{-1}, respectively, for the textile preparation surfactant.
- For the textile dye precursor PB, on the other hand, relative activated sludge inhibition rates always remained below 20Percent and hence no EC values could be calculated for this model pollutant. Conclusively, PB did not exhibit any toxic effect on activated sludge microorganisms in the studied concentration range.
- DOS, a relatively simply structured, aliphatic compound, could be effectively treated via both H_2O_2/UV-C and Photo-Fenton processes. The difference in the overall removal efficiencies and reaction rate coefficient was not significant and hence it appeared that mere H_2O_2/UV-C treatment in the absence of Fe^{2+} photocatalyst was sufficient for complete DOS and associated organic carbon content abatements. The main advantage of the H_2O_2/UV-C treatment process over Photo-Fenton treatment is its robustness towards reaction pH below pH = 11 hence not requiring a pH adjustment to acidic values.
- For aromatic compounds that strongly absorp UV-C light at 254 nm thus hindering its efficient use by H_2O_2 for HO• production, Photo-Fenton treatment improves the degradation rate considerably. PB abatement via H_2O_2/UV-C treatment process resulted in relatively slow and incomplete degradation rates. In the presence of Fe^{2+}, however, PB degradation was enhanced by a factor of 10 and its removal was complete after 30-40 min Photo-Fenton treatment. These results suggest that the H_2O_2/UV-C process is not capable of effectively degrading the complicated aromatic structure of PB and under these circumstances; Photo-Fenton treatment is the treatment process of choice.

According to the activated sludge inhibition tests no toxic/bioinhibitory degradation intermediates and/or oxidation end products were formed during and after photochemical treatment of DOS (60 min H_2O_2/UV-C treatment at an initial pH of 10.5 with 42 mM H_2O_2) and PB (Photo-Fenton treatment for 80 min at pH 3; in the presence of 0.72 mM $Fe^{2+}$and 40 mM H_2O_2). The obtained findings indicated that photochemical advanced oxidation processes are potential, promising options for the effective treatment of refractory and toxic industrial pollutants.

PILOT YARN YARN PLANT EXPERIENCES: TREATMENT STEPS FOR THE BIODEGRADATION OF YARN

Considering both the volume and the effluent composition, the textile industry wastewater is rated as the most polluting among all industrial sectors. Important pollutants are present in textile effluents; they are mainly recalcitrant organics, colour, toxicants and inhibitory compounds (Khelifi et al., 2008). Textile industries however, have caused serious environmental problems because of the wastewater produced. Most textile industries produce wastewater with relatively high BOD, COD, suspended solids and colour. The wastewater may also contain heavy metals depending on the type of colouring substances used. In general, the objective of textile industry wastewater treatment to reduce the level of organic pollutants, heavy metal, suspended solids and colour before discharge into the river. Colouring substances are used for dyeing and printing processes.

The wastewater from these two processes is the most polluted liquid waste in a textile industry. Biological, chemical, physical or the combination of the three treatment technologies can be used to treat textile industry liquid waste (Suwardiyno and Wenten, 2005). It has been proven that some of these dyes and/or products are carcinogens and mutagens (Manu and Chaudhari 2003). A part from the aesthetic deterioration of the natural water bodies, dyes also cause harm to the flora and fauna in the natural environment (Kornaros and Lyberatos 2006). So, textile wastewater containing dyes must be treated before their discharge into the environment (Forgas et al., 2004). Numerous processes have been proposed for the treatment of coloured waste water *e.g.*, precipitation, flocculation, coagulation, adsorption and wet oxidation (Hongman et al., 2004; Thomas et al., 2006). All these methods have different colour removal capabilities, capital costs and operating speed. Among these methods coagulation and adsorption are the commonly used; however, they create huge amounts of sludge which become a pollutant on its own creating disposal problems (Nyanhongo et al., 2002). Among low cost, viable alternatives, available for effluent treatment and decolourization, the biological systems are recognised, by their capacity to reduce biochemical oxygen yarn yarn demand (BOD) and chemical oxygen yarn yarn demand (COD) by conventional aerobic biodegradation (Forgas et al., 2004; Kornaros and Lyberatos 2006; Balan and Monteiro, 2001). Work on the use of combined bacterial process to treat textile wastewater has been carried out over the years by many research groups. Recent study has used the combination of anaerobic and aerobic steps in an attempt to achieve not only decolourization but also mineralization of dyes (Forgas et al., 2004; Ong et al., 2005). Aerobic processes have been recently used for the treatment of textile wastewater as standalone processes (Khelifi et al., 2008) and it is confirmed that they are efficient, costeffective for smaller molecules and that the aerobic reactor is an effective technique to treat industrial wastewater (Coughlin et al., 2002; Coughlin et al., 2003; Buitron et al., 2004; Ge et al., 2004; Sandhaya et al., 2005; Steffan et al., 2005; Sudarjanto et al., 2006). The aerobic reactor has the advantage of being a closed and comparatively homogeneous and stable ecosystem. Since little is known about this ecosystem, a molecular inventory is the first step to describe this dynamic bacterial community without cultivation (Godon et al., 1997). In order to better understand the functions of the bacterial community, a yarn yarn full description of the bacterial ecosystem is required (Bouallagui et al., 2004). Acquisition of DNA sequences is now a fundamental component of most phylogenetic, phylogeographic and molecular ecological studies. Single-strand conformation polymorphism (SSCP) offer a simple, inexpensive and sensitive method for detecting whether or not DNA fragments are identical in sequence, and so can greatly reduce the amount of sequencing necessary (Sunnucks et al., 2000). SSCP can be applied without any a priori information on the species and then can give a more objective view of the bacterial community. SSCP has been applied to study microbial communities from several habits including water, compost and anaerobic digesters (Duthoit et al., 2003; Bouallagui et al., 2004). In this research, we used the mixture design in the experimental design (Minitab 14.0) to optimize the formulation of the predominant strains isolated from textile waste water yarn yarn plant. After biodegradation, the Chemical Oxygen Yarn yarn demand (COD) and percentage of decolourization were measured. The relationships between the different combinations and products were analyzed by Minitab to select the optimal bacterial combination and to investigate the aerobic degradability of a textile industry wastewater in Tunisia by an aerobic Stirred Bed Reactor (SBR).

MATERIALS AND METHODS

MATERIALS

The microbial strains were microcapsules of *Sphingomonas paucimobilis* (14×10^7 cfu), *Bacillus sp*. (4.2×10^8 cfu) and *Staphylococcus epidermidis* (7×109 cfu), which were isolated from textile Waste Water yarn yarn plant in KsarHellal, Tunisia. *Sphingomonas paucimobilis*, *Staphylococcus epidermidis* and *Bacillus sp*. were isolated in previous works of Ayed et al.2009a,b and Ayed et al 2010a,b,c with the ability of degrading azo and triphenylmethane dyes (Congo Red, Methyl Red, Methyl Orange, Malachite Green, Phenol Red, Fushin, Methyl Green and Crystal Violet). All chemicals used were of the highest purity available and of analytical grade.

NUTRIENT AGAR PREPARATION

Nutrient agar was used as the growth medium for microbial isolation. For this purpose, 28 g of nutrient agar was dissolved in 1 l of distilled water, and was then autoclaved at 121°C for 20 min. After autoclaving, the agar was left to cool at room temperature for 15 min, and it was then poured out into Sterilin © disposable Petri dishes.

MICROBIAL STRAIN

The culture was cultivated and maintained by weekly transfers on to nutrient agar slants. For production experiments, the culture was revived in nutrient broth (pH 7.0) and freshly prepared 3 hold culture (λ600 nm= 1) prepared in Mineral Salt Medium (MSM) at 37°C, 150 rpm (New Brunswick Scientific Shaker, Edison, NJ) was used as the inoculum. The used medium was composed in 1000 ml of distilled water: glucose (1250 mg/l), yeast extract (3000 mg/l), $MgSO_4$ (100 mg/l); $(NH_4)_2SO_4$ (600 mg/l); NaCl (500 mg/l); K_2HPO_4 (1360 mg/l); $CaCl_2$ (20 mg/l); $MnSO_4$ (1.1 mg/l); $ZnSO_4$ (0.2 mg/l); $CuSO_4$ (0.2 mg/l); $FeSO_4$ (0.14 mg/l) and it was maintained at a constant pH of 7 by the addition of phosphate buffer (Ayed et al., 2010a,b,c).

ACCLIMATIZATION

The acclimatization was performed by gradually exposing *Sphingomonas paucimobilis*, *Bacillus sp*. and *Staphylococcus epidermidis* to the higher concentrations of effluent (Kalme et al., 2006). This bacteria were grown for 24 h at 30°C in 250 ml Erlenmeyer flasks containing in g/l yeast extract (3.0) and glucose (1.25) (pH 7.0). During the investigation, nutrient broth concentration was decreased from 90Percent (w/v) to 0Percent (w/v) and finally the organism was provided with effluent as sole source of nutrient. Acclimatization experiments were carried out at optimum temperature (Kalme et al., 2006).

OPERATIONAL CONDITIONS OF LABORATORY BIOREACTORS

A laboratory scale aerobic bioprocess was used in this study. The aerobic system used was SBR bioreactor. The system was operated continuously at a constant temperature of 30°C using an external water bath. A continuous stirred tank reactor with a 500 ml working volume was used. Mixing was assured by the continuous rotation of the magnetic stirrer. The system was first inoculated with a microbial consortia (*Sphingomonas paucimobilis*, *Bacillus sp*. and *Staphylococcus epidermidis*) obtained from a textile wastewater treatment yarn yarn plant. These inocula were selected because of the large variety of microorganisms that could be found in the biomass degrading dyes in textile wastewater, and because mixed cultures offer considerable advantages over the use of pure culture. In fact, individual strains may attack the dye molecules at different position or may use decomposition products produced by another strains for further decomposition. In fact, it is mentioned that adaptation is important for successful decolourization, and as acclimation occurred, the decolourization time becomes constant (Buitron and Quezada Moreno, 2004).

The system was fed by a peristaltic pump with the textile effluent obtained from textile wastewater yarn yarn plant in Ksar Hellal (Tunisia), and its pH was maintained at approximately 7. Air was provided from the bottom of the aeration of the combined bacterial process using diffusers and an air pump. Bioreactors operating conditions were (COD: 1700 (mg O_2/l); BOD_5: 400 (mg O_2/l); Colour: 3600 (U.C); pH: 7; MES: 810 (mg/l)).

ANALYTICAL METHODS

The effluent from each bioreactor was collected daily, centrifuged at 6000 rpm for 10 min and analysed for colour, COD, pH, volatile suspended solids (VSS) and colonies forming units (cfu). COD and colour measurements were carried out on the clear supernatant. Colour was measured by an UV–vis spectrophotometer (Spectro UV-Vis Double Beam PC Scnning spectrophotomètre UVD-2960) at a wavelength of 275 nm in which maximum absorbance spectra was obtained. The decolourization and COD removal were calculated according to the following formulation (Eq 1and Eq 2) (Ayed et al., 2009a,b). In this study, *Sphingomonas paucimobilis, Bacillus sp*. and Filamentous bacteria were used as mixture starters, with different proportions ranging from 0 to 100Percent. Decolourization experiments were taken according to the ratio given by the experimental design, and 10Percent of mixed culture were inoculated into the effluent (3.0 g/l yeast extract and 1.25 g/l glucose) at 37°C for 10 hin shaking conditions (150 rpm) (Ayed et al., 2010a,b,c).

$$\% \text{ Decolorization} = \frac{(I - F)}{I} \times 100$$

Where I was the initial absorbance and F the absorbance at incubation time t

$$\text{COD removal}(\%) = \frac{\text{initial COD}(0\,h) - \text{observed COD}(t)}{\text{initial COD}(0\,h)} \times 100$$

The pH was measured using a digital calibrated pH-meter (Inolab, D-82362 Weilheim Germany). All assays were carried in triplicate.

Table. Mixture Design Matrix with the Experimental Analysis

Assay	Sphingomonas Paucimobilis	Bacillus sp.	Staphylococcus Epidermidis	Total	COD Removal (%)	Decolorization (%)
1	0.66667	0.16667	0.16667	1.00000	60	70
2	0.50000	0.50000	0.00000	1.00000	76	77
3	0.50000	0.00000	0.50000	1.00000	77	88
4	0.33333	0.33333	0.33333	1.00000	75	80
5	0.00000	1.00000	0.00000	1.00000	70	77
6	1.00000	0.00000	0.00000	1.00000	81	89
7	0.16667	0.16667	0.66667	1.00000	53	63
8	0.00000	0.00000	1.00000	1.00000	49	55
9	0.00000	0.50000	0.50000	1.00000	45	44
10	0.16667	0.66667	0.16667	1.00000	60	64

PILOT YARN YARN PLANT DESIGN

As described earlier, the pilot yarn yarn plant comprised several treatment steps.

RESULTS AND DISCUSSION

MODEL ESTABLISHMENT

Through linear regression fitting, the regression models of tow responses (COD Percent and decolourization Percent) were established. The regression model equations are as follows:

$$Y_{decolourizationPercent} = 85.34\ S1 + 67.70\ S2 + 55.43\ S3 + (-21.77)\ S1{*}S2 + (67.69)\ S1{*}S3 + (-73.59)\ S2{*}S3 \quad R^2 = 84.82\text{Percent};\ P=0.09$$

$$Y_{CODPercent} = 77.11\ S1 + 70.11\ S2 + 49.02\ S3 + (-1.55)\ (S1{*}S2) + (44.27)\ (S1{*}S3) + (-57.73)\ (S2{*}S3) \quad R^2 = 75.27\text{Percent};\ P = 0.2$$

Where S1: Sphingomonas paucimobilis; S2: Bacillus sp. and S3: Staphylococcus epidermidis

EFFECT OF FORMULATION ON THE PERCENTAGE OF DECOLOURIZATION AND COD REMOVAL OF EFFLUENT

The mixture design is now used widely in the formulation experiment of food, chemicals, fertilizer, pesticides, and other products. It can estimate the relationship between formulation and performance through regression analysis in fewer experiment times (Zhang et al., 2006).

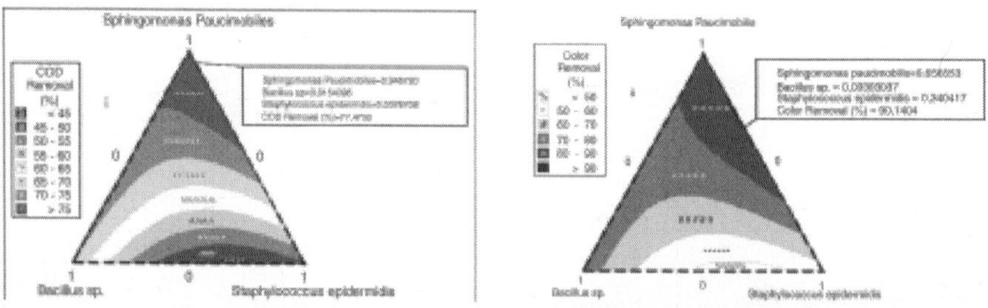

Fig. Mixture Contour Plots between the Variables (*Sphingomonas Paucimobilis*, *Bacillus sp* and *Staphylococcus Epidermidis*.) for i COD Removal (Percent), ii Colour Removal (Percent).

Zhang et al. (2006) studied the formulation of yarn yarn plant protein beverage using the mixture design, obtaining the optimized combination of walnut milk, peanut milk, and soy milk. In the mixture design, the effect of the change of variables on the responses can be observed on the ternary contour map. Figure shows the effect of the interaction of *Sphingomonas paucimobilis*, *Bacillus sp*. and *Staphylococcus epidermidis* on the decolourization of effluent; Figure shows the effect of the interaction of *Sphingomonas paucimobilis*, *Bacillus sp*. and *Staphylococcus epidermidis* on the variation of COD. The statistical significance of the ratio of mean square variation due to regression and mean square residual error was tested using analysis of variance. ANOVA is a statistical technique which subdivides the total variation in a set of data into component parts associated with specific sources of variation for the purpose of testing hypotheses on the parameters of the model. Only results obtained for decolourization and COD removal were presented herein for clarity of purpose. The regression adjusted average squares were (305.8) and (231), the linear regression adjusted average squares were (1529.3) and (1115.02) allowed the calculation of the Fisher ratios (*F*-value) for assessing the statistical significance. The model *F* -value (4.33) and (2.43) implies that most of the variation in the response can be explained by the regression equation.

Table. Analysis of Variance of Percent Decolourization (ANOVA) for the Selected Linear and Interactions Model for Effluent Textile Wastewater

Source	Degrees of Freedom	Sum of Square	Sum of Adjusted Squares	Adjusted Average Squares	*F*-ratio	*P*-value (Significance)
Regression	5	1529,3	1529,303	305,861	4,33	0,090
Linear regression	2	996,33	521,279	260,639	3,69	0,123
Quadratic regression	3	532,97	532,970	177,657	2,52	0,197
Residual error	4	282,30	282,297	70,57		
Total	**9**	**1811,60**				

The *P*–value for the regression obtained R2= 84.82Percent; P=0.09 for decolourization was less than 0.1 and means consequently that at least one of the term in the regression equation has significant correlation with the response variable. The associated *P*–value is used to judge whether *F*–ratio is large enough to indicate statistical significance. A *P*-value is more than 0.1 (*i.e.* α =0.05 or 95Percent confidence) indicates that the model is not to be considered statistically significant. The non-significant value of lack of fit (>0.05) revealed that the quadratic model is statistically significant for the response and therefore it can be used for further analysis (Zhou et al., 2007). The ANOVA test also shows a term for residual error, which measures the amount of variation in the response data left unexplained by the model (Xudong and Rong, 2008).The collected data were analyzed by using Minitab® 14 Statistical Software for the evaluation of the effect of each parameter on the optimization criteria. In order to determine the effective parameters and their confidence levels on the colour removal process, an analysis of variance was performed. A statistical analysis of variance (ANOVA) was performed to see which process parameters were statistically significant. *F*-test is a tool to see which process parameters have a significant effect on the dye removal value. The *F*-value for each process parameter is simply a ratio of the mean of the squared deviations to the mean of the squared error. The colour removal from the real textile wastewater was investigated in different experimental conditions.

Table. Analysis of Variance of CODPercent (ANOVA) for the Selected Linear and Interactions Model for Effluent Textile Wastewater

Source	Degrees of Freedom	Sum of Square	Sum of Adjusted Squares	Adjusted Average Squares	*F*-ratio	*P*-value (Significance)
Regression	5	1155,02	1155,023	231,005	2,43	0,205
Linear regression	2	885,44	470,406	235,203	2,48	0,199
Quadratic regression	3	269,58	269,578	89,859	0,95	0,498
Residual error	4	379,48	379,477	94,869		
Total	**9**	**1534,50**				

The mixture surface plots, which are a three-dimensional graph, was represented using COD and colour removal were represented based on the simultaneous variation of *Sphingomonas paucimobilis*, *Bacillus sp*. and *Staphylococcus epidermidis* in the consortium composition ranging from 0 to 100 Percent for each strain. The mixture surface plot also describing individual and cumulative effect of these three variables and their subsequent effect on the response (Liu et al., 2009; Ayed et al., 2010b,c). The mixture contour plots between the variables such as *Sphingomonas paucimobilis*, *Bacillus sp*. and *Staphylococcus epidermidis* are given in Figure. The lines of contour plots predict the values of each response at different proportion of *Sphingomonas paucimobilis*, *Bacillus sp*. and *Staphylococcus epidermidis*. These values are more or less same to the experimental values.

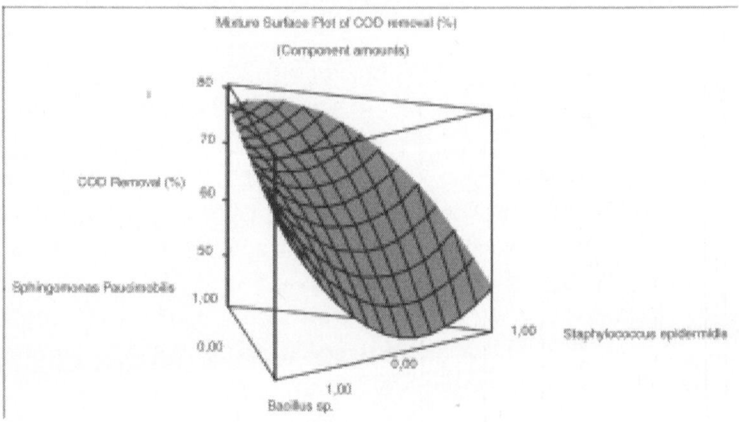

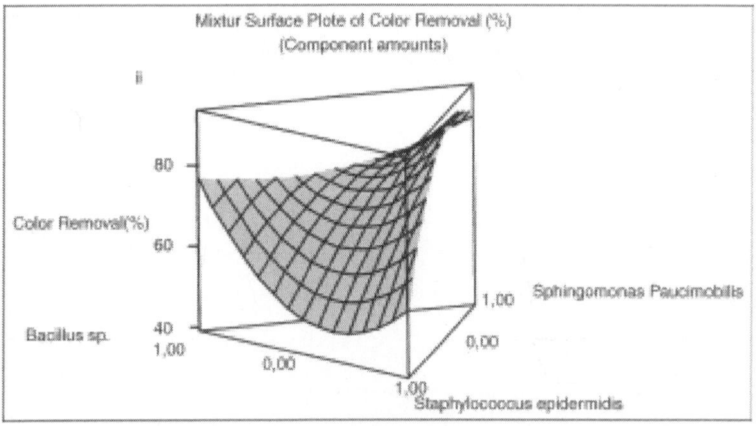

Fig. Mixture Surface Plots between the Variables (*Sphingomonas Paucimobilis*, *Bacillus sp* and *Staphylococcus Epidermidis*.) for i COD Removal (Percent), ii Colour Removal (Percent).

CONCLUSIONS

The developed consortium showed a better decolourization yields as compared to pure cultures, which proved a complementary interaction among various isolated bacteria. The consortium achieved significantly a higher reduction in colour (90.14Percent) and COD removal (77.47Percent) in less time (96h). The biodegradation of the effluent textile wastewater was achieved by the developed consortium using *Sphingomonas paucimobilis*, *Bacillus sp.* and *Staphylococcus epidermidis*.

6

Surface and Bulk Modification of Synthetic Textiles to Improve Dyeability

INTRODUCTION

Synthetic fibres, mainly polyethylene terephthalate (PET), polyamide (PA), polyacrylonitrile (PAN) and polypropylene (PP), are the most widely used polymers in the textile industry. These fibres surpass the production of natural fibres with a market share of 54.4Percent. The advantages of these fibres are their high modulus and strength, stiffness, stretch or elasticity, wrinkle and abrasion resistances, relatively low cost, convenient processing, tailorable performance and easy recycling. The downside to synthetic fibres use are reduced wearing comfort, build-up of electrostatic charge, the tendency to pill, difficulties in finishing, poor soil release properties and low dyeability. These disadvantages are largely associated with their hydrophobic nature. To render their surfaces hydrophilic, various physical, chemical and bulk modification methods are employed to mimic the advantageous properties of their natural counterparts. This review is focused on the application of recent methods for the modification of synthetic textiles using physical methods (corona discharge, plasma, laser, electron beam and neutron irradiations), chemical methods (ozone-gas treatment, supercritical carbon dioxide technique, vapor deposition, surface grafting, enzymatic modification, sol-gel technique, layer-by-layer deposition of nano-materials, micro-encapsulation method and treatment with different reagents) and bulk modification methods by blending polymers with different compounds in extrusion to absorb different colourants. Nowadays, the bulk and surface functionalization of synthetic fibres for various applications is considered as one of the best methods for modern textile finishing processes. This last stage of textile processing has employed new routes to demonstrate the great potential of nano-science and technology for this industry. Combination of physical technologies and nano-science enhances the durability of textile materials against washing, ultraviolet radiation, friction, abrasion, tension and fading. European methods for application of new functional finishing materials must meet high ethical demands for environmental-friendly processing. For this purpose the process of textile finishing is optimized by different researchers in new findings. Application of inorganic and organic nano-particles have enhanced synthetic fibres attributes, such as:

- Softness,
- Durability,
- Breathability,
- Water repellency,
- Fire retardancy and
- Antimicrobial properties.

This review article gives an application overview of various physical and chemical methods of inorganic and organic structured material as potential modifying agents of textiles with emphasis on dyeability enhancements. The composition of synthetic fibres includes:

- Polypropylene (PP),
- Polyethylene terephthalate (PET),
- Polyamides (PA) or
- Polyacrylonitrile (PAN).

Synthetic fibres already hold a 54Percent market share in the fibre market. Of this market share, PET alone accounts for almost 50Percent of all fibre materials in 2008. Polypropylene, a major component for the non-wovens market accounts for 10Percent of the market share of both natural and synthetic fibres worldwide. It is apparent that synthetic polymers have unique properties, such as high uniformity, mechanical strength and resistance to chemicals or abrasion. However, high hydrophobicity, the build-up of static charges, poor breathability, and resistant to finishing are undesirable properties of synthetic materials. Synthetic textile fibres typically undergo a variety of pre-treatments before dyeing and printing is feasible. Compared to their cotton counterparts, fabrics made from synthetic fibres undergo mild scouring before dyeing. Non-etheless, these treatments still create undesirable process conditions which can result in increased waste production, unpleasant working conditions and higher energy consumption. Therefore reducing pollution in textile production is becoming of utmost importance for manufacturers worldwide. In coming years, the textile industry must implement sustainable technologies and develop environmentally safer methods for textiles processing to remain competitive.

FIBRE-FORMING SYNTHETIC POLYMERS

Fibres comprising at least 85Percent by weight of a substituted aromatic carboxylic acid ester are termed polyesters. The most important representative of this category is polyethylene terephthalate or PET. PET is a hydrophobic fibre with maximum moisture regain of only 1Percent at 100Percent relative humidity. Until the development of disperse dyes, dyeing of polyester was difficult. Disperse dyes with very low water solubility can sublime into PET fibres by heat through Thermosol and/or thermofixation processes. It can be applied with heat, pressure or via carriers by an exhaust process. Alkali treatment can be used to etch the PET surface to increase the hydrophilicity of the fibre resulting in better dyeability. However, the rate of hydrolysis is very low without a catalyst and surface-limited. In generic terms, aliphatic polyamides (PA) are called nylons and aromatic PAs are called aramids. The first important PA was Nylon 66 as produced by the reaction of adipic acid and hexamethylene diamine monomer. Several structural modifications with differing temperature capabilities have become commercially available, including Nylon 46, 610, 612, 6, 11, and so on. Polyamides have applications in many areas, the most important being in the production of fibre-based materials. Nylons are dyeable with disperse dyes or with acid dyes under mild acidic conditions. Aqueous acids as well as bases cause the rupture of the polymer backbone. In the case of acid dyeing, dye molecules only attach to available amino end groups, thus shade depth is determined by the ratio of negatively charged groups of the dye molecule to positively charged end groups in the fibre.

CLASSIFICATION OF METHODS USED FOR SYNTHETIC TEXTILES MODIFICATION

Synthetic fibres have relatively high levels of orientation and crystallinity that impart the desired properties. These same characteristics contribute to their structural resistance to colouration by dye compounds and finishing with various materials necessitating enhancement of the fibre surface for improved dye receptivity. In this sense, a large body exists to improve fibre dyeability via physical methods, chemical methods (enzymatic modification grafting of different monomers, utilization of supercritical carbon dioxide as the solvent carrier for disperse dyes, sol-gel technique, layer-by-layer deposition and treatment with different reagents) and bulk modification methods using various additives during fibre processing. In last decade, traditional methods that consume high amounts of energy and water are under pressure for replacement due to high manufacturing costs and negative environmental impact.

In addition, some processing negates the bulk properties of fibres, require harsh process conditions, and produce undesirable side effects and Dor waste disposal problems. Recent methods address these challenges and deficiencies of traditional techniques and will be discussed in this chapter.

PHYSICAL METHODS

Traditional transformation of a hydrophobic polymer such as poly (PDMS) to a hydrophilic state has been achieved via techniques such as corona and plasma treatments. Corona, plasma, irradiation, and laser technologies are ideal for textile surface modifications due to the energy-efficient dry-state processing, continuous on-line applicability, and minimal precursor quantity requirements. In addition, surface modification does not ingress to the bulk fibre's mechanical properties although physical alterations are generally realized on the fibre's outermost surface layer. Depending on the treatment duration, the changes can propagate several microns below the surface. X-ray photoelectron spectroscopy (XPS) analysis on oxygen plasma treated samples demonstrated a rapid substitution of carbon atoms by oxygen atoms, which led to the formation of hydrophilic surfaces. This treatment will propagate several hundred nanometers below the surface with irreversible chemical changes at the near-surface region. The various physical methods all have their merits with some processes, such as corona, more applicable to simple on-line installation. This section will highlight each technique with some examples on their resultant properties.

Corona Discharge

Corona discharge is the breakdown of a gas between two electrodes. When the gas is anything but air, it is termed plasma modification to be discussed in the next section. The generation of the initial hydrophilic surface is the same in both modifications with just different outcomes based on the introduced gas stream. A typical schematic for dielectric discharge modification. When the high voltage or electric field is applied in a gas layer exposed to a polymer surface, the gas molecules, in this case air, breakdown to ions to conduct electricity. It is the bluish air glow from the electric source that is termed corona. This phenomenon starts with a few stray electrons colliding with other gas molecules. The collision rapidly generates several multiple electrons, positive ions, and excited molecules.

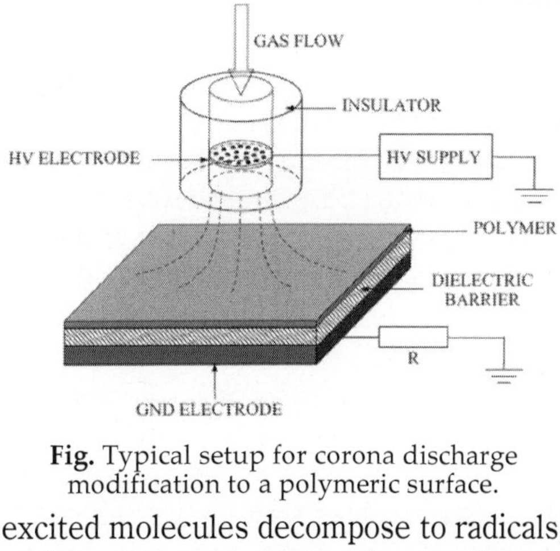

Fig. Typical setup for corona discharge modification to a polymeric surface.

The unstable excited molecules decompose to radicals, ions and photons, *i.e.* reactive species. When the gas is oxygen, the reactive species are elemental oxygen (O•), ozone (O_3) and activated oxygen (O_2*). Scheme 1 shows the general reactions of polyethylene terephthalate (PET) in the presence of either UV or corona as the mechanisms are believed to be the same. The terminal phenols and hydroxy phenolics are rapidly formed on the surface of corona-exposed PET between the reactive oxygen species and moisture in the air. Other polar moieties that are formed during corona treatment include carbonyls (-CO) and carboxyls (-COO). This change in surface polarity has been widely studied for increases in adhesion, wettability, printing, and as the subject of this chapter, dyeing. Several excellent reviews are referenced here that detail the specific and mechanisms of the treatment.

A recent study evaluated not just how corona discharge affects conversion to hydrophilicity in terms of enhanced dyeability of synthetic fibres but also in how the fibre topography changes as a result of physical treatment ultimately influencing dyeability as well. The study postulated that without fully understanding topography, specifically fibre roughness, correlating the varying results of the many conflicting studies of fibre dyeing with respect to corona treatment is too convoluted. To this extent, three synthetic fibres in separate woven mats were evaluated:

- Polyester (PET),
- Polyamide (PA6) and 3)
- Polypropylene (PP).

The corona generator was unique in that its construction afforded a high degree and uniform surface modification to the fibres. This was by designing a multi-segmented electrode system where continuous low energy doses equated the larger required total energy dose for adequate surface modification. The process conditions previously determined for each fibre resulted in total activation energies *Ej* of 75.6 J/cm^2 for PET, *Ej* of 18.9 J/cm^2 for PA6, and *Ej* of 22.7 J/cm^2 for PP fabrics. The difference via atomic force microscopy (AFM) before and after modification for the PP fabric sample where clear disruption of the fibre surface is apparent. In addition to increased roughness, the change in free surface energy was approximately 10, 4, and 30 J/cm^2 for PET, PA6, and PP respectively. Dyeability for the modified samples was assessed with two techniques; 1) exhaustion and 2) the preferred Thermosol continuous method. The results between unmodified and modified samples showed little difference in the dyeing attributes via the exhaustion methods, as measured by degree of dye exhaustion, ΔE or colour difference, and dyeing fastness. This was attributed to the fact that corona discharge does not modify the entire fibre depth but rather only the first 200nm from the surface of the approximately 6500nm diameter fibre. Therefore in the dyeing techniques, such as exhaustion, where the entire fabric is immersed into the bath batch-wise, the dye pick-up rate is minimal. In contrast, dyeing methods that are applied only to the surface of the fibre, as in the case with the Thermosol method, surface hydrophilicity and roughness are far greater parameters to gauge the receptivity of the dye. In this case, the dye intensity for the corona-modified samples was significantly enhanced as measured by the colour difference ΔE for the two dyes studied, C.I. Disperse Blue 73 and C.I. Disperse Red 54.

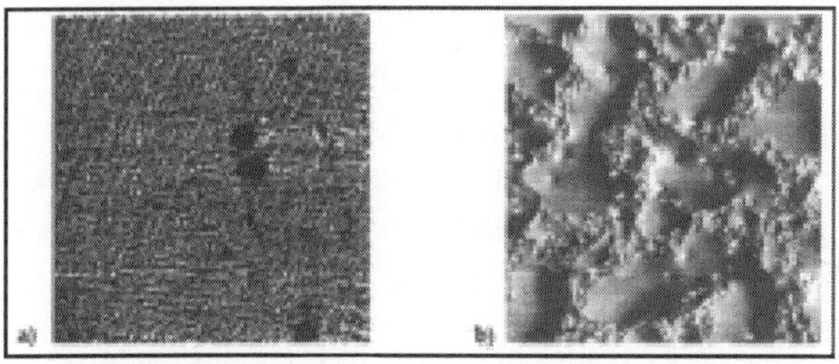

Fig. AFM images of polypropylene woven fabric b) before and b) after corona discharge treatment.

Plasma functionalization

As mentioned, when the dielectric discharge occurs in environments besides air, the technology is referred to as plasma treatment; it has been studied at both vacuum and atmospheric pressures. The various gases employed include oxygen, nitrogen, argon, ammonia and reactive monomers. The hydrophilicity of poly(ethylene terephthalate) (PET) was greatly improved with the plasma method where the discharge barrier occurred in argon, nitrogen and air. The surface wettability enhancements were due to two reactions;

- Direct reaction (*i.e.* oxidation) of reactive gases (oxygen plasma) and
- Free radical formation and their subsequent reactions such as degradation and crosslinking.

In this study, it was determined that a nitrogen atmosphere with a glow discharge of 30W provided the most durable and wettable surface finish for PET. The optimum power level of 30W was chosen from experiments in air at power levels ranging from 10-30W. Figure illustrates the differences of wettability at the optimum power level.

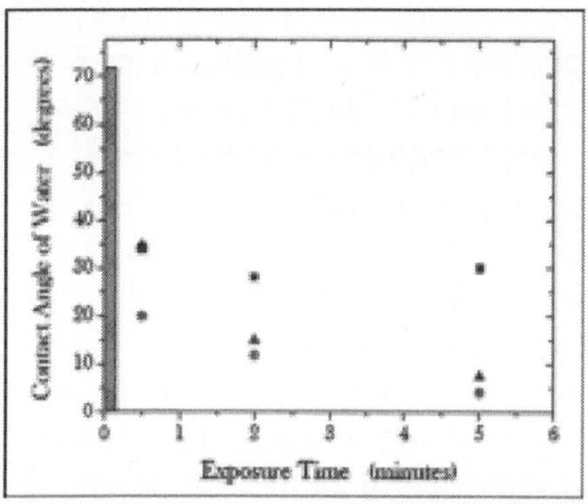

Fig. Advancing water contact angles for PET modified as a function of plasma exposure time in air (■), argon (▲), and nitrogen(•). The bar at time zero is the contact angle for virgin PET.

Plasma treatment has also been performed on PET fibre prior to coating with a PDMS surfactant. In this case the efficacy of the surfactant was greatly enhanced with plasma activation at 1kW, 10 kHz, in the presence of air. It was determined by scanning electron microscopy that the plasma treated fibres were rougher allowing increased PDMS deposition via observation of a smoother coating, less moisture regain after coating with PDMS which is hydrophobic in nature, better drape recovery and more wrinkle resistance

Laser treatment

Laser induced surface modification of polymers provides a unique and powe.rful method for the surface modification without any changes in their bulk properties. The smooth surface of synthetic fibres can be modified by this technique to a regular, roll-like structure, which has a striking effect on adhesion of particles and coatings, wetting properties and optical appearance. Yip et al. applied a 193 nm argon fluoride excimer laser on polyamide (nylon 6) fabrics. Micrometer-sized ripple like structures were developed on the surface of irradiated fabric and chemical analysis indicates that carbonization has occurred. It is believed that the laser treatment breaks the long chain molecules of nylon, thus increasing the number of amine end-groups. Kan stated that properties such as wettability and air permeability of polyester were positively affected by laser while fibre weight and diameter, tensile strength, yarn abrasion and bending were adversely affected. In this study, laser irradiation was not found to affect the bulk properties of polymer due to its low penetration depth.

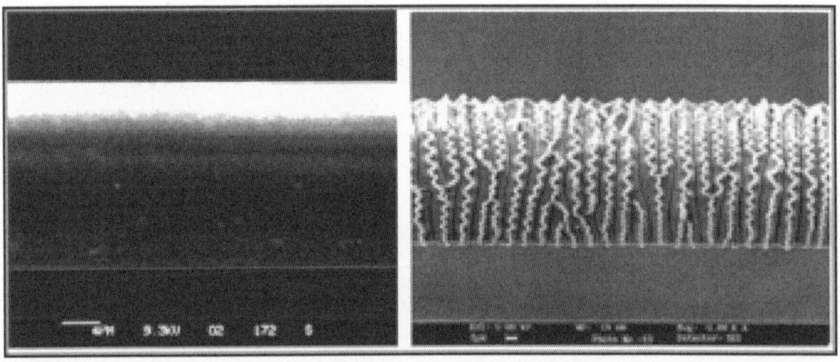

Fig. Surface structure of polyester fibre before and after laser treatment.

Other physical methods

Exposing fibres to a stream of high-energy electrons is another method for surface modification. The dyeability of hydrophobic polypropylene fibres was enhanced by Kim and Bae using electron beam irradiation and sulfonic acid incorporation. The colour strength of polypropylene fibres after irradiation was examined according to the dyeing conditions including the pH of the dye bath, absorbed doses, and the introduction of a functional group to the fibre substrate. The best dyeing result was obtained with cationic dyes at alkaline conditions. Neutron irradiation significantly changes the material properties by displacement of lattice atoms and the generation of helium and hydrogen by nuclear transmutation. Mallick et al. considered the shift in some of the Raman peak positions to a higher value with the development of micro-stresses due to neutron irradiation of synthetic fibres. The defects due to irradiation were confirmed by SEM micrographs of virgin and irradiated fibres. Critical temperature and pressure describes a fluid at temperatures and pressures higher than those at which the liquid and gaseous states of the given substance would have the same density. Prorokova et al. showed modification of the surface of PET fabrics by application of a water-repellent coating in supercritical carbon dioxide medium. It was believed that in treating PET fibre materials with a solution of low-molecular-weight poly(tetrafluoroethylene) in supercritical carbon dioxide, an ultrathin layer of fluoropolymer is formed on the surface. This layer gives the fabric a high degree of water repellency. Another potentially attractive approach would take advantage of UV light's ability to transform chemistry of the polymer surface. Zhu and Kelley modified the surface of PET by deep UV (172 nm) irradiation. The main effects were increased conversion of the C=O moiety to carboxylic acid with negligible change to fibre topography. Other studies revealed that surface chemical composition, morphology, adhesion, thermomechanics, and stiffness/modulus of PET are strongly influenced by UV irradiation in the presence of bifunctional media.

CHEMICAL METHODS

The methods discussed in the following sections include ozonolysis, supercritical carbon dioxide, chemical vapor deposition, "grafting to", enzymatic modification, sol-gel deposition, layer-by-layer, micro-surface modification by alkaline or acidic means, and macro-encapsulation of dye molecules. In many instances, the best performance is obtained from a combination of physical and chemical methods, especially for poly(propylene).

Ozone gas treatment

Hydrophilicity of synthetic polymer surfaces can be achieved with functionalization of peroxide molecules via exposure to ozone (O_3). An advantage of ozone treatment is the uniform coverage of the reactive molecules on 3-D structures. It is also well-known that ozone will treat not only the surface but diffuse through the polymer bulk. The process is often done in the gas phase but aqueous treatment has also been employed with good results. In an early study by Fujimoto et al., the surface oxidation of polyurethane (PU) and polyethylene (PE) films were evaluated by both ozone and plasma treatment. The process conditions for ozone treatment followed typical gas reactor set-up as illustrated in Figure.

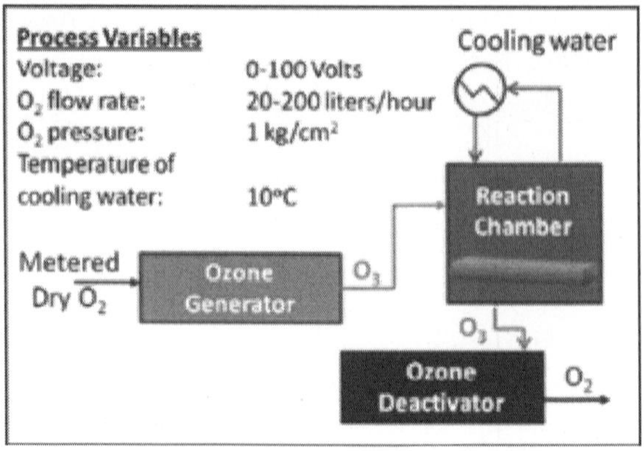

Fig. Typical set-up for ozonolysis of substrates.

Of the process variables to control, gas mass flow rate, voltage and time are the most common to control ozone concentration. The formed polymeric peroxides on the surface and in the bulk were quantified by iodide, DPPH, and peroxidase spectrophotometric methods. Additional surface analysis to confirm reactivity of the ozone included wettability measurements, Fourier infrared spectroscopy in the attenuated reflectance mode (FTIR-ATR), and X-Ray Photoelectro Spectroscopy (XPS/ ESCA). Subsequent to oxidation was the graft polymerization of acrylamide which will be discussed in Section 2.2.4. The surface oxidation efficiency was evaluated based on water contact angle and peroxide concentration. Of interest is that the wettability of the plasma exposed films is much higher than the ozonated films but the peroxide concentration in the plasma treated polyurethane film (Power = 24W, exposure time = 20 seconds) is lower than its ozone counterpart (Voltage =100V, exposure time =20 minutes, O_2 flow = 50 L· hr-1). It should also be noted the concentration of peroxide decreases with polymer rigidity and non-polar constituents (C-F).

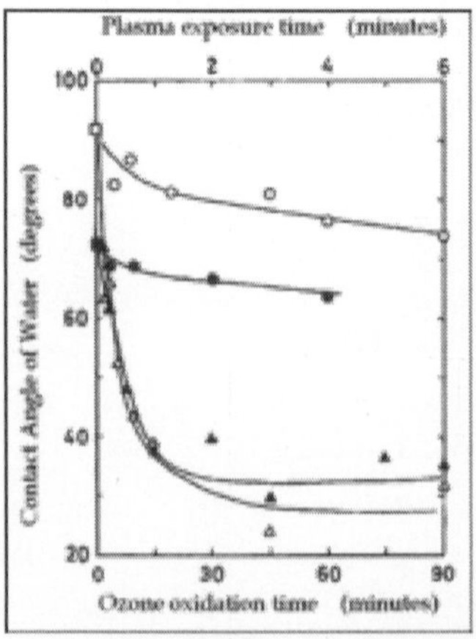

Fig. Decrease in contact angle of PE and PU films by ozone oxidation and plasma exposure. Ozone (60V, 50Lhr-1):(o)PE, (•)PU. Plasma (Ar, 24W): (Δ) PE, (▲)PU. Virgin:(□)PE, (■)PU.

Ozone self-decomposes rapidly in water producing free radicals, a stronger oxidant than ozone itself. This property was utilized to produce hydrophilic and highly reactive high-density polyethylene (HDPE) films. The O_2 gas rate in this study was about twelve times higher with exit ozone concentrations ranging between 1-3.7 weight percent. Figure illustrates the difference of ozone treatment in the aqueous phase versus the gas phase. While initially there is no apparent difference in the media treatment, the peroxide generation is greater and faster for ozonation in the aqueous phase. For all samples, it was demonstrated that the stability of the generated peroxides lasted at least 15 days with no change in concentration. The subsequent grafting of acrylamide on the aqueous ozone treated samples was successful but its performance in terms of surface energy was not contrasted to acrylamide graft polymerization on gas-phase ozone treated samples.

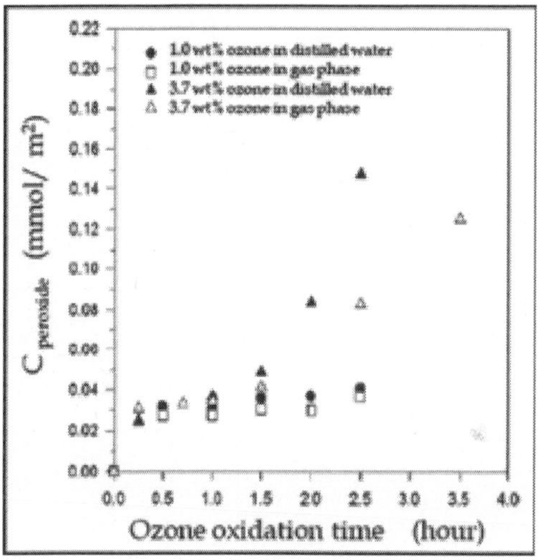

Fig. Peroxides generated after diffrent Ozonation times in different media.

Specific to fabrics, chemical modification via gas-phase ozone treatment was performed on nylon 6 taffeta, polyester taffetas, cationic dyeable polyester (CDP) and poly(butylene terephthalate) fibres. The surface analysis via ESCA was very comparable to the plasma studies by others, the reader is referred to previous methods for process conditions. Notable is that the processing time was only 10 minutes with similar surface chemistries for operations at atmospheric pressure and 0.1 MPa. The ozone-modified fabrics were treated with Disperse Red 60 and Disperse Blue 56 dyes in batch immersion at 100°C for up to 120 hours. The authors found that the internal structure of the fibres increased in crystallinity (as measured by a density gradient column and X-ray diffraction), wettability and moisture uptake upon ozone treatment. These characteristics of the modified fibres were attributed to the increase in dye uptake rate, especially for polyester fibres. The equilibrium dye uptake increased for PBT fibres, polyester taffeta and nylon 6 taffeta but remained unchanged for CDP fibres.

Supercritical carbon dioxide technique

In recent years, waterless dyeing in supercritical carbon dioxide ($scCO_2$) fluid has been gaining much interest to textile chemists. This process is deemed an environmentally safe solvent as opposed to the traditional solvent of choice, water. Application of these techniques can result in reducing waste and cost for the entire dyeing process of synthetic textiles. The possible advantages of this process are:

- Plasticizing effect due to CO_2 on textiles decreasing the glass transition temperature.
- Elimination of contaminated waste water streams.
- Elimination of chemicals, such as leveling agents, pH regulations and dispersants, to solubilize disperse dyes in medium.
- Controllable solubilities of disperse dyes via pressure.
- Diffusion of dyes and penetration of voids within the fibre structure in the fluid is higher/ faster due to zero surface tension between air and $scCO_2$ carbon dioxide.
- Generation of effluents due to dyeing or recycling of contaminated gas streams does not exist.
- Energy consumption is low for heating up dyeing liquor.
- For polyester, elimination of reduction clearing process, short dyeing times, and high diffusivities resulting in high extraction/ reaction rates.

Fabric and dye are put in the container before starting the process. The apparatus is then sealed and heated to a pre-selected dyeing temperature and CO_2 is pumped simultaneously to the set pressure. The dyeing is carried out on textile depending on the type of fibre and then the pressure is slowly reduced to atmospheric at isothermal conditions.

Bach et al. dyed PP fibres in $scCO_2$ with different disperse dyes. They showed that disperse azo dyes with a naphthalene moiety gave much deeper colours on PP versus benzo-azo or anthraquinone dyes. They stated that improvement in dyeability is due to the changes in the crystal network of PP by the treatment in CO_2 as contrasted with PP dyed in water or air. FTIR and NMR results obtained from Nylon 6,6 fabric samples that underwent $scCO_2$ dyeing of with a disperse-reactive dye confirmed a covalent bond with the fibres.

Wash and light fastness of the fabrics showed satisfactory results. Their results indicated that fabric immersed in $scCO_2$ does not undergo any fibre damage. Shim et al. studied sorption of disperse dyes in PET and PTT textiles in the presence of $scCO_2$. They found that the dyeing rate increased monotonically with pressure at isothermal conditions and increased with temperature isobaric conditions. Generally, this method plasticizes the polymeric fibrous chains enhancing dye diffusion rates and increasing the ease of solvent removal. Moreover, it replaces water in dyeing processes, overcoming the problem of wastewater treatment.

Textile surface functionalization by vapor deposition methods (VDM)

Sputter coating is a significant technique producing functional nanostructured fibres. These functionalized fibres are essential for realizing their applications in microelectronic elements, photonics devices, and medical implants.

Wei et al. used magnetron sputter coatings to generate functional nanostructures on polymer fibre surfaces. Conducting aluminum (Al) film, piezoelectric aluminum nitride (AlN) film, and ceramic film of aluminum oxide (Al_2O_3) were deposited onto PET fibres at low temperature.

These nanostructured fibres have great potential for applications ranging from conductive shields, packing, and protective materials to electronic sensors.

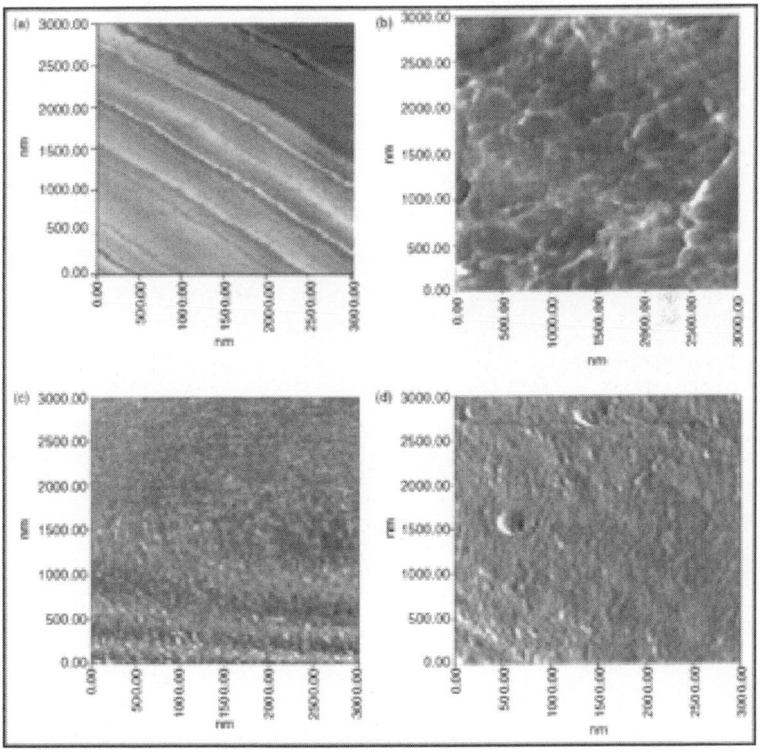

Fig. Surface morphology of textile fibre: (a) original polypropylene fibre; (b) 20nm copper coated polypropylene fibre; (c) 50nm copper coated polypropylene fibre; and (d) 100nm copper coated polypropylene fibre.

Copper (Cu) nanocomposite textiles were prepared by magnetron sputter coating as discussed by other researchers. The surface conductivity of the textiles coated with Cu nanostructures showed a significant increase compared to the uncoated ones. The increased coating thickness improved electrical conductivity. The coated textiles also showed considerable improvement in UV and visible light shielding as examined by UV/Vis spectrometer.

Surface grafting

Covalent bonding of a molecule to most non-reactive synthetic surfaces such as polyolefins typically requires surface activation via one the physical methods described earlier. To this extent, graft peroxide-initiated polymerization of acrylamide (AAm) proceeded successfully on the ozonated surfaces. FTIR confirmed the presence of the amide groups with peak intensity at 1667 cm-1 increasing for the samples exposed longer to ozone treatment. Finally contact angle of the samples proceed from 74.92° for virgin HDPE to 38.55° for the AAmgrafted HDPE. Grafting of AAm to the PU and PE films by Fujimoto et al. was favourable although the methodology was different. In this case, thermal activation of the peroxide for AAm grafting occurred at lower temperature (60°C versus 85°C) for less time (3 hr versus 24 hr). The end result was still an outermost layer of polyacrylamide on the PU film as measured by FTIR, optical microscopy, and graft density determination by the ninhydrin method (150 μg cm^{-1}). It was also demonstrated, however, that the grafting efficiency was reduced when the procedure was performed on PE film. This is attributed to lower levels of peroxides incorporated to the more chemicallyresistant PE film as compared to PU film. These two surface modifications are categorized as *"grafting from"* methods as the peroxide initiator was tethered to the fibre surface prior to the polymerization reaction. When polymer chains are absorbed (and then subsequently reacted) to a solid surface, the correct term is *"grafting to"*. A thorough overview of this subject is provided by Minko.

Fig. SEM images of the morphology of the film surfaces (×5000:a) virginHDPE film,b) HDPE film after ozonation in distilled water at 3.7wtPercent for 1 hour, and c) AAm-grafted HDPE fim ozonated in distilled water at 3.7wtPercent for 1hour: adapted from.

Polymerization processing parameters for the grafting of 4-vinyl pyridine to PET fibre was done to increase wettability and heavy-metal capture from aqueous media. Similar to previous work, benzoyl peroxide was used as the initiator. The researchers pre-swelled the fibres in dichloroethane to aid in the absorption of initiator and monomer prior to polymerization. They evaluated monomer concentration, initiator concentration, reaction temperature and time to determine the optimum grafting conditions for a maximum yield of 70Percent. The incorporation of 4-vinyl pyridine resulted in a 40Percent increase of water absorption at the maximum graft yield. The chemical initiator immersion technique for benzoyl peroxide and ammonium persulphate was implemented for grafting acrylic acid onto polylactide fibres. Both techniques for grafting were inefficient to produce significant improvement in dyeability. They do note that layer-by-layer deposition provided surprising results in that the alternative barrier layer impeded the diffusion of oxygen thus prevention of polymer (PLA) degradation fibres. Dyeability as observed by K/ S values was also the highest for the layer-by-layer modification technique. PET modification by glycolysis was determined most effective as measured by percent concentration of hydroxyl groups. The quantity of hydroxyl groups was assessed by X-ray photoelectron spectroscopy (XPS), dynamic contact angle (DCA) and labeling reactions. The labeling reactions combined with DCA are key results to conclude the increase effectiveness of glycolysis over hydrolysis and reduction of PET surfaces for incorporating hydrophilicity. Although XPS is surface sensitive to the first 100 angstroms, it is also performed in vacuum. Functionalizing the surface with non-polar moieties (fluorine) insure that their preferred orientation state is at the non-polar interface (air then vacuum) versus buried beneath the polymer surface. One final chemical reaction to be discussed for PET modification is aminolysis. This reaction forms amide groups through the reaction of polyester's carbonyl with a primary amine. This reaction was done in the early 1960's to improve wettability, reduce static electricity and increase dyeability. This reaction can be quite severe to the fibres with complete degradation as a result. To temper fibre degradation, long chain multifunctional amines such as tetraethylenepentamine (TTEPA) has been employed with favourable results. While XPS showed no difference in Percent nitrogen content with reaction time, titration and DCA methods show an optimum wettability or amide content at 180 minutes. This indicated that the reaction was proceeding through the depth of the fibre which can also positively impact the diffusion of a dye throughout the fibre.

Enzyme surface modification of textiles

Processing with enzymes is one of the best environmental friendly applications of biotechnology in textile industry. Enzymes are biological catalysts that mediate virtually all of the biochemical reactions that constitute metabolism in living systems. They accelerate the rate of chemical reaction without themselves undergoing any permanent chemical change. All known enzymes are proteins and consist of one or more polypeptide chains. The influence of many chemical and physical parameters such as salt concentration, temperature and pH on the rate of enzyme catalysis can be explained by their influence on protein structure. Some enzymes require small non-protein molecules, known as cofactors, in order to function as catalysts. Enzymes differ from chemical catalysts in at least two ways. Enzymes have far greater reaction specificity than chemically catalyzed reactions rarely forming by-products. In contrast to chemical catalysis, enzymes catalyze reactions under milder reaction conditions, at atmospheric pressures and at neutral pHs. In this section, we summarize how enzymes can be used to increase the hydrophilicity of selected synthetic polymers. Increase in hydrophilicity often leads to improved dyeability of the textile products. Enzymes are also used for the synthesis, surface functionalization and grafting of polymers that are used as textile fibres, however it is beyond the scope of this book chapter. We will discuss both the enzymes for hydrolysis of synthetic fibres as well as the test methods utilized for their characterization and performance.

Enzymatic hydrolysis of polyesters / PET

Enzymes are potential tools for PET hydrolysis. PET hydrolyzing enzymes belong to the hydrolase class such as cutinases, lipases and esterases. PET was hydrolyzed by cutinases from organism *F. solani, F. oxysporum* and from *Pencillium citrinum*. Other PET-hydrolyzing enzymes are lipases, such as those from *Humicola sp., Candida sp., Pseudomonas sp.* and *Thermomyces lanuginosus*. In addition to enzymatic hydrolysis, the simple adsorption of enzyme protein to the polymer can also increase the hydrophilicity of PET owing to the hydrophilicity of the protein. High crystallinity of PET polymers negatively affects the ability of the enzymes to hydrolyze which has already been shown for enzymes from *F. solani* and from *T. fusca*.

Enzymatic hydrolysis of polyacrylonitrile (PAN)

PAN is a collective name for all polymers that consist of at least 85Percent acrylonitrile monomer. The homopolymer is difficult to process and dye thus is only for industrial applications. The co-monomers in acrylic fibres are selected for fibre specific properties, such as dyeability with sodium methallyl sulfonate, sodium sulfophenyl methallyl ether, etc. Acrylic fibres comprising negative groups can be dyed with basic dyes under carefully controlled conditions. Dyeing is usually performed in the presence of a retarder, which decreases the dyeing process rate for uniform shade reproduction. Finishing processes for PAN are limited since desirable properties can be more easily incorporated by copolymerization or by modification on the fibre level. For example, highly absorbent fibres are made by inclusion of a hydrophilic co-monomer which is subsequently removed by hydrolysis. It has been shown that bacterial strains, such as *Micrococcus luteus,* can degrade PAN fibres. During this process, poly(acrylic acid) is released from PAN as confirmed by NMR analysis. The release of poly(acrylic acid) from PAN, together with the formation of ammonia, was also shown for commercial nitrilases. Several researchers converted the nitrile groups of PAN to the corresponding acids or amides by nitrilases or by an enzyme system comprising nitrile hydratase and amidase, respectively resulting in major increases in hydrophobicity. These changes in surface properties corresponded to an 80Percent increase in the surface oxygen-to-carbon (O/C) ratio attributed to enzymatic hydrolysis of the nitrile groups. Commercial PAN-based materials usually contain around 7Percent vinyl acetate to reduce rigidity of the polymer. The vinyl acetate moieties in PAN can be hydrolyzed by cutinases and lipases, making this approach applicable to most commercially available PANs.

Enzymatic hydrolysis of polyamides (PA)

Enzymes such as proteases, amidases and cutinase can hydrolyse PA. A model substrate (adipic acid bishexyl-amide) has been developed for screening PA hydrolysis activity of given enzymes. It has been found that protease from *Beauveria sp.*, an amidase from *Nocardia sp.* and a cutinase from *F. solani pisi* can degrade the model substrate and correlated with PA hydrolysis activity. For actual PA substrates; it has been demonstrated by Parvinzadeh that protease treated Nylon 66 fabrics shows higher dye bath exhaustion with reactive and acid dyes. The intensity of major peaks in FTIR spectra of the protease treated samples is in favour of chemical changes of the polypeptide functional groups in the fabric. The results of colour measurements showed that there is a direct co-relation between the concentrations of enzyme against the darker shade of the dyed fabric. In a separate study performed by the same researchers, it was confirmed that acid and disperse dyes showed higher exhaustion on the protease and lipase treated Nylon 6 samples. To conclude, in coming years, the textile industry will go towards sustainable technologies and developing environmentally safer methods for textiles processing. One way is the processing with enzymatic system, rather than conventional chemical methods. There are several ways to improve the dyeability of any synthetic fibres. However, enhancement of the hydrophilicity of synthetic polymers is a key requirement for better dyeability. Enzymes have proved to be environmental friendly tools for hydrolysis of synthetic polymers, specifically on the polymer surface, without compromises in polymer bulk properties. In general hydrolases class of enzymes are used for improving hydrophilicity of synthetic fibres such as PET, PAN and PA respectively.

Modification of textile surfaces using sol-gel technique

A sol-gel technology is probably one of the most important developments in material science during the last decades. The sol-gel technique offers far reaching possibilities for creating new surface properties. Scientific literature demonstrates a wide array of functionalities that have been achieved by application of sol-gel coatings on textile surfaces. Its inorganic nature makes sol–gel layers very strong with nanometer-thick layers. Sol gel applications for textiles includes manipulation or changing

- Key textile properties *e.g.* stiffness, handle, absorbency, permeability etc.
- Surface properties *e.g.* hydrophobicity, hydrophilicity, abrasion resistance, photocatalytic activity, other barrier functions etc.
- Optical properties *e.g.* improving dyeability, photochromic effect, UV-absorption properties.
- Bio-active systems such as biocidal coatings, controlled release systems, immobilization of biological materials etc.
- And other physical properties *e.g.* heat resistance, conductivity etc.

Sol-gel technology offers the possibility of tailoring surface properties to a certain extent, and combining different functionalities into a single material. The added advantage is that the application of sols can be carried out with techniques commonly used in the textile industry such as, a simple dip or padding process followed by a thermal treatment in a stenter frame.

The sol gel principle

The preparatory material or precursor used to produce the "sol" usually consists of inorganic metal salts or metal organic components, such as metal alkoxides. These precursors are subjected to a series of hydrolysis and polymerization reactions to create a colloidal suspension or "sol". This sol is deposited on the surface of materials, transferred into a gel and finally into a layer of oxide by heat treatment. In production and for research purposes, SiO_2 and TiO_2 layers are deposited most often. Also, layers of many other compositions containing Al_2O_3, B_2O_3, ZrO_2, PbO and other oxides are often prepared. Next to the clearly inorganic layers, hybrid inorganic-organic layers have also been developed, which contain both chemical bonding of organic substances and functional groups next to silicon, titanium, zirconium and oxygen.

Dyeing of textiles based materials

Currently, we are observing an intensive increase in the production of textiles made of synthetic polymers. Synthetic fibres have high mechanical properties and are extremely rigid apart from other properties such as having low porosity and lower swelling etc. These properties are directly related to low dyeability with the standard dyeing technology. Synthetic fibres normally have a high glass transition temperature which makes it impossible for the dye molecule to penetrate into the fibres, especially when using water as a solvent. In some cases, it is possible to use other solvents or other fibre-swelling compounds to reduce the glass transition temperature of fibres. Some key examples of sol gel method applied on textile based materials to improve it dyeability. In the surface treatment of polymer structures, it is necessary to use hybrid layers based on a mixture of inorganic and organic polymer compounds, which are connected at the end of the process to a single macromolecular network. The inorganic part is linked with chemical, mechanical and thermal stability. The application of sol gel technology for improving dyeability is still at relatively early stage of development. However, looking at the trends, it is expected that sol gel technology will play important role towards improving dyeability and bringing new functionality together.

Nano-modifications of textiles surfaces using layer-by-layer deposition methods

A variety of functional thin films can be produced using the layer-by-layer assembly technique. LbL-based thin films are currently being evaluated for properties that include antimicrobial, anti-reflection, electrical conductivity, anti-flammable, gas barrier, and UV resistance. These films, typically < 1μm thick, are created by alternately exposing a substrate to positively- and negatively-charged molecules, polymer electrolytes, or particles. Steps 1 – 4 are continuously repeated until the desired number of "bilayers" is achieved. Figure provides an illustration of a film deposited with cationic and anionic polymers. Individual layers may be 1 – 100+ nm thick depending on chemistry, molecular weight, charge density, temperature, deposition time, counterion, and pH of species being deposited. The ability to control coating thickness down to the nm-level, easily insert variable thin layers without altering the process, avoid disturbing intrinsic mechanical behaviour of the substrate, and process under ambient conditions are some of the key advantages of this deposition technique. In non-wovens, each thread can be individually coated with a uniform LbL nanocoating and still remain soft and flexible.

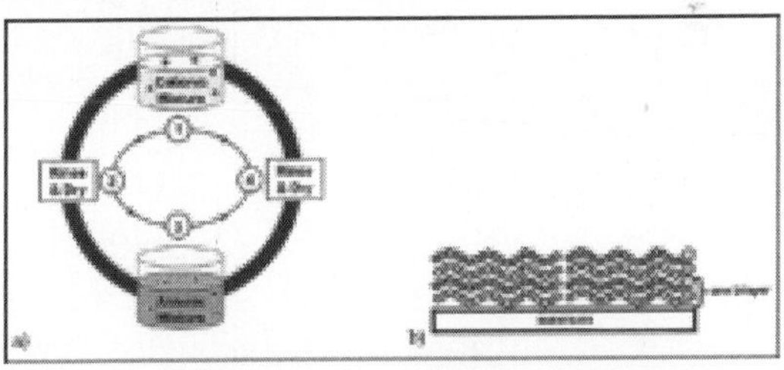

Fig. a) Schematic of layer-by-layer deposition process used to preapre functional thin films and b) steps 1-4 are repeated until the desired number of bilayers are generated on a substrate.

Surface roughness of materials has been intensely studied over the last decade. When this aspect is coupled with low surface energy components, "rough" materials become superhydrophobic. Olephobicity also comes into play when the material becomes nanoporous, minimizing void volume between molecular substituents, preventing wetting of low surface tension liquids in addition to polar liquid. Designing superhydrophobic surfaces via the layerby-layer assembly method have included covalently bonded interlayers, integrated organic and inorganic components, and induced micro-roughness from the underlying substrate to mimic the back of the Stenocara beetle where the hydrophilic/ superhydrophobic regions allow self-cleaning surfaces. Interesting alternating layers of the anionic and cationic mixtures have allowed for facile fabrication of the resultant surface chemistries. To further enhance the viability of the LbL technique for organic polymers, pH-amplified exponential growth LbL self-assembly was implemented for poly(ethylenimine) (PEI) and poly(acrylic acid) PAA as the polycation and polyanion respectively. This technique takes advantage of the synergistic effect of the pH-dependent tunable charge density and weak polyelectrolyte diffusivity. The end result is fast LbL layer formation in a limited number of deposition cycles. This research proved that only three bilayers were necessary to achieve a lotus-like superhydrophobic surface. LbL deposition of polyelectrolytes can also be used to impart a hydrophilic surface to polyolefins. In the case of polyethylene, this would be useful for high performance fibres that might eventually be used for athletic clothing to wick away moisture as well as increase dye receptivity. With a contact angles (θdH_2O) around 70°, polyethylene (PE) surfaces could actually be called slightly hydrophilic. In reality, contact angles below 45o are typically needed for a surface to exhibit facile wetting by a water droplet or hydrophilic behaviour. A matrix of polymer polyelectrolytes that are used for imparting various finishes to substrates via LbL deposition. Specifically implemented by the Grunlan laboratory to determine the possible hydrophilic coatings onto PE, 2.5 bilayers were needed to achieve a θdH_2O of 22° when using the system of poly(diallyldimethylammonium chloride) (PDDA-PAA) as compared to other systems comprising 6 or more bilayers. The elevated pH of PAA (pH = 5) created an increased negative charge density and thinner deposition relative to unmodified PAA (pH < 3). Additionally, stopping deposition at half bilayers, where PDDA was at the surface, proved much more hydrophilic than full bilayers due to its high charge density. They also noticed upon further evaluation, full bilayers of PDDA-PAA (pH = 5) had a contact angle of 11 degrees. This near superhydrophilic value is much lower than the initial screening attributed to a more pristine PE film cleaning and controlled polyelectrolyte deposition. The half bilayer protocol, with PDDA at the film's surface, had contact angles that could no longer be measured, suggesting a superhydrophilic surface. This was shown to be reproducible and unchanging as the assembly was built from 2.5 to 6.5 bilayers. Another unique utilization of LbL deposition was demonstrated by Cho and researchers. They formed a nanoporous substrate with microscale roughness through alternating electrostatic deposition of poly(allylamine hydrochloride) (PAH) as the polycation and SiO_2 nanoparticles as the polyanion for a substrate with reversible wetting properties. The exposed SiO_2 layer was treated with 3- (aminopropyltrichlorosilane) providing reactive binding sites (-NH_2) for photo-switchable moieties. The CF_3AZO moieties in vis (440nm) light are in a hydrophobic *trans* orientation but orient to a *cis* state upon exposure to UV (365 nm) light. After the LbL-CF_3AZO fabrication, the surface measured contact angles up to 1560, as dictated by the number of bilayers. In contrast, the contact angle of water for a CF_3AZO-functionalized flat film measured 76° (θdH_2O). The flat surface after exposure to UV light for 10 minutes, demonstrated a contact angle change of 50, whereas the LbL functionalized surface could switch between being superhydrophobic at 1560 to superhydrophilic at < 50, for nine bilayers. This reversibility was repeated up to 5 cycles with essentially zero hysteresis.

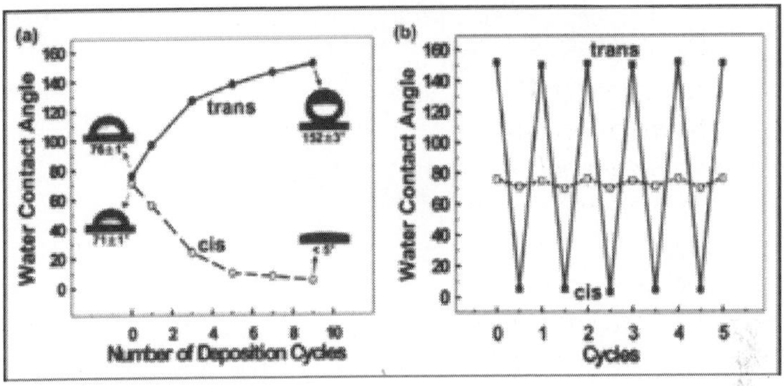

Fig. a) The number of LBL bilayers necessary to induce super-hydrophobicity/ super hydrophilicity on CF_3AZO-LBL surfaces and b) corrosponding reversiblity for the run nine bilayer surface after UV/ Vis exposure.

To conclude, LbL deposition is a unique surface modifying technique that gives ultimate flexibility for the design of the surface. Challenges for this technique include cost-effective commercial implementation at high on-machine line speeds for continuous operations. Engineering creativity to meet this challenge holds promise for a facile surface treatment technology for textiles.

Micro-encapsulation method to enhance dyeing process

Microencapsulation is actually a micro-packaging process involving the production of microcapsules. These materials act as barrier walls of different solids or liquids as cores. The wall has the ability to protect the core from hazardous environments, *i.e.* oxidization, heat, acidity, alkalinity, moisture or evaporation. They are produced by depositing a thin polymer coating on small solid particles or liquid droplets, or on dispersions of solids in liquids. The core contents are released under controlled conditions to suit a specific purpose. The most commonly methods used for preparation of microencapsules are complex coacervation, polymer-polymer incompatibility, interfacial polymerization and in situ polymerization, spray drying, centrifugal extrusion, air suspension coating, pan coating and emulsion hardening method. In dyeing of synthetic fibres, the major interest in microencapsulation is currently in the application of dyes as core and liposome as shell. Liposomes are artificially prepared vesicles made of lipid bilayer that can be filled with various materials. They comprise naturally-derived phospholipids with mixed lipid chains.

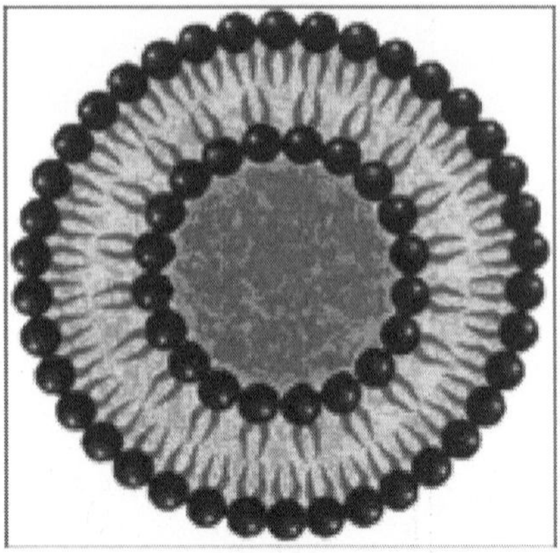

Fig. Structure of liposome

Gomez and Baptista studied microencapsulation of the dye in liposomes with lecithin from soy as an alternative to retarding and leveling agents. Liposomes were prepared with soy lecithin at different concentrations, containing the commercial acid dye C.I. Acid Blue 113. The effect on the dyeing rate of the microencapsulated dyes was compared with that from common retarding and leveling agents. The influence of surfactants on the stability of the liposomes and hence on the exhaustion curves of the dyeing was also evaluated. Interesting results obtained from exhaustion curves of anionic and non-ionic surfactants compared with commercial retarding and leveling agent. Marti et al. used phosphatidylcholine liposomes instead of synthetic surfactants as dispersing agent for disperse dyeing. They calculated the turbidity ratio to assess the dispersion behaviour of different liposome-dispersed dye preparations compared with commercial dye forms. Results indicated that liposomic preparations diminish the aggregation of dye molecules that normally occurs at high temperatures.

They also found the potential efficacy of liposomes as natural surfactants which can be applied to disperse dye formulations to dye polyester fibres with good dye exhaustion and washing fastness. This environmentally friendly biological surfactant, phosphatidylcholine, duly structured as liposomes, can substitute high amounts of synthetic dispersing agents in disperse dye formulations. Yan suggested microencapsulated disperse dyes to dye PET in the absence of auxiliaries and without reduction clearing. They studied the dyeing behaviours and dyeing kinetic parameters of microencapsulated disperse dye on PET compared with those of commercial disperse dyes with auxiliary solubilization.

Their results showed that the dyeing behaviours of disperse dye are influenced greatly by microencapsulation. The diffusion of disperse dyes from microcapsule onto fibres can be adjusted by the reactivity of shell materials and mass ratios of core to shell. The disparity of diffusion between two disperse dyes can be reduced by microencapsulation. In addition, the microencapsulation improves the utilization of disperse dyes due to no auxiliary solubilization.

Micro surface modification of textiles by aqueous solutions

Alkaline, acidic and solvents hydrolysis is another method to improve various physical and chemical properties of synthetic fibres. Alkaline hydrolysis has been studied extensively to overcome some problems of low water absorption properties and softness as alkaline hydrolysis improves the water absorption properties and softness of the PET fibre to give it a character similar to that of natural fibres. The alkaline hydrolysis of PET fibres is usually carried out with an aqueous alkaline solution, such as sodium hydroxide. In the alkaline hydrolysis process, PET undergoes a nucleophilic substitution. Chain scission of PET occurs, resulting in a considerable weight loss and the formation of hydroxyl and carboxylate end groups, which improves the handling, moisture absorption and dyeability of the fabric with enhanced softness. The effects of pretreatment reagents on the hydrolysis and physical properties of PET fabrics were investigated under various alkaline hydrolysis treatment and pretreatment conditions by Kim and his colleagues. Solvents used for pretreatment included benzyl alcohol and 2- phenyl ethanol. Results indicated that fabric weight loss, crystallinity, the initial and maximum water absorption increased with increasing hydrolysis time. Jain et al. reduced multi-filamentous polyacrylonitrile (PAN) fibres to amino groups using lithium aluminum hydride for immobilization of antibodies and detection of analyte. 24 h reduced fibres gave the most stable and reproducible results on immobilization of antibodies. Modified PAN fibres had a strong potential to be used as matrix for the detection of pathogenic bacteria and medical diagnostics. Another approach carried out by Cui and Yoon to modify the surface of PET film by treatment with ethoxylated hexylaminoanthraquinones synthesized by the reaction of 1-aminoanthraquinone with poly(ethylene glycol)s via hexamethylene spacer. The ethoxylated hexylaminoanthraquinones were adsorbed only onto the extreme surface of PET and water contact angle was decreased by the adsorption. A comprehensive collection of wet-chemical analyses of oxidized surfaces of poly(ethylene terephthalate) or polyolefin was presented by Knittel and Schollmeyer. Advanced oxidation of textile samples has been done using ozone and UV functionalization. They claimed that method presented uses inexpensive equipment and can be done quickly in a normal lab even as a process control.

BULK MODIFICATION OF SYNTHETIC TEXTILES USING NANOMATERIALS

In recent years, synthetic polymer-nanometric filler composites have generated significant attention in diverse applications such as transportation vehicles, construction materials, electronics, sporting goods, packaging, household and textile industries. The aim is to enhance a wide range of properties including mechanical properties, barrier, flame retardancy, solvent and heat resistance, biodegradability, chemical and thermal stability as well as improvement in dyeability relative to a virgin polymer. In order to obtain these specifications, fillers such as cellulose, clay, calcium carbonate, carbon, metal oxides and various forms of silica have been developed by different researchers. In this regard, the geometrical shape of the particles plays an important role in determining the properties of composites. Processing of such polymeric nanocomposites are more difficult compared to the corresponding pure polymers since such inorganic nanoparticles have strong tendencies to agglomerate. To overcome such difficulties, the sol-gel method, LbL deposition, in situ polymerization and melt processing are put into practice. The last method is still the most cost effective, simple, feasible and environmentally benign process for the mass production of polymeric nanocomposite. Considerable efforts have been devoted to improve various physical, mechanical and barrier properties of PET through mixing it with nanoclays. The layered clays used are mica, fluoro-mica, hectorite, saponite, etc., but one of the most commercially interesting clay is bentonite belonging to a structural family known as the 2:1 phyllosilicates. It is well known that the clay minerals have also been used as adsorbent for removal of acid, reactive, disperse and basic dyes from aqueous solutions due to the fact that they are globally abundant and inexpensive. Their inner layers comprise an octahedral sheet, which is situated between two tetrahedral sheets. The substitutions of Al^{3+} for Si^{4+} in the tetrahedral layer and Mg^{2+} or Zn^{2+} for Al^{3+} in the octahedral layers result in a net negative surface charge ion in water which cause the repulsion interaction with anionic dyes. The dyeability of synthetic fibres depends on their physical and chemical structure. Dyeing process consist of three steps including the diffusion of dye through the aqueous dye bath on to the fibre, the adsorption of dye into the outer layer of the fibre and the diffusion of dye from the adsorbed surface into the fibre interior. It was shown by researchers that functional groups of PET and water molecules play a great role in this process. The terminal carboxylic and hydroxyl groups in PET chains interact with water molecules. This makes a swelled fibre resulting to increase the attraction of disperse dye by these functional groups of fibre. The proportion of crystalline and amorphous regions of polymer is another factor influencing the dyeability. Researchers are concerned with the development and implementation of new techniques in order to fulfill improvement in dyeability of various polymers. Blending of polymeric fibres with nanoclays as inexpensive materials is still claimed as cost effective method to enhance dyeability. Up to now, only two research articles are focused on dyeing properties of polypropylene- and polyamide 6- layered clay incorporated nanocomposites prepared by melt compounding. Toshniwal et al. suggested that polypropylene fibres could be made dyeable with disperse dyes by addition of nanoclay particles in polymer matrix. Another research work done by Razafimahefa and her colleagues showed that the introduction of the nanoclay improves the dyeing ability of nylon with disperse dyes. Nevertheless, because of the interactions between the anions in montmorillonite and the amino groups on the polyamide, the dyeing sites are occupied with the nanoclay. This led to inferior dyeing with acid or metal complex dyes than in the case of the unfilled polymer. Our previous study on dyeability of PET/clay nanocomposites stated the following type of interactions between the disperse dye and clay surfaces:

- Hydrogen bonding between OH groups of modified clays and the NH_2 and CO groups of disperse dye molecules.
- Electrostatic bonding between the negatively charged oxygen atom of carbonyl groups in disperse dye molecule and positively charged nitrogen atom of quaternary ammonium salt in modified clays.
- Direct π interactions and van der Waals forces between methyl and ethyl groups of modified clays on one hand and methoxy group and benzene rings of disperse dye molecule on the other hand.

The second reason for improving disperse dye absorption of PET/clay nanocomposites could be the relatively large voids between clay platelets after modification with quaternary ammonium salts. It was shown that the surface morphology of PET/clay nanocomposites has great influence on water contact angle of the resultant nanocomposite.

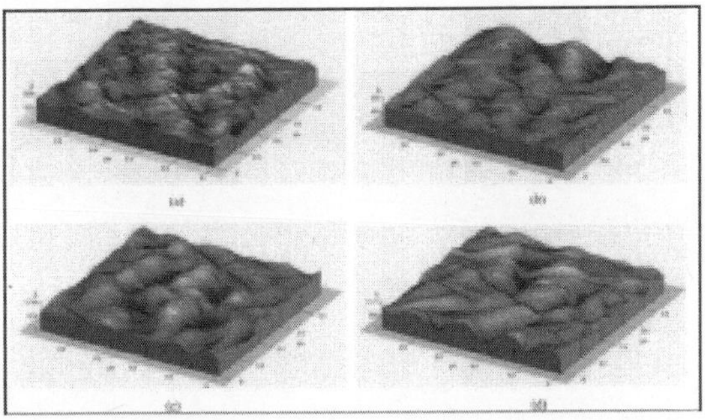

Fig. topographic images of atomic force microscopy for various composites: (a) Pure PET, (b) PET=15A, (c) PET=30B, (d) PET=Na+

Another applicable nanofiller is silica nanoparticle which impacts high stiffness, tensile strength, modulus, toughening, crystallinity, viscosity, creep resistance, coefficient of friction, wear resistance, toughness and interfacial adhesion in various polymer nanocomposites. Recently, a research programme to explain dyeability of PET/silica nanocomposite was done by Yang and Gu. They used *in situ* polymerization to prepare PET/SiO_2 nanocomposites. According to the results, the colour strength of the dyeing increased with increasing SiO_2 content in polymer. Over the last 20 years polymer chemistry has created a non-linear polymeric structure coined dendritic polymers. Their architecture arises from the introduction of a large number of branches with many functional end groups.

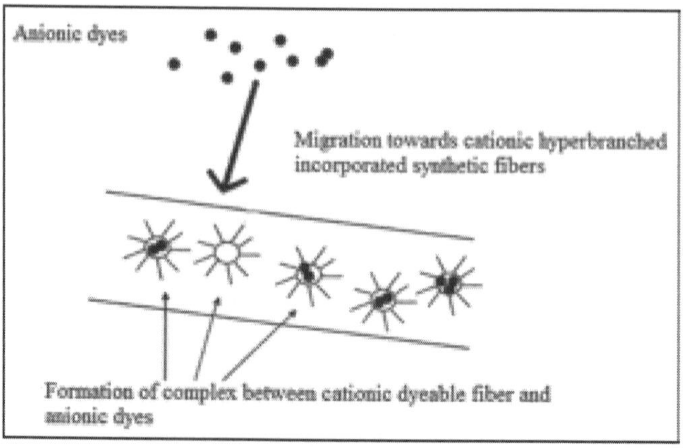

Fig. Incorporation of hyperbranched polymers into the fibre structure

Two classes of these polymers are dendrimers with a perfectly branched uniform structure and hyperbranched polymers with non-uniform ones. It was already suggested by different authors that highly branched molecules should be able to act as a host for the encapsulation of guest molecules of dyes or a dendritic box. This structure can lead to industrial development of dyeable poly(propylene) fibre. Other authors suggested that an improvement in dyeability of the polymeric- hyperbranched additive nanocomposites can be attributed to the decrease in glass transition temperature and the lower crystallinity of the polymers.

REMARKS AND OUTLOOK

Various types of physical and chemical finishing methods have been described in this chapter. Most of them are developed to solve problems with synthetic fibres to expand their usefulness. Examples of such problems are their insufficient fabric softness, low absorbency of water, flammability, and pilling, low dyeability, slipping and static problems during production and usage. New finishing processes using physical and chemical methods can solve these problems and restrictions. Both the improved and the newly developed finishes based on nano-science are valuable tools that can project an enhanced image of the finish producers for the next textile industry revolution.

Bibliography

Ainley, Albert: *Woolen and worsted loomfixing; a book for loomfixers and all who are interested in the production of plain and fancy worsteds and woolens*, Lawrence Mass, 1900.

Alford, Marianne Margaret Compton Cust: *Needlework as art*, London: Sampson Low Marston Searle and Rivington, 1886.

Alfred Jenks & Son: *Illustrated Catalogue of Machines Built by Alfred Jenks & Son, Manufacturers of Every Variety of Cotton and Wool Carding, Spinning, and Weaving Machinery, in All Its Departments*, Bridesburg, Pa.: Alfred Jenks & Son, 1853.

Allen, Frederick James: *The shoe industry*, Boston: The Vocation bureau of Boston, 1916.

Allen, Nellie B: *Cotton and other useful fibers*, Boston: Ginn, 1929.

Allington, Sara May: *Practical sewing and dressmaking*, Boston: Page Co., 1913.

Allinson, May: *Dressmaking as a trade for women in Massachusetts*, Washington: Govt. print. off., 1916.

Atkinson, Edward: *Cotton and Cotton Manufactures in the United States*, Boston: Franklin Press, 1880.

Baird, Robert H: *The American Cotton Spinner, and Managers' and Carders' Guide: A Practical Treatise on Cotton Spinning*, Philadelphia: H. C. Baird, 1869.

Batchelder, Samuel: *Introduction and Early Progress of the Cotton Manufacture in the United States*, Boston: Little, Brown and Company, 1863.

Brown Brothers & Co: *Brown Brothers & Co., Successors to Butler, Brown & Co.: Manufacturers of Shaw's U. S. Standard Ring Travelers, Belt Hooks, Wire Goods, Etc., and Dealers in Supplies for Cotton, Woolen, Silk, Jute & Flax Mills, Providence*, R.I.: Brown Brothers & Co., c1881.

E. Jenckes Manufacturing Co: *Illustrated Price List: E. Jenckes Mfg. Co., Sole Manufacturers of Hicks' U. S. Standard Ring Travelers, Manufacturers' Supplies of Every Description*, Pawtucket, R.I.: E. Jenckes Mfg. Co., 1883.

Lowell Machine Shop: *The Lowell Machine Shop, Lowell, Mass.: Builders of Cotton Machinery of Every Description*, Lowell, Mass.: Stone, Bacheller & Livingston, 1882.

Mason Machine Works: *The Mason Machine Works, Taunton Massachusetts, U.S.A.: Inventors and Builders of Cotton Machinery*, Taunton, Mass.: The Company, 1898.

Merrill, Gilbert R: *American Cotton Handbook: A Practical Reference Book for the Entire Cotton Industry*, New York: American Cotton Handbook Co., c1941.

Scott, Robert: *The Practical Cotton Spinner, and Manufacturer: The Managers', Overlookers' and Mechanics' Companion*, Corr. and enl. Philadelphia: H. C. Baird, 1851.

Webber, Samuel: *Manual of Power for Machines, Shafts, and Belts, with the History of Cotton Manufacture in the United States*, New York: D. Appleton and Company, 1879

Index